ESI Lectures in Mathematics and Physics

Erwin Schrödinger International Institute for Mathematical Physics
Boltzmanngasse 9
A-1090 Wien
Austria

The Erwin Schrödinger International Institute for Mathematical Phyiscs is a meeting place for leading experts in mathematical physics and mathematics, nurturing the development and exchange of ideas in the international community, particularly stimulating intellectual exchange between scientists from Eastern Europe and the rest of the world.

The purpose of the series *ESI Lectures in Mathematics and Physics* is to make selected texts arising from its research programme better known to a wider community and easily available to a worldwide audience. It publishes lecture notes on courses given by internationally renowned experts on highly active research topics. In order to make the series attractive to graduate students as well as researchers, special emphasis is given to concise and lively presentations with a level and focus appropriate to a student's background and at prices commensurate with a student's means.

Previously published in this series:

Arkady L. Onishchik, *Lectures on Real Semisimple Lie Algebras and Their Representations*
Werner Ballmann, *Lectures on Kähler Manifolds*
Christian Bär, Nicolas Ginoux, Frank Pfäffle, *Wave Equations on Lorentzian Manifolds and Quantization*
Recent Developments in Pseudo-Riemannian Geometry, Dmitri V. Alekseevsky and Helga Baum (Eds.)
Boltzmann's Legacy, Giovanni Gallavotti, Wolfgang L. Reiter and Jakob Yngvason (Eds.)

Hans Ringström

The Cauchy Problem in General Relativity

European Mathematical Society

Author:

Hans Ringström
Department of Mathematics
KTH
100 44 Stockholm
Sweden

2000 Mathematics Subject Classification (primary; secondary): 83-02; 83C75

Key words: General relativity, Cauchy problem, strong cosmic censorship, non-linear wave equations

ISBN 978-3-03719-053-1

The Swiss National Library lists this publication in The Swiss Book, the Swiss national bibliography, and the detailed bibliographic data are available on the Internet at http://www.helveticat.ch.

Contact address:

European Mathematical Society Publishing House
Seminar for Applied Mathematics
ETH-Zentrum FLI C4
CH-8092 Zürich
Switzerland

Phone: +41 (0)44 632 34 36
Email: info@ems-ph.org
Homepage: www.ems-ph.org

Typeset using the authors' TEX files: I. Zimmermann, Freiburg
Printed in Germany

9 8 7 6 5 4 3 2 1

Les philosophes font subir à la réalité, pour pouvoir l'étudier pure, à peu près les mêmes transformations que le feu ou le pilon font subir aux corps: rien d'un être ou d'un fait, tels que nous l'avons connu, ne paraît subsister dans ces cristaux ou dans cette cendre.

Marguerite Yourcenar, *Mémoires d'Hadrien*

Preface

Given initial data to Einstein's equations, there is a maximal globally hyperbolic development. This is a fundamental fact concerning the Cauchy problem in general relativity, and these notes are the consequence of a desire to understand the details of the proof of it. Concerning this result, there is material of an overview character available, as well as the original research articles. However, reconstructing a complete, detailed proof with these articles as a starting point is something that requires an effort, even for someone familiar with the fields involved. Since the Cauchy problem has received an increasing amount of attention the last 15 years or so, it seems natural to write down all the details in a coherent fashion. Those potentially interested in understanding the proof come from many different fields of mathematics, and for this reason, the goal of these notes has been to keep the prerequisites at a minimum.

Stockholm, April 2009 Hans Ringström

Contents

1 Introduction

1.1 Historical overview

Even though Einstein introduced his equations in 1915, cf. [35] and [36], it was not until 1952 that it was firmly established that they allow a formulation as an initial value problem. The seminal paper was written by Yvonne Choquet-Bruhat, cf. [39], and it contains a proof of local existence of solutions. Due to the diffeomorphism invariance of the equations, the step from local existence to the existence of a maximal development is non-trivial. The reason is that even if there is a maximal development in the sense that it cannot be extended, there is no reason to expect this development to be unique. In the end, one does in fact need to restrict one's attention to a special class of developments in order to get an element which is maximal and unique in the given class. Partly as a consequence of this, it was not until 1969 that Choquet-Bruhat and Robert Geroch, cf. [10], demonstrated that, given initial data, there is a unique maximal globally hyperbolic development (MGHD). The existence of a MGHD does not say anything about the "global" properties of solutions, but it is nevertheless a fundamental theoretical starting point for any analysis of solutions to Einstein's equations. To take but one example, the question of predictability in general relativity, as embodied in the strong cosmic censorship conjecture, is phrased in terms of the MGHD.

The initial value point of view illustrates the strong connection between Einstein's general theory of relativity and the theory of hyperbolic partial differential equations (PDE's). Historically, this is a connection which has not received as much attention as one might have expected. In the beginning, most people working in the field focused on writing down different explicit solutions. Obviously, this was a natural starting point, and some of the solutions written down in this period will with all probability always be of fundamental importance. However, it is possible to work with such spacetimes without even being aware of the hyperbolic PDE character of the equations.

The question of singularities in general relativity has always been important. At an early stage, it was suggested that some of the singularities present in the model solutions of the universe and of isolated systems would not be present in less symmetric solutions. Remarkably, the existence of singularities in a more general situation was demonstrated by methods which avoid the PDE aspect of the problem altogether. The singularity theorems of Hawking and Penrose, cf. [66], [46], [47] for some of the original papers and [48], [87], [65] for textbook presentations, demonstrate that singularities are generic, assuming one is prepared to equate the existence of a singularity with the existence of an incomplete causal geodesic. These results had an important impact on the field, not least because they illustrated the importance of the geometry, and partly as a consequence of them, Lorentz geometry has become an important subject in its own right. Surprisingly enough, not much information concerning Einstein's equations is required in order to obtain these theorems; some general energy conditions concerning the matter present suffice. On the other hand, one does not obtain that much information. It would of course be of interest to know if the gravitational fields become arbitrarily strong as one approaches a singularity, i.e., in the incomplete directions of causal geodesics. To address this question, it would seem that it is necessary to take the PDE aspect into account.

In the formulation of Einstein's equations as an initial value problem, the initial data cannot be chosen freely; they have to satisfy certain constraint equations. The various methods introduced for solving these equations in the end lead to non-linear elliptic PDE's on manifolds, so that even before coming to the evolutionary aspect of the equations, one is faced with a difficult analysis problem. Given initial data, the problem of local existence has been solved by the work of Yvonne Choquet-Bruhat as mentioned above. However, in the end one is interested in the global structure. In terms of analysis, this often makes it necessary to prove global existence of solutions to a non-linear hyperbolic PDE, which is usually very difficult. In view of these complications, it is not surprising that it has taken such a long time for the hyperbolic PDE aspect of the equations to become the focus of attention.

1.2 Some global results, recent developments

To our knowledge, the first global results in the absence of symmetries are due to Helmut Friedrich, cf. [40]. In this paper, he demonstrates, among other things, the stability of de Sitter space. Furthermore, he proves that initial data close to those induced on a hyperboloid in Minkowski space yield developments that are future null geodesically complete. In addition to these results, he derives detailed information concerning the asymptotics. However, it is interesting to note that the methods he uses are very geometric in nature. The proof is based on a very intelligent and geometric choice of equations which makes it possible to reduce a global problem, from the geometric point of view, to a problem of local stability in the PDE setting.

The proof of the global non-linear stability of Minkowski space, see [20], by Demetrios Christodoulou and Sergiu Klainerman is based on very different methods. In [20], the authors solve a problem which is global in nature also from the non-linear hyperbolic PDE point of view, and for this reason more work on the analysis part of the problem is required. Beyond being an important result in its own right, the work of Christodoulou and Klainerman seems to have had the effect of bringing Einstein's equations to the attention of the non-linear hyperbolic PDE community, as illustrated by e.g. [57] and [58]. Furthermore, the interest in the question of local existence has been revived.

In the case of non-linear hyperbolic PDE's, it is of interest to prove well posedness for as low a regularity of the data as possible. One reason is that a local existence result is usually associated with a continuation criterion; e.g. if the maximal existence time to the future is T_+, then there is a statement of the form: either $T_+ = \infty$ or a suitable norm of the solution is unbounded as $t \to T_+-$. Proving local existence in a lower degree of regularity usually results in a continuation criterion involving a weaker norm. In the optimal situation, it is possible to prove local existence in such a weak regularity that the norm appearing in the continuation criterion is controlled by a quantity which is preserved by the evolution. In such a situation, one is allowed to conclude that solutions do not blow up in a finite time simply due to the local existence result. A striking illustration of this perspective is given by the work of Klainerman and Matei Machedon, cf. [51], which proves that solutions to the Yang–Mills equations in $3 + 1$ dimensional Minkowski space do not blow up in finite time due to such an argument. However, it should be noted that the latter conclusion had already been obtained by the work of Douglas Eardley and Vincent Moncrief in [33] and [34]. In the case of Einstein's equations, there is no hope for a similar result, but the question of local existence for as rough data as possible is nevertheless of interest. Since the result of Choquet-Bruhat, [39], there have been many papers concerned with the question of local existence. One classical reference is [38]. More recently, Klainerman and Igor Rodnianski have undertaken an ambitious programme in order to obtain the optimal regularity result, cf. [52] and the references cited therein.

The results [40] and [20] are the first examples of results that are global in nature, concern a situation without symmetry and have been obtained by PDE methods. Some examples of more recent results in this category are [3], [57] and [77]. The number of results that are global in nature, concern a situation with symmetry and have been obtained by PDE methods is, due to the work carried out the last twenty years or so, quite large.

1.3 Purpose

The above sections were intended to illustrate that in recent years, the perspective on Einstein's equations taken by researchers has tended toward the initial value point of view. Unfortunately, it seems that the main references available relevant to this perspective are of one of the following types: textbooks on geometric aspects, i.e., books on Lorentz geometry, textbooks on non-linear hyperbolic PDE's, research papers and overview papers. The reason this is unfortunate is that the books on Lorentz geometry typically ignore the hyperbolic PDE aspects (or, as in the case of [48] and [87], give an outline rather than a detailed exposition) whereas the books on non-linear hyperbolic PDE's typically ignore the geometry. That geometry is important in the context of non-linear hyperbolic PDE's even outside the field of general relativity can be illustrated in the following way. Consider a non-linear wave equation of the form

$$g^{\mu\nu}(u, \partial u)\partial_\mu \partial_\nu u = f(u, \partial u)$$

on $\mathbb{R}^{n+1}$, where ∂u is used to represent the first derivatives and we use Einstein's summation convention, cf. Section A.1. Then $g^{\mu\nu}(u, \partial u)$ are the components of the inverse of a Lorentz metric, and such fundamental aspects of the equations as uniqueness are most naturally expressed in terms of the causal structure of this metric. The research papers are of course worth reading, but at some stage a coherent presentation is preferable. The overview papers fill an important function in that they provide the intuition behind the results, but in the end one would like to have complete proofs.

The purpose of these notes is, among other things, to give a complete proof of the existence of a maximal globally hyperbolic development. For each matter model one couples to Einstein's equations, it is necessary to give a proof for the resulting combination of equations. Since it is not the ambition of these notes to give an exhaustive list of matter models which can be coupled to Einstein's equations, we shall restrict our attention to the non-linear scalar field case. This is a matter model which allows vacuum with a cosmological constant as a special case. The reason we have chosen it is that it has attracted attention in recent years in the cosmological setting as a model which produces different types of accelerated expansion, something recent observational data indicate that our universe is undergoing. Beyond this main goal, there are, however, several other results that are of fundamental importance, but seem to fall outside of the scope of most textbooks. For instance, uniqueness of solutions to linear wave equations expressed in terms of the causal structure of the Lorentz metric involved is something standard textbooks on PDE's typically avoid. Cauchy stability in the context of general relativity is another result which is of fundamental importance, but detailed proofs are rarely written down in an accessible way. Finally, there are recent results concerning the structure of globally hyperbolic spacetimes, cf. [4], [5], [6], which deserve to be presented in the present context.

It is not the ambition of these notes to give an up-to-date overview of the methods developed to prove local existence in low regularity. In fact, the methods presented here for proving local existence are quite old. The main purpose is rather to present some elementary theory for hyperbolic PDE's and some elementary Lorentz geometry in a unified way, the hope being that this will be of use to those working in the field. The reason for doing so at this point in time is that the importance of the combination has become apparent in the last fifteen years or so.

As we mentioned above, researchers working on problems related to Einstein's equations come from many different backgrounds. In particular, there are those with a geometric background but not such a strong background in PDE's, and there are those who come from the non-linear hyperbolic PDE community. For this reason, these notes have been written with the ambition of not presupposing more of the reader than a knowledge of measure and integration theory, some very elementary functional analysis, basic Lorentz geometry and elementary differential geometry. To be more specific, it is, from a logical point of view, possible to read these notes given a knowledge of analysis corresponding to *Principles of Mathematical Analysis* and *Real and Complex Analysis* by Walter Rudin, cf. [78] and [79]. In fact, only the first eight chapters of the latter book are needed. In particular, no previous familiarity with PDE's is required. Concerning Lorentz geometry, we assume the reader is familiar with the material contained in Barrett O'Neill's book *Semi-Riemannian Geometry*, [65], though no mention of manifolds will be made until Chapter 10. Concerning differential geometry, we presuppose the basics, including familiarity with Stokes' theorem, as contained in, e.g., [25], and, in the last part of these notes, a familiarity with Lie groups corresponding to the material given in, e.g., Lee's book [55]. Assuming less than the material contained in these references seems unreasonable, since they are all very well written and much more fundamental than the results presented in these notes. However, assuming more also seems unwarranted, given the goal already mentioned. Let us mention that, recently, two related books were published; see [19] and [9]. Finally, let us recommend the book [69], treating similar material from a different perspective, as a good companion to these notes.

Acknowledgments. The author would like to express his gratitude to the Erwin Schrödinger International Institute for Mathematical Physics, Vienna, for the hospitality afforded him during the programmes "Singularity formation in non-linear evolution equations", organized by P. C. Aichelburg, P. Bizon and S. Klainerman in 2004, and "Mathematical Cosmology", organized by P. C. Aichelburg, G. F. R. Ellis, V. Moncrief and J. Wainwright in 2001. Both events were related to the subject of these notes. Parts of the material contained in the notes were presented in seminars during these programmes. The author is also grateful to the Mittag-Leffler Institute for hosting the programme entitled "Geometry, Analysis, and General Relativity" during which these notes were completed. Finally, thanks are due to Lars Andersson, Mattias Dahl and Alan Rendall for comments on earlier drafts of these notes. This work was supported by the Swedish Research Council and the Göran Gustafsson Foundation. The author is a Royal Swedish Academy of Sciences Research Fellow supported by a grant from the Knut and Alice Wallenberg Foundation.

2 Outline

2.1 PDE theory

Since the main purpose of these notes is to prove the existence of a maximal globally hyperbolic development, given initial data, a natural starting point is to prove local existence of solutions to the type of non-linear PDE's that result after making suitable gauge choices. One way of doing so is to first prove local existence of solutions to linear wave equations. Thus, let us start by considering equations of the form

$$g^{\mu\nu}\partial_\mu\partial_\nu u + a^\mu\partial_\mu u + bu = f, \tag{2.1}$$

$$u(0,\,\cdot\,) = u_0, \tag{2.2}$$

$$\partial_t u(0,\,\cdot\,) = u_1, \tag{2.3}$$

where $g^{\mu\nu}$ are the components of an $(n+1)\times(n+1)$ real matrix-valued function on $\mathbb{R}^{n+1}$ which is such that each element in its range is symmetric, has one negative eigenvalue and n positive eigenvalues. Furthermore, a^μ, b and f are functions on $\mathbb{R}^{n+1}$, u_0 and u_1 are functions on $\mathbb{R}^n$, and we use Einstein's summation convention and the convention concerning coordinates described in Section A.1 (precise conditions that ensure the existence of solutions are specified in Section 8.2). In practice, we prove local existence of solutions to linear *symmetric hyperbolic systems*, a problem that of solving (2.1)–(2.3) can be reduced to. These are equations of the form

$$A^\mu\partial_\mu u + Bu = f, \tag{2.4}$$

$$u(0,\,\cdot\,) = u_0, \tag{2.5}$$

where A^μ, $\mu = 0,\dots,n$, and B are $N\times N$ real matrix-valued functions on $\mathbb{R}^{n+1}$ and f and u_0 are $\mathbb{R}^N$-valued functions on $\mathbb{R}^{n+1}$ and $\mathbb{R}^n$ respectively. The conditions of symmetric hyperbolicity are that A^μ, $\mu = 0,\dots,n$, be *symmetric* and that A^0 be *positive definite*. We shall use the notation Lu for the left-hand side of (2.4), i.e.,

$$Lu = A^\mu\partial_\mu u + Bu. \tag{2.6}$$

Both the equations (2.1) and (2.4) have energies associated with them. In the case of (2.1), one choice of energy is

$$\mathcal{E} = \frac{1}{2}\int_{\mathbb{R}^n}[-g^{00}|u_t|^2 + g^{ij}\partial_i u\cdot\partial_j u + |u|^2]\,dx, \tag{2.7}$$

assuming that $g^{00}(t,x) < 0$ and that $g^{ij}(t,x)$, $i,j = 1,\dots,n$, are the components of a positive definite matrix for all $(t,x)\in\mathbb{R}^{n+1}$. In the case of (2.4), a natural energy is

$$E = \frac{1}{2}\int_{\mathbb{R}^n} u^t A^0 u\,dx, \tag{2.8}$$

where the t denotes transpose. Given the energies (2.7) and (2.8), the standard procedure is to differentiate with respect to time, to integrate by parts in the term inside the integral involving spatial derivatives and to use the equation. This yields inequalities that form the basis of the proof of existence of solutions to linear equations as well as non-linear ones (in Section 7.2 and the beginning of Chapter 8 we write down, in detail, the arguments needed to derive such inequalities in the symmetric hyperbolic and linear wave equation setting respectively). The resulting inequalities also imply uniqueness of solutions to (2.1)–(2.3) and of solutions to (2.4)–(2.5). Loosely, estimates arising from these types of arguments are referred to as energy estimates. In some situations it is of interest to consider an energy which, as opposed to (2.7), is geometrically defined. To this end, let us define a stress energy tensor $T_{\mu\nu}$ by

$$T_{\mu\nu} := \nabla_\mu u \cdot \nabla_\nu u - \frac{1}{2} g_{\mu\nu}(g^{\alpha\beta}\nabla_\alpha u \cdot \nabla_\beta u),$$

where ∇ is the Levi-Civita connection associated with g viewed as a Lorentz metric on $\mathbb{R}^{n+1}$. Given a suitable spacelike hypersurface, one can then contract $T_{\alpha\beta}$ with the timelike unit normal twice and integrate the result over the hypersurface in order to obtain a geometrically defined energy. It should, however, be mentioned that in certain circumstances it is quite useful to contract the stress energy tensor with other types of vector fields and to integrate over surfaces that are not spacelike. Nevertheless, we shall use the expression (2.7) which, even though it is not geometric, is quite useful from a practical point of view.

The idea of how to prove existence of solutions to (2.4)–(2.5) is taken from [50]. Let us give a rough outline. Let $T > 0$ and let L^* be the L^2-adjoint of the operator defined in (2.6). First, one proves that the definition

$$F(L^*\phi) = \int_0^T (\phi(t), f(t))_{L^2}\, dt \tag{2.9}$$

makes sense for smooth functions ϕ with compact support such that $\phi(t,x) = 0$ for $t \geq T$. In fact, this is a consequence of the energy estimates. Here $(\,\cdot\,,\,\cdot\,)_{L^2}$ is the L^2 inner product; to start with, we assume f to be smooth and to have compact support, and for this reason we can consider f and ϕ to be maps from $\mathbb{R}$ into the set of L^2-functions on $\mathbb{R}^n$. The energy estimates also imply

$$|F(L^*\phi)| \leq C \int_0^T \|(L^*\phi)(t,\,\cdot\,)\|_X\, dt,$$

where X is the norm of some suitable Hilbert space. Due to the Hahn–Banach theorem, one can extend F to be a bounded linear functional on $L^1([0,T],X)$. The existence of a solution then follows by a duality saying that the dual of $L^1([0,T],X)$ is $L^\infty([0,T],Y)$ for some suitable space Y and that

$$F(v) = \int_0^T (v(t), u(t))_{L^2}\, dt \tag{2.10}$$

for some $u \in L^\infty([0,T],Y)$ (in order for this to make sense, one of course has to interpret the right-hand side of (2.10) in a suitable way). Combining the above observations, one obtains that

$$\int_0^T (L^*\phi(t), u(t))_{L^2}\,dt = \int_0^T (\phi(t), f(t))_{L^2}\,dt \tag{2.11}$$

for suitable smooth functions ϕ. This is a weak statement of the fact that u is a solution to the equation.

2.1.1 The Fourier transform and Sobolev spaces. In order for the above argument to make sense, some preliminary observations are necessary. First of all, one needs to define $L^p([0,T],X)$ for a Banach space X, and thus, in particular, to define the concept of measurability in this context. Furthermore, one needs to prove the relevant dualities. Chapter 3 provides the necessary background. The Hilbert space X we shall use in the above argument is a Sobolev space corresponding to a negative number, say $-s$, of derivatives and Y is a Sobolev space corresponding to s derivatives. The greater the s, the more regular the weak solution. For s large enough, one obtains a classical solution. Furthermore, making strong enough assumptions concerning the initial data and the coefficients, one obtains smooth solutions due to uniqueness. By the above, it is clear that we need to introduce the concept of a negative number of derivatives. If k is a non-negative integer, then the Fourier transform of $(1-\Delta)^k f$, where Δ is the ordinary Laplacian, is $(1+|\xi|^2)^k \hat{f}$, where $\hat{f}$ is the Fourier transform of f. Thus, we can define $(1-\Delta)^s f$ for any real s to be the inverse Fourier transform of $(1+|\xi|^2)^s \hat{f}$. In particular, we can speak of a negative number of derivatives. However, it is clear that in order to be able to justify the above arguments rigorously, we need to present some elementary properties of the Fourier transform, and this is the subject of Chapter 4. The Fourier inversion formula will also be used to prove Sobolev embedding, which relates the C^k norm of a function with the Sobolev norm.

In Chapter 5, we introduce Sobolev spaces. The starting point in the definition of these spaces is the concept of a weak derivative; u is said to be k times weakly differentiable if, for every multiindex α such that $|\alpha| \le k$, there is a function u_α such that

$$\int_{\mathbb{R}^n} u\partial^\alpha \phi\, dx = (-1)^{|\alpha|} \int_{\mathbb{R}^n} u_\alpha \phi\, dx$$

for all $\phi \in C_0^\infty(\mathbb{R}^n)$. Needless to say, it is necessary to demand measurability of u and u_α as well as some degree of integrability (they should be in $L^1_{\text{loc}}(\mathbb{R}^n)$). The Sobolev spaces $W^{k,p}(\mathbb{R}^n)$ are then defined as the set of k times weakly differentiable functions all of whose weak derivatives are in $L^p(\mathbb{R}^n)$. The corresponding norm can be defined as the sum of the L^p-norms of the weak derivatives. In the end, we shall only use the special case $p = 2$, which we shall denote $H^k(\mathbb{R}^n)$.

The idea of how to prove local existence of solutions to non-linear wave equations is to set up an iteration. To get convergence, one needs a complete space in which to carry out the iteration, and, furthermore, it is essential that the equation preserve the degree of regularity of the initial data. That a (high enough) Sobolev degree of regularity is preserved by the equations is related to the existence of the energies (2.7) and (2.8) and their behaviour under the evolution. Furthermore, one basic property of the Sobolev spaces is that they are complete. In other words, the basic properties of the equations themselves indicate that it is natural to consider initial data in Sobolev spaces. As we mentioned above, in the proof of existence of solutions to linear wave equations, we shall need to use Sobolev spaces corresponding to s derivatives in $L^2(\mathbb{R}^n)$, where s is a real number. We shall denote these spaces by $H_{(s)}(\mathbb{R}^n)$ and they are defined using the Fourier transform as indicated above, with a norm given by (2.13) below. Note, however, that if s is negative, these spaces do not consist of functions but of tempered distributions, a concept we define in Chapter 5. We end the chapter by proving some elementary properties of $H_{(s)}(\mathbb{R}^n)$ as well as the duality statement which will constitute the basis of the existence proof of solutions to symmetric hyperbolic systems.

In Chapter 6 we relate Sobolev spaces to C^k spaces. In particular, we prove the estimate

$$\|f\|_{C_b^k(\mathbb{R}^n,\mathbb{C})} \leq C\|f\|_{(s)} \tag{2.12}$$

for $s > k + n/2$, where $\|\cdot\|_{C_b^k}$ is the C^k-norm and $\|\cdot\|_{(s)}$ is the $H_{(s)}$ norm, defined by

$$\|f\|_{(s)} = \left[\frac{1}{(2\pi)^n}\int_{\mathbb{R}^n} |\hat{f}(\xi)|^2(1+|\xi|^2)^s\,d\xi\right]^{1/2}, \tag{2.13}$$

where $\hat{f}$ is the Fourier transform of f. The estimate (2.12) is a special case of Sobolev embedding, and in this particular form, it is an immediate consequence of the Fourier inversion formula. The reason such a result is important is that, as we mentioned above, in order to prove local existence of solutions to non-linear wave equations, we need to work with Sobolev spaces. In order to go from the element we have constructed in the complete space to a classical (smooth or C^k) solution to the equation, we need to relate regularity from the Sobolev point of view to regularity from the C^k point of view, and this achieved by (2.12).

When proving convergence of the iteration used to construct a local solution in the non-linear case, we need estimates for the non-linearity. In that context the estimate

$$\|\partial^{\alpha_1}\phi_1\ldots\partial^{\alpha_l}\phi_l\|_2 \leq C\sum_{i=1}^{l}\sum_{|\alpha|=k}\|\partial^{\alpha}\phi_i\|_2\prod_{j\neq i}\|\phi_j\|_\infty \tag{2.14}$$

is useful, where $\|\cdot\|_p$ denotes the L^p norm and $k = |\alpha_1| + \cdots + |\alpha_2|$. This is a consequence of estimates that are due to Gagliardo and Nirenberg, and we prove (2.14) in Chapter 6. In the beginning of Chapter 6 we give examples of the use of this inequality.

2.1.2 Symmetric hyperbolic systems. Chapter 7 begins with a proof of Grönwall's lemma, which states that if f, k and G are non-negative functions such that G is non-decreasing and

$$f(t) \leq G(t) + \int_0^t k(s)f(s)\,ds$$

for all $t \in [0, T]$, then

$$f(t) \leq G(t) \exp\left(\int_0^t k(s)\, ds\right)$$

for all $t \in [0, T]$. To prove this in the case that f and k are continuous is straightforward. However, the context of interest to us is when $k \in L^1([0, T])$ and $f \in L^\infty([0, T])$. Grönwall's lemma is a basic technical tool which is used frequently in the proof of existence of solutions to symmetric hyperbolic systems as well as local existence of solutions to non-linear wave equations.

The proof of Grönwall's lemma is followed by a section on energy inequalities. To prove that E, defined by (2.8), satisfies an estimate of the form

$$\partial_t E \leq CE + C\|f\|_2 E^{1/2},$$

where the constant C depends on the bounds for A^μ and B, is straightforward, but due to its central importance we give a detailed proof in Section 7.2. To prove similar estimates for derivatives of u does also not require much of an effort. However, in the end, we shall need estimates for a negative number of derivatives in order to prove local existence of solutions to linear equations.

Before proceeding to the question of existence, we provide a primitive uniqueness statement. The argument is based on the idea of integrating an expression of the form

$$\partial_\alpha[e^{-kt} u^t A^\alpha u]$$

over a suitable region and using the equations. By a suitable choice of region $\mathcal{D}$, a suitable choice of k and the assumption that the initial data vanish, one obtains an equality in which one side is non-positive and smaller than

$$-c_1 \int_{\mathcal{D}} e^{-kt} u^t u\, dx,$$

where $c_1 > 0$, and the other side is non-negative. The conclusion is then that $u = 0$ in $\mathcal{D}$.

After these preliminaries we turn to the question of existence. The outline of the argument for proving the existence of a weak solution, cf. (2.11), has already been given. In order to go from a weak solution to a smooth solution, technical arguments are required, some of which are relegated to Appendix A.7. The chapter ends with a proof of the existence of unique smooth solutions to equations of the form (2.4)–(2.5) on all of $\mathbb{R}^{n+1}$ under quite general circumstances.

2.1.3 Linear and non-linear wave equations. Chapter 8 deals with existence of solutions to linear wave equations. We begin by giving a detailed proof of the basic energy inequality. The first section is then concerned with the linear algebra needed. We restrict our attention to symmetric matrix-valued functions with components $g_{\mu\nu}$, $\mu, \nu = 0, \dots, n$, such that $g_{00} < 0$ and such that g_{ij}, $i, j = 1, \dots, n$, are the components of a positive definite matrix. Under these circumstances, we demonstrate that g

is invertible, and if we denote the components of the inverse by $g^{\mu\nu}$, then $g^{00} < 0$ and g^{ij}, $i, j = 1, \dots, n$, are the components of a positive definite matrix. In other words, the energy written down in (2.7) makes sense. After these preliminary observations, we write down an existence and uniqueness result of solutions to linear wave equations. Since linear wave equations of the form we are interested in can be reformulated to symmetric hyperbolic equations, this is immediate from the results presented in Chapter 7.

Chapter 9 is concerned with the question of local existence of solutions to non-linear wave equations. Since symmetric hyperbolic systems are more general, and since some aspects of the arguments involved are easier to carry out in that case, it might seem natural to work in that setting when considering non-linear equations as well. However, in our experience, it is advantageous to work with second order equations when dealing with questions that are global in nature. For this reason, we wish to acquaint the reader with the associated methods. The equations of interest are

$$g^{\mu\nu}\partial_\mu\partial_\nu u = f$$

on $\mathbb{R}^{n+1}$ (though, in the end, we shall only get existence on a subset of the form $(T_-, T_+)\times\mathbb{R}^n$). Here $g^{\mu\nu}$ and f depend on $(t, x) \in \mathbb{R}^{n+1}$, u and ∂u. The first section is concerned with introducing the necessary terminology and proving a coarse uniqueness result. After that, we set up the basic iteration and prove local existence. It is of interest to note that in the case where the coefficients of the highest order derivatives depend on the solution and its first derivatives, the natural spaces for proving convergence of the iteration is

$$C_w\{[0, T], H_{(k+1)}(\mathbb{R}^n, \mathbb{R}^N)\}, \quad C_w\{[0, T], H_{(k)}(\mathbb{R}^n, \mathbb{R}^N)\},$$

for the function and time derivative of the function respectively, where C_w signifies continuity in the weak topology, given that the initial data are such that

$$[u(0, \cdot), \partial_t u(0, \cdot)] \in H_{(k+1)}(\mathbb{R}^n, \mathbb{R}^N) \times H_{(k)}(\mathbb{R}^n, \mathbb{R}^N). \tag{2.15}$$

If the coefficients of the highest order derivatives were instead, say, those of Minkowski space, the natural spaces would be

$$C\{[0, T], H_{(k+1)}(\mathbb{R}^n, \mathbb{R}^N)\}, \quad C\{[0, T], H_{(k)}(\mathbb{R}^n, \mathbb{R}^N)\},$$

and this would simplify the analysis in many respects. It is also of interest to note that the first existence result we state does not guarantee the existence of local smooth solutions. The reason is that the existence time produced by the argument depends on the degree of regularity of the data. Thus, the existence time could very well shrink to zero as the degree of regularity tends to infinity. It is consequently of interest to prove that the maximal existence time of solutions with a given degree of regularity depends on a quantity which is independent of the degree of regularity of the initial data. In the end, we prove that if $T_{k,+}$ is the maximal future existence time corresponding to specific initial data with a regularity of the form (2.15), then either $T_{k,+} = \infty$ or

$$\lim_{t\to T_{k,+}-} \sup_{0\le\tau\le t} m[u](\tau) = \infty,$$

where

$$m[u](t) = \sum_{|\alpha|+j\leq 2} \sup_{x\in\mathbb{R}^n} |\partial^\alpha \partial_t^j u(t,x)|.$$

For this to make sense, we of course need to require that $k > n/2+1$ in view of Sobolev embedding, cf. (2.12). This is an example of a continuation criterion of the type we mentioned in Section 1.2. As a consequence of this and uniqueness, the existence of smooth solutions given smooth initial data is immediate.

We end the chapter by proving a local stability result. The specific form is as follows. Let

$$U_0, U_1, U_{0,l}, U_{1,l} \in C_0^\infty(\mathbb{R}^n, \mathbb{R}^N),$$

assume that $(U_{0,l}, U_{1,l})$ converges to (U_0, U_1) with respect to the space appearing in (2.15) and let the corresponding smooth solutions be u_l and u respectively. Assuming that the maximal existence intervals for u_l and u are $(T_{-,l}, T_{+,l})$ and (T_-, T_+) respectively, the statement is that if $T_0 \in (T_-, T_+)$, then $T_0 \in (T_{-,l}, T_{+,l})$ for l large enough and

$$\lim_{l\to\infty} \left[\|u_l(T_0,\cdot) - u(T_0,\cdot)\|_{H_{(k+1)}} + \|\partial_t u_l(T_0,\cdot) - \partial_t u(T_0,\cdot)\|_{H_{(k)}}\right] = 0.$$

2.2 Geometry, global hyperbolicity and uniqueness

The main point of the part of these notes which concerns geometry, global hyperbolicity and uniqueness is to prove a uniqueness statement for wave equations on Lorentz manifolds phrased in terms of the geometry. However, the different characterizations of global hyperbolicity provided have independent value.

2.2.1 Geometry and global hyperbolicity. In Chapter 10 we remind the reader of basic differential geometry and Lorentz geometry. In particular, we write down Stokes' theorem, derive a divergence theorem that will be of use in proving uniqueness, fix our conventions concerning tensors and curvature, and write down an expression for the Ricci curvature in local coordinates which will be the basic starting point for proving local existence of solutions to Einstein's equations. Furthermore, we recall the basic causality concepts such as $I^\pm(A)$, $J^\pm(A)$, $D^\pm(A)$, etc., as well as the strong causality condition and the concepts of global hyperbolicity, Cauchy hypersurface, etc. In particular, we say that a Lorentz manifold is globally hyperbolic if the strong causality condition holds at each of its points and if, for each pair $p < q$, the set $J(p,q) = J^+(p) \cap J^-(q)$ is compact. This definition may seem a bit technical. However, Lorentz manifolds satisfying this definition have an important property; if there is a causal curve from p to q, then there is a length maximizing causal geodesic from p to q.

In Chapter 11 we relate global hyperbolicity to the existence of a smooth spacelike Cauchy hypersurface, and draw some related conclusions. We start by proving that if there is a smooth spacelike Cauchy hypersurface Σ in a Lorentz manifold (M,g), then

M is diffeomorphic to $\mathbb{R} \times \Sigma$. The reason for this is that there is a complete timelike vector field on M, assuming that M is time orientable. Using the flow of this vector field and the properties of Cauchy hypersurfaces, we deduce the desired conclusion. We then give a proof of a result of Geroch stating that a globally hyperbolic Lorentz manifold has a Cauchy hypersurface. The idea of the proof is as follows. First one defines a "nice" measure on the spacetime, say μ, such that the total volume of the spacetime is 1. Then one defines

$$f_{\pm}(p) = \mu[J^{\pm}(p)], \quad t(p) = \ln \frac{f_{-}(p)}{f_{+}(p)}.$$

Using the global hyperbolicity of the spacetime, one can argue that t is a continuous function such that its range is $\mathbb{R}$ along any inextendible causal curve. Furthermore, it is strictly increasing along any causal curve. As a consequence, all the level sets of t are acausal topological Cauchy hypersurfaces.

After the presentation of the proof of the theorem of Geroch, we describe some of the results of [4], [5], [6]. In particular, we prove that if (M, g) is a globally hyperbolic Lorentz manifold, there is a smooth function τ on M whose gradient is everywhere past directed timelike and which has the property that $\tau \circ \gamma(s) \to \pm\infty$ as $s \to t_{\pm}$, assuming that $\gamma\colon (t_{-}, t_{+}) \to M$ is a future directed inextendible causal curve. Thus, each level set of τ is a smooth spacelike Cauchy hypersurface. By the above observation, one can then draw the conclusion that M is diffeomorphic to $\mathbb{R}\times\Sigma$ where Σ is one of the smooth spacelike Cauchy hypersurfaces. We also give a proof, following the papers mentioned above, of the fact that given a smooth spacelike Cauchy hypersurface Σ in M, there is a function τ which, in addition to the properties mentioned above, has the property that $\tau^{-1}(0) = \Sigma$. We end the chapter by making some auxiliary observations that are of interest in the discussion of constant mean curvature hypersurfaces, cf. Chapter 18.

2.2.2 Uniqueness. In Chapter 12 we consider, among other things, equations of the form

$$\Box_g u + Xu + \kappa u = 0 \tag{2.16}$$

on a globally hyperbolic Lorentz manifold. Here $\Box_g$ is the wave operator associated with the metric g, X is an $N \times N$ matrix of vector fields on M and κ is an $N \times N$ matrix of smooth functions on M. Assuming Ω to be a subset of a spacelike Cauchy hypersurface Σ, u to solve the equation (2.16), and u and its normal derivative to vanish in Ω, we prove that $u = 0$ in $D(\Omega)$. This is the geometrically natural formulation of uniqueness. Let us give an outline of the argument in the case of Minkowski space. Let $p \in D^{+}(\Omega)$. Then $D = J^{-}(p) \cap J^{+}(\Sigma)$ is given by the triangle, depicted in Figure 2.1, and its interior, in the case of 2-dimensional Minkowski space (assuming $p = (1, 0)$ and Σ to be the Cauchy hypersurface defined by $t = 0$). By assumption, the initial data are zero at the base of the triangle, and we wish to prove that the solution vanishes inside the triangle. One way of doing so is to consider the set of points q to the past of p such that $q - p$ is a timelike vector of squared length $c < 0$. Let us call this set Q_c and let $D_c = J^{-}(Q_c) \cap J^{+}(\Sigma)$. One such set D_c is depicted

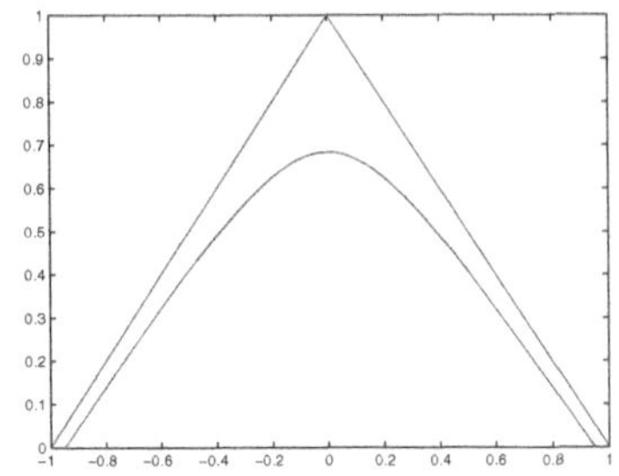

Figure 2.1. The geometric setup for proving uniqueness.

in Figure 2.1; it is the region below the hyperbola and above the base of the triangle. The point is that the closure of the union of the D_c for $c < 0$ coincides with D. In other words, it is enough to prove that the solution vanishes in D_c for every $c < 0$. To achieve this goal, one constructs a suitable vector field, defined using the solution, and integrates the divergence of this vector field over D_c. One then argues that the part of the boundary at the base does not yield any contribution to the resulting boundary integral (since the data are zero at the base), that the contribution from the part of the boundary associated with $Q_c \cap J^+(\Sigma)$, i.e. the hyperbola depicted in Figure 2.1, is non-positive and that the divergence of the vector field is bounded from below by a positive constant times $|u|^2$. This leads to the conclusion that both the boundary and the bulk terms have to vanish, so that $u = 0$ in D_c. In the case of a general Lorentz manifold, the situation is of course somewhat more complicated, but the basic idea is the same if one restricts one's attention to suitable convex neighbourhoods. However, there are several technical observations needed in order to make sense of the ingredients, and this is the starting point of Chapter 12. To go from these local observations, relevant in a convex neighbourhood, to global results, requires some additional technical observations. To prove existence of solutions to tensor wave equations is then, given the uniqueness result and the existence results concerning linear wave equations, quite straightforward.

2.3 General relativity

The general relativity part of these notes contains a proof of the fact that there is a maximal globally hyperbolic development of initial data to Einstein's equations. This, of course, constitutes the main goal, but we also include a detailed proof of Cauchy stability in the case of Einstein's equations. The latter result is of interest in various contexts. Say, for the sake of argument, that we wish to prove future global non-linear stability of a certain class of spatially homogeneous spacetimes. Say, furthermore, that all the spacetimes in this class have future asymptotics such that we have a stability result which is valid starting at a late enough hypersurface of spatial homogeneity. The only thing missing in order for us to be allowed to go from the statement that small neighbourhoods of the initial data at a late enough hypersurface of spatial homogeneity yield future causally geodesically complete spacetimes, to the statement that small neighbourhoods of the initial data at an arbitrary hypersurface of spatial homogeneity yield future causally geodesically complete spacetimes is Cauchy stability. Another example of the use of Cauchy stability is given by an elementary application to the so-called Bianchi IX vacuum spacetimes, cf. the last part of these notes. These are the maximal globally hyperbolic developments that result by specifying left invariant vacuum initial data on the 3-dimensional unimodular Lie group SU(2). Due to a result of Lin and Wald, cf. [56], all these spacetimes recollapse, i.e., they are future and past timelike geodesically incomplete (we shall devote Chapter 21 to a proof of this fact). Furthermore, in each of them, there is one Cauchy hypersurface on which the trace of the second fundamental form has one sign and another on which it has the opposite sign. By Cauchy stability, one concludes that there is an open set of initial data such that the same holds for the corresponding maximal globally hyperbolic developments. As a consequence of Hawking's theorem, cf. Theorem 55A, p. 431 of [65], this implies that these developments recollapse. Thus, given Bianchi IX vacuum initial data, there is an open neighbourhood of these data that have recollapsing maximal globally hyperbolic developments (for a detailed proof, see Chapter 21). However, there is one point that should be kept in mind in this context, and that is the fact that for the above arguments to make sense, one has to prove the existence of initial data close to the specified Bianchi IX vacuum initial data.

2.3.1 Constraint equations and local existence. We begin Chapter 13 by writing down Einstein's equations. As mentioned above, we shall restrict our attention to matter models of non-linear scalar field type. The idea of the initial value formulation of Einstein's equation is that the initial data should correspond to the metric, second fundamental form, scalar field and normal derivative of the scalar field induced on a spatial hypersurface in the spacetime one wishes to construct. This perspective immediately leads to the problem that the initial data cannot be specified freely; they have to satisfy certain equations that are referred to as the *constraint equations*. In Section 13.2, we derive these equations from the Gauß and Codazzi equations, cf. [65]. After that, we proceed to the question of gauge choices. In coordinates, the Ricci tensor takes the form

$$R_{\mu\rho} = -\frac{1}{2}g^{\alpha\beta}\partial_\alpha\partial_\beta g_{\mu\rho} + \nabla_{(\mu}\Gamma_{\rho)} + g^{\alpha\beta}g^{\gamma\delta}[\Gamma_{\alpha\gamma\mu}\Gamma_{\beta\delta\rho} + \Gamma_{\alpha\gamma\mu}\Gamma_{\beta\rho\delta} + \Gamma_{\alpha\gamma\rho}\Gamma_{\beta\mu\delta}],$$

cf. (10.13). Here,

$$\Gamma_{\alpha\beta\gamma} = \frac{1}{2}[\partial_\alpha g_{\gamma\beta} + \partial_\gamma g_{\alpha\beta} - \partial_\beta g_{\alpha\gamma}], \quad \Gamma^\mu_{\alpha\gamma} = g^{\mu\beta}\Gamma_{\alpha\beta\gamma}$$

are the Christoffel symbols and we have used the notation

$$\nabla_\mu\Gamma_\rho = \partial_\mu\Gamma_\rho - \Gamma^\alpha_{\mu\rho}\Gamma_\alpha, \quad \nabla_{(\mu}\Gamma_{\rho)} = \frac{1}{2}(\nabla_\mu\Gamma_\rho + \nabla_\rho\Gamma_\mu)$$

and

$$\Gamma_\alpha = g^{\mu\nu}\Gamma_{\mu\alpha\nu}.$$

Unfortunately, in coordinates, the Ricci tensor, considered as a differential operator acting on the metric, is not hyperbolic. Due to the diffeomorphism invariance, hyperbolicity is, however, of course not to be expected. To overcome this problem, we shall use an idea from [41]. The problematic term in the above expression for the Ricci tensor is the second one, $\nabla_{(\mu}\Gamma_{\rho)}$. The reason for this is that the first term is a perfectly good hyperbolic differential operator acting on the metric and the last term, consisting of the sum of the products of the Christoffel symbols, only involves first derivatives of the metric. In other words, if the second term were absent, we would not have any problems. For this reason, the idea is to replace the Ricci tensor in the equations with

$$\hat{R}_{\mu\nu} = R_{\mu\nu} + \nabla_{(\mu}\mathcal{D}_{\nu)},$$

where

$$\mathcal{D}_{\mu} = F_{\mu} - \Gamma_{\mu},$$

cf. (14.4). The object F_{μ} is referred to as a gauge source function and it is allowed to depend on the metric and the spacetime coordinates but not on the derivatives of the metric. Then $\hat{R}_{\mu\nu}$, considered as a differential operator acting on the metric, is hyperbolic. As far as local existence is concerned, the particular choice of gauge source function is not of any fundamental importance, but some choices are more convenient than others. In particular, there are choices of F_{μ} such that $\mathcal{D}_{\mu}$ becomes the components of a tensor, which is quite convenient. The resulting equations are then

$$\hat{R}_{\mu\nu} - \nabla_{\mu}\phi\nabla_{\nu}\phi - \frac{2}{n-1}V(\phi)g_{\mu\nu} = 0, \tag{2.17}$$

$$\nabla^{\mu}\nabla_{\mu}\phi - V'(\phi) = 0. \tag{2.18}$$

Note that only certain components of the metric and the time derivative of the metric are specified by the initial data. When setting up initial data for the modified system, there is thus an additional freedom. However, there is also a problem. We now have a system which we can solve, but the relation of this system to the one we are interested in is not so clear. The point is that (2.17)–(2.18) imply a wave equation for $\mathcal{D}_{\mu}$, so that if we can set up the initial data in such a way that the initial data for $\mathcal{D}_{\mu}$ are zero, we obtain a solution to the original problem. In part this is achieved by the above mentioned freedom; it leads to the possibility of ensuring that $\mathcal{D}_{\mu} = 0$ originally. However, to ensure that the normal derivative of $\mathcal{D}_{\mu}$ vanishes originally, the constraint equations are needed. To sum up, the problem of local existence is reduced to the problem of solving (2.17)–(2.18) for correctly set up initial data, and this is a problem which essentially reduces to the type of non-linear wave equations we have already considered in these notes.

It is of interest to note that this argument is essentially the same as the original one by Choquet-Bruhat [39]; the only difference is that she used the gauge source function $F_{\mu} = 0$. This choice is referred to as wave coordinates and is sometimes useful even for global problems, cf. [57], [58]. It is also of interest to note that the same kind of problem that arose above appears in the context of the Ricci flow and the context of Cheeger–Gromov convergence/collapse theory. In both these settings it would be convenient if the Ricci tensor were an elliptic operator acting on the metric. However, this is not true, and in the former case this problem is overcome by considering a modified system and in the latter it is overcome by the introduction of harmonic coordinates.

In view of the above observations, the proof of local existence of solutions to Einstein's equation is straightforward. However, in the proof of the existence of a maximal globally hyperbolic development of initial data, there is one fact which is needed and which in the end turns out to be more complicated to prove than one might expect. We need to know that two developments are extensions of a common development. In order to be able to prove this, one has to construct the common development, which is not so complicated, and to prove the existence of an isometry from the common development into the two developments with which one started. The problem is that the construction of the isometry is rather technical.

2.3.2 Cauchy stability. We begin the chapter on Cauchy stability by introducing Sobolev spaces on manifolds. After that, we introduce the class of background solutions we shall be considering and prove Cauchy stability. Note that comparing two developments of different initial data by necessity has a non-canonical element to it. The reason for this is that in order to be able to make the comparison, one needs to choose a diffeomorphism from a subset of one development into the other and pull back the metric and scalar field using this diffeomorphism. However, as far as the type of practical applications mentioned above are concerned, this is not a major problem.

The actual proof of Cauchy stability is quite long. The reason for this is that setting up the initial data, describing the geometric set up, and writing down all the technical details requires some time. The length is thus not caused by any fundamental difficulty.

2.3.3 Existence of a maximal globally hyperbolic development. In Chapter 16 we prove the theorem that motivates these notes. We begin by introducing the material from set theory that is needed and then define the concept of a maximal globally hyperbolic development. The proof, by and large, follows the arguments of [10] and is in part based on a Zorn's lemma type argument. However, it should be noted that the Zorn's lemma part only gives a maximal element. The hard and important part of the proof is to show that this maximal element is an extension of every globally hyperbolic development.

2.4 Pathologies, strong cosmic censorship

The last part of these notes consists of a study of the maximal globally hyperbolic developments (MGHD's) corresponding to left invariant vacuum initial data on unimodular Lie groups. However, in the process we also prove some general results, such as uniqueness of constant mean curvature hypersurfaces in globally hyperbolic and spatially compact Lorentz manifolds satisfying the timelike convergence condition. One motivation for the study is the following question: are there initial data for Einstein's equations such that the MGHD is extendible? The answer to this question is yes. In fact, as we shall show, there are left invariant vacuum initial data on SU(2) (a unimodular Lie group) such that the corresponding MGHD has two extensions which are inextendible, and in this sense maximal but not isometric. The fact that there are inequivalent maximal extensions means that the initial data do not uniquely determine a maximal development. In this sense, the general theory of relativity is not deterministic. Nevertheless, the known examples are such that one is naturally led to the so-called strong cosmic censorship conjecture (SCC), stating that for generic initial data, the MGHD is inextendible. To hope to prove this conjecture in all generality is not realistic at this time. A more realistic problem would be to consider the same question in classes of initial data satisfying a given symmetry condition. In Section 17.2 we briefly describe some such results, the motivation being to illustrate some of the techniques and different notions of inextendibility and genericity that have been used. In later chapters we consider the question of extendibility of the MGHD in the class of vacuum initial data on unimodular Lie groups. In fact, we characterize the initial data for which the Kretschmann scalar, $R_{\alpha\beta\gamma\delta} R^{\alpha\beta\gamma\delta}$, is unbounded in the incomplete directions of causal geodesics in the corresponding MGHD (in which case the MGHD is inextendible) and, in the remaining cases, explicitly construct extensions. However, in the case of the Lie groups SU(2) and $\widetilde{\mathrm{SL}}(2,\mathbb{R})$, the universal covering group of $\mathrm{SL}(2,\mathbb{R})$, we shall in part refer to the research papers for the analysis of the asymptotics of the relevant ODE's.

In Chapter 17 we sketch the proof of the existence of inequivalent extensions and give a more complete outline of the contents of the last part of these notes.

Part I

Background from the theory of partial differential equations

3 Functional analysis

The goal of the PDE part of these notes is to prove that there is uniqueness and local existence of solutions to non-linear wave equations, given initial data. The proof of local existence is based on the construction of a sequence of approximate solutions and a proof of the fact that the sequence converges to a solution to the equation of interest. The approximate solutions are obtained by solving linear wave equations, and consequently, it is necessary to be able to solve such equations. As we mentioned in the outline, we shall prove the existence of solutions to linear wave equations by reducing the problem to that of solving linear symmetric hyperbolic equations. The essential step in the resolution of the latter problem is to use energy estimates together with the Hahn–Banach theorem in order to obtain a solution, in a weak sense, in the dual of a function space of the form $L^1([0,T],X)$, where X is a Hilbert space. In the end, X will be a Sobolev space corresponding to a negative number of derivatives. In order for this information to be useful, it is necessary to identify what the dual of $L^1([0,T],X)$ is when X is a complex Hilbert space, and that is the purpose of the present chapter. Before characterizing the dual, it is necessary to give a definition of what is meant by $L^1([0,T],X)$, something we provide in Section 3.1. In Section 3.2, we then prove that the dual of $L^1([0,T],X)$ is $L^\infty([0,T],X)$ when X is a separable complex Hilbert space.

3.1 Measurability

We shall be interested in function spaces of the form $L^p(S,H)$, where H is a Hilbert space. We thus need to define the concept of measurability for a function $f : S \to H$. We shall use the terminology of [86], p. 130.

Definition 3.1. Let $(S,\mathcal{A},m)$ be a measure space and $x : S \to X$, where X is a Banach space. Then x is said to be *weakly measurable* if $f[x(s)]$ defines an $\mathcal{A}$-measurable function for every $f \in X^*$. The function x is said to be *finitely-valued* if it is constant $\neq 0$ on a finite number of disjoint $A_j \in \mathcal{A}$ with $m(A_j) < \infty$ and $x = 0$ on $S - \bigcup_j A_j$. The function x is said to be *strongly measurable* if there exists a sequence of finitely-valued functions converging pointwise to x m-almost everywhere on S.

Remark 3.2. We shall take for granted that all measure spaces are *complete*, meaning that all subsets of measurable subsets with measure zero are measurable.

We have two different concepts of measurability, but in the end we shall only be interested in σ-finite measure spaces and separable Banach spaces, for which the two concepts are equivalent, cf. p. 131 of [86]. For the sake of completeness, we include a proof in the Appendix, cf. Theorem A.1. Recall that a Banach space is said to be *separable* if it contains a countable dense subset. A measure space $(S,\mathcal{A},m)$ is said to be σ-finite if there is a sequence $A_n \in \mathcal{A}$, $n = 1,2,\dots$, such that $m(A_n) < \infty$ and $\bigcup_n A_n = S$.

It is of interest to note that, under the above assumptions, if $\{x_n\}$ is a sequence of strongly measurable functions converging a.e. to a function x, then x is strongly measurable. The reason is that if $f \in X^*$, then $f \circ x_n$ is measurable and converges a.e. to $f \circ x$. By standard measure and integration theory, $f \circ x$ is measurable, so that x is weakly and thus strongly measurable. Note that if $x\colon S \to X$ is strongly measurable, then $\|x\|_X$ is a measurable function from S to $\mathbb{R}$. Thus the following definition makes sense.

Definition 3.3. Let $(S, \mathcal{A}, m)$ be a σ-finite measure space, let X be a separable Banach space and let $x\colon S \to X$ be a strongly measurable function. For $1 \le p < \infty$, we say that $x \in \mathcal{L}^p(S, X)$ if $\|x\|_X^p$ is integrable and we write

$$\|x\|_p = \left(\int_S \|x(s)\|_X^p \, dm\right)^{1/p}.$$

If $\|x\|_X$ is essentially bounded, we say that $x \in \mathcal{L}^\infty(S, X)$ and write

$$\|x\|_\infty = \operatorname*{ess\,sup}_{s \in S} \|x(s)\|_X.$$

We define $L^p(S, X)$ to be the set of equivalence classes of elements of $\mathcal{L}^p(S, X)$, two elements being equivalent if the set on which they differ has measure zero.

Remark 3.4. The objects $\|\cdot\|_p$ are norms on $L^p(S, X)$, the argument being the same as when the functions are real-valued. That $L^p(S, X)$ are Banach spaces with these norms for $1 \le p \le \infty$ follows by an argument that is identical to the proof of Theorem 3.11 of [79]; all one needs to do is to replace $|\cdot|$ by $\|\cdot\|_X$.

3.2 Dualities

Proposition 3.5. *Let $T > 0$ and*

$$X_n = L^1\{[0, T], \mathbb{C}^n\}, \quad Y_n = L^\infty\{[0, T], \mathbb{C}^n\},$$

where we take the standard norm for $\mathbb{C}^n$ in the definition of the norm for these spaces. Furthermore, for $f \in X_n$, $g \in Y_n$, let

$$\langle f, g\rangle = \int_0^T f(t) \cdot \bar{g}(t)\, dt.$$

Given $F \in X_n^*$, there is a $g \in Y_n$ such that

$$F(f) = \langle f, g\rangle \tag{3.1}$$

for all $f \in X_n$, and

$$\|F\|_{X_n^*} = \|g\|_{Y_n}. \tag{3.2}$$

Proof. Let $e_j \in \mathbb{C}^n$ be the vector whose jth component is 1 and all of whose other components are zero. Let $Z = L^1([0,T],\mathbb{C})$. Note that for every $j = 1,\dots,n$,

$$F_j(f) = F(fe_j)$$

defines an element of Z^*. By Theorem 6.16 of [79], there is a $g_j \in L^\infty([0,T],\mathbb{C})$ such that for all $f \in Z$,

$$F_j(f) = \int_0^T fg_j\,dt.$$

Let

$$g = \sum_{j=1}^n \bar{g}_j e_j.$$

Then (3.1) holds. However, we do not get the optimal result (3.2) as far as the norms are concerned. The equality (3.1) immediately gives the inequality $\|F\|_{X_n^*} \le \|g\|_{Y_n}$. If $\|g\|_{Y_n} = 0$, we are thus done, so assume not. For $0 < \varepsilon < \|g\|_{Y_n}$, let $\mathcal{A}_\varepsilon$ be the subset of $[0,T]$ such that $|g(t)| \ge \|g\|_{Y_n} - \varepsilon$. Then $\mu(\mathcal{A}_\varepsilon) > 0$ and we can define

$$f_\varepsilon = \frac{1}{\mu(\mathcal{A}_\varepsilon)}\frac{g(t)}{|g(t)|}\chi_{\mathcal{A}_\varepsilon}.$$

Note that this function is measurable and that

$$\|f_\varepsilon\|_{X_n} = 1.$$

Adding up the above observations, (3.1) yields that

$$\|F\|_{X_n^*} \ge |\langle f_\varepsilon, g\rangle| \ge \|g\|_{Y_n} - \varepsilon.$$

Since this is true for any $\varepsilon > 0$, we conclude that (3.2) holds. □

Proposition 3.6. *Let H be an infinite dimensional, separable complex Hilbert space with inner product $(\,\cdot\,,\,\cdot\,)$. Let $T > 0$ and*

$$X = L^1\{[0,T],H\}, \quad Y = L^\infty\{[0,T],H\},$$

where we take the standard norm for H in the definition of the norm for these spaces, and define, for $f \in X$, $g \in Y$,

$$\langle f,g\rangle = \int_0^T (f(t),g(t))\,dt.$$

Given $F \in X^*$, there is a $g \in Y$ such that $F(f) = \langle f,g\rangle$ holds for all $f \in X$. Furthermore

$$\|F\|_{X^*} = \|g\|_Y. \tag{3.3}$$

Remark 3.7. As a consequence of the strong measurability of f and g, it is clear that $(f(t), g(t))$ defines a measurable function, which is furthermore bounded by an integrable function.

Proof. Since H is separable, it has a countable ON basis $\{e_i\}$, $i = 1, 2, \ldots$, see Lemma A.7. We can define an element F_N of X_N^*, cf. the terminology of the previous proposition, by

$$F_N(f) = F\Big(\sum_{i=1}^{N} f_i e_i\Big).$$

We obtain an $h_N \in Y_N$ such that

$$F_N(f) = \int_0^T f \cdot \bar{h}_N \, dt$$

for all $f \in X_N$. If we let

$$g_N = \sum_{i=1}^{N} h_{N,i} e_i,$$

this equality reads

$$F(f) = \langle f, g_N \rangle \tag{3.4}$$

for all $f \in Z_N$, where

$$Z_N = \big\{ \textstyle\sum_{j=1}^{N} f_j e_j : f \in X_N \big\},$$

and we take it to be understood that f_j are the components of f. Furthermore,

$$\|g_N\|_Y = \Big\| \Big(\sum_{i=1}^{N} |h_{N,i}|^2 \Big)^{1/2} \Big\|_\infty \le \|F\|_{X^*}. \tag{3.5}$$

Note that if $i \le N_1 \le N_2$, then $h_{N_1,i} = h_{N_2,i}$ a.e. due to (3.4) and an argument similar to the one given at the end of the previous proposition. Thus we can define $h_i = h_{N,i}$ for any $i \ge N$. What remains to be proved is that the definition

$$g = \sum_{i=1}^{\infty} h_i e_i$$

makes sense, that $F(f) = \langle f, g \rangle$ and that (3.3) holds. Since we have (3.5), we can redefine h_i on a set of measure zero in order that

$$\sum_{i=1}^{N} |h_i(t)|^2 \le \|F\|_{X^*}^2$$

for all $t \in [0, T]$ and all N. For a fixed t, the left-hand side is an increasing sequence of numbers. Consequently, for every $t \in [0, T]$,

$$\sum_{i=1}^{\infty} |h_i(t)|^2 \leq \|F\|_{X^*}^2 \tag{3.6}$$

and the left-hand side is an absolutely convergent series. The sequence of functions

$$g_N(t) = \sum_{i=1}^{N} h_i(t) e_i$$

is consequently a Cauchy sequence in H for every fixed t. Let us call the limit $g(t)$. Note that it is strongly measurable, that $\|g\|_Y \leq \|F\|_{X^*}$ due to (3.6) and that $(g(t), e_i) = h_i(t)$. We need to prove that $F(f) = \langle f, g\rangle$ holds. If $f \in Z_N$, we have

$$\langle f, g\rangle = \langle f, g_N\rangle = F(f). \tag{3.7}$$

For $f \in X$, let us define

$$f_N(t) = \sum_{i=1}^{N} (f(t), e_i) e_i.$$

Note that $f_N \to f$ pointwise and that $\|f_N\|_X \leq \|f\|_X$, cf. Lemma A.7 and its proof. Let us define

$$r_N(t) = \|(f - f_N)(t)\|_H,$$

which converges to zero pointwise and is bounded by $2\|f(t)\|_H$. By Lebesgue's dominated convergence theorem, we conclude that r_N converges to zero in the space $L^1([0, T])$, so that $f_N \to f$ in X. Since both the left- and right-hand sides of (3.7) define elements of X^* and the equality holds for f_N, we conclude that equality holds for f. Since the equality proves that $\|F\|_{X^*} \leq \|g\|_Y$ and we already have the opposite inequality, the last statement of the proposition follows. □

4 The Fourier transform

Consider a solution u to the standard linear wave equation

$$-\partial_t^2 u + \Delta u = 0 \tag{4.1}$$

on $\mathbb{R}^{n+1}$, where Δ is the standard Laplacian on $\mathbb{R}^n$. Assuming u to be of a high enough degree of regularity and to decay fast enough, one can define

$$E_k = \frac{1}{2} \sum_{|\alpha| \le k} \int_{\mathbb{R}^n} [(\partial^\alpha \partial_t u)^2 + |\nabla \partial^\alpha u|^2] \, dx,$$

where ∇ is the ordinary gradient on $\mathbb{R}^n$. Furthermore, using (4.1) and integration by parts, one obtains the conclusion that the time derivative of E_k is zero. As a consequence, it is clear that a regularity of the initial data corresponding to $k+1$ spatial derivatives of u in L^2 and k spatial derivatives of $\partial_t u$ in L^2 is preserved by the evolution. Consequently, it is natural to consider initial data with this degree of regularity. In the case of a non-linear wave equation, E_k need of course not be conserved. However, it is possible to bound it in a neighbourhood of the starting time if it is bounded originally, assuming k to be large enough. It is natural to ask what happens if one specifies initial data such that, say, u is $k+1$ times continuously differentiable and $\partial_t u$ is k times continuously differentiable originally. That this degree of regularity need not be preserved is illustrated by (1.6) and (1.7) on p. 5 of [82] and by Theorem 1.1 on p. 6 of [82]. In other words, due to the properties of the wave equation, it is more natural to consider functions with the property that up to k derivatives of it are bounded in L^2 than functions with the property that up to k derivatives of it are bounded in L^∞.

The proof of local existence of solutions to non-linear wave equations is based on the construction of a Cauchy sequence of approximate solutions, and so, in order to prove convergence (and thereby existence of solutions to the non-linear problem), it is necessary to have complete function spaces. In other words, instead of considering k times continuously differentiable functions u with the property that

$$\|u\|_{H^k} = \Big(\sum_{|\alpha| \le k} \int_{\mathbb{R}^n} (\partial^\alpha u)^2 \, dx \Big)^{1/2} \tag{4.2}$$

is bounded, it is necessary to take the completion of this space with respect to the norm $\|\cdot\|_{H^k}$. This leads to the introduction of the Sobolev space $H^k(\mathbb{R}^n)$, something we shall discuss in detail in the following chapter.

The above discussion illustrates that in connection with the study of solutions to (4.1) and to related equations, Sobolev spaces appear naturally. However, one is often interested in obtaining solutions that are k times continuously differentiable or even smooth. In that context, it is of interest to ask if it is possible to relate Sobolev space regularity to classical differentiability. It turns out that this is possible; there is an inequality which goes under the name of Sobolev embedding which says that if

$u \in H^k(\mathbb{R}^n)$, then u is l times continuously differentiable assuming $k > l + n/2$, cf. Theorem 6.5. There are many ways of proving this inequality, but the easiest one is via the Fourier transform; using the Fourier inversion formula (4.9) and Hölder's inequality, one obtains

$$\begin{aligned}|u(x)| &= \left|\frac{1}{(2\pi)^n}\int_{\mathbb{R}^n} e^{ix\cdot\xi}\hat{u}(\xi)\,d\xi\right| \\ &\leq \frac{1}{(2\pi)^n}\left(\int_{\mathbb{R}^n}(1+|\xi|^2)^{-k}\,d\xi\right)^{1/2}\left(\int_{\mathbb{R}^n}(1+|\xi|^2)^k|\hat{u}(\xi)|^2\,d\xi\right)^{1/2} \\ &\leq C_k\left(\int_{\mathbb{R}^n}(1+|\xi|^2)^k|\hat{u}(\xi)|^2\,d\xi\right)^{1/2}\end{aligned}$$

for some constant C_k, assuming that $k > n/2$. Using the basic properties of the Fourier transform, such as Parseval's identity and that fact that, disregarding constants, multiplication by ξ^j on the Fourier side corresponds to differentiation with respect to x^j on the function side, one can conclude that

$$\left(\int_{\mathbb{R}^n}(1+|\xi|^2)^k|\hat{u}(\xi)|^2\,d\xi\right)^{1/2} \leq C_k\|u\|_{H^k}$$

for some constant $C_k > 0$. Combining the above two inequalities, one obtains the simplest version of Sobolev embedding. In the proof of Theorem 6.5, we provide the details of the argument.

The main purpose of the present chapter is to provide enough information concerning the Fourier transform to make it possible to carry out the above argument. In Section 4.1, we introduce the class of Schwartz functions, a class which is particularly convenient to use; the class of smooth functions with compact support is, for instance, inappropriate since it is not preserved by the Fourier transform. In the end it turns out that the Fourier transform is a homeomorphism of the class of Schwartz functions to itself. In Section 4.1, we also define the Fourier transform and prove that it is a continuous map from the Schwartz class to itself. In Section 4.2, we then prove the Fourier inversion formula as well as Parseval's formula.

The main references for this chapter are [45] and [49].

4.1 Schwartz functions, the Fourier transform

Before defining the Fourier transform and writing down its basic properties, it is natural to define the set of Schwartz functions.

Definition 4.1. The Schwartz class $\mathcal{S}(\mathbb{R}^n)$ is the set of $f \in C^\infty(\mathbb{R}^n, \mathbb{C})$ (i.e., smooth, complex-valued functions) such that for every pair of multiindices α and β, there is a real constant $C_{\alpha,\beta}$ such that

$$|x^\alpha \partial^\beta f(x)| \leq C_{\alpha,\beta} \tag{4.3}$$

for all $x \in \mathbb{R}^n$.

Remark 4.2. Concerning the notation and the concept of a multiindex we refer the reader to Section A.1.

Note that in the above definition, we use the notation

$$x^\alpha = (x^1)^{\alpha_1} \dots (x^n)^{\alpha_n}$$

for $x \in \mathbb{R}^n$ and a multiindex α. For multiindices α, β and $f \in \mathcal{S}(\mathbb{R}^n)$, let us define

$$p_{\alpha,\beta}(f) = \sup_{x \in \mathbb{R}^n} |x^\alpha \partial^\beta f(x)|.$$

These objects are *seminorms* meaning that they are real-valued and that

$$\begin{aligned} p_{\alpha,\beta}(f + g) &\leq p_{\alpha,\beta}(f) + p_{\alpha,\beta}(g) \\ p_{\alpha,\beta}(zf) &= |z| p_{\alpha,\beta}(f) \end{aligned}$$

for all $f, g \in \mathcal{S}(\mathbb{R}^n)$ and all $z \in \mathbb{C}$. Since the set of pairs of multiindices is countable, we can think of these seminorms as being indexed by the positive integers, and we shall write p_i. For $f, g \in \mathcal{S}(\mathbb{R}^n)$, let us define

$$d(f, g) = \sum_{k=1}^{\infty} 2^{-k} \frac{p_k(f - g)}{1 + p_k(f - g)}. \tag{4.4}$$

Lemma 4.3. *The function d given by* (4.4) *defines a metric on $\mathcal{S}(\mathbb{R}^n)$.*

Remark 4.4. In fact, $\mathcal{S}(\mathbb{R}^n)$ with the metric d is a *Fréchet* space, meaning that it is a topological vector space, cf. [80], that the metric is complete, translation invariant, i.e. $d(f + h, g + h) = d(f, g)$, and that there is a local base for the topology whose members are convex. A *local base* is a collection $\mathcal{B}$ of neighbourhoods of the origin such that every neighbourhood of the origin has to contain a member of $\mathcal{B}$ (in our setting the origin is the function 0). These facts will however not be of any relevance to us. We refer the reader interested in a more complete discussion of these aspects to [80].

Proof. Note that $p_{0,0}(f - g) = 0$ implies $f = g$. The only non-trivial aspect of the proof that d defines a metric is the triangle inequality. This follows from the fact that $x/(1 + x)$ is an increasing function for $x > 0$ and the inequality

$$\frac{x + y}{1 + x + y} \leq \frac{x}{1 + x} + \frac{y}{1 + y}$$

which holds for all non-negative real numbers x and y. □

Let us define the Fourier transform.

Definition 4.5. Let $f \in \mathcal{S}(\mathbb{R}^n)$. Define the Fourier transform of f, $\hat{f}$, by

$$\hat{f}(\xi) = \int_{\mathbb{R}^n} e^{-ix \cdot \xi} f(x)\, dx. \tag{4.5}$$

Since f is a Schwartz function, there is for every k a constant C_k such that

$$|f(x)| \leq C_k(1 + |x|^2)^{-k},$$

cf. (4.3). Consequently (4.5) makes sense. If $f \in \mathcal{S}(\mathbb{R}^n)$, then $\hat{f}$ is a smooth function and one can differentiate under the integral sign. As a consequence,

$$\partial_\xi^\alpha \hat{f}(\xi) = \int_{\mathbb{R}^n} (-ix)^\alpha e^{-ix\cdot\xi} f(x)\, dx. \tag{4.6}$$

By integration by parts, we also obtain

$$\xi^\alpha \hat{f}(\xi) = \int_{\mathbb{R}^n} i^{|\alpha|} \partial_x^\alpha (e^{-ix\cdot\xi}) f(x)\, dx = \int_{\mathbb{R}^n} e^{-ix\cdot\xi} (-i)^{|\alpha|} \partial_x^\alpha f(x)\, dx. \tag{4.7}$$

Combining these two observations, we have the following conclusion.

Lemma 4.6. *If $f \in \mathcal{S}(\mathbb{R}^n)$, then $\hat{f} \in \mathcal{S}(\mathbb{R}^n)$ and the function $\mathcal{F} : \mathcal{S}(\mathbb{R}^n) \to \mathcal{S}(\mathbb{R}^n)$ defined by $\mathcal{F}(f) = \hat{f}$ is continuous.*

Proof. Let α and β be multiindices and consider

$$\xi^\alpha \partial_\xi^\beta \hat{f} = \int_{\mathbb{R}^n} e^{-ix\cdot\xi} (-i)^{|\alpha|} \partial_x^\alpha [(-i)^{|\beta|} x^\beta f]\, dx,$$

where we have used (4.6) and (4.7). We obtain

$$\sup_{\xi \in \mathbb{R}^n} |\xi^\alpha \partial_\xi^\beta \hat{f}| \leq \int_{\mathbb{R}^n} (1 + |x|^2)^{-n}\, dx \sup_{x \in \mathbb{R}^n} |(1 + |x|^2)^n \partial_x^\alpha [x^\beta f]|. \tag{4.8}$$

The right-hand side is bounded since $f \in \mathcal{S}(\mathbb{R}^n)$. Thus $\hat{f} \in \mathcal{S}(\mathbb{R}^n)$. In order to prove continuity, note that $f_n \to f$ with respect to d defined in (4.4) if and only if $p_l(f_n - f) \to 0$ for all seminorms p_l. By considering (4.8) with f replaced by $f - f_n$, we see that $\hat{f}_n \to \hat{f}$. □

4.2 The Fourier inversion formula

Our next goal is to prove that it is possible to invert the Fourier transform. In fact, we wish to prove that for $f \in \mathcal{S}(\mathbb{R}^n)$,

$$f(x) = \frac{1}{(2\pi)^n} \int_{\mathbb{R}^n} e^{ix\cdot\xi} \hat{f}(\xi)\, d\xi. \tag{4.9}$$

Note that by arguments similar to ones given above, this relation defines a continuous map from $\mathcal{S}(\mathbb{R}^n)$ to itself. After having proven (4.9), we can thus conclude that the Fourier transform is a homeomorphism of $\mathcal{S}(\mathbb{R}^n)$ to itself. As a preparation for the proof of (4.9), let us prove the following lemma.

Lemma 4.7. *Let f_0 be defined by*

$$f_0(x) = \exp\Big(-\frac{1}{2}|x|^2\Big). \tag{4.10}$$

Then

$$\hat{f}_0(\xi) = (2\pi)^{n/2} \exp\Big(-\frac{1}{2}|\xi|^2\Big).$$

Proof. Note first that

$$\int_{\mathbb{R}^n} f_0(x)\,dx = (2\pi)^{n/2}. \tag{4.11}$$

Compute

$$\begin{aligned}\hat{f}_0(\xi) &= \int_{\mathbb{R}^n} e^{-ix\cdot\xi} \exp\Big(-\frac{1}{2}|x|^2\Big)\,dx \\ &= \exp\Big(-\frac{1}{2}|\xi|^2\Big) \int_{\mathbb{R}^n} \exp\Big[-\frac{1}{2}(x+i\xi)\cdot(x+i\xi)\Big]\,dx\end{aligned}$$

Note that

$$\begin{aligned}&\int_{\mathbb{R}^n} \exp\Big[-\frac{1}{2}(x+i\xi)\cdot(x+i\xi)\Big]\,dx \\ &\qquad = \prod_{j=1}^{n} \int_{-\infty}^{\infty} \exp\Big[-\frac{1}{2}(x^j+i\xi^j)\cdot(x^j+i\xi^j)\Big]\,dx^j.\end{aligned}$$

Each factor on the right-hand side is an integral in the complex plane, and by standard methods of complex analysis, we are allowed to shift the contour $t + i\xi^j$, $t \in \mathbb{R}$, to the real axis. By (4.11), we obtain the result. □

Due to Lemma 4.7, we conclude that (4.9) holds for $f = f_0$ (in order to obtain this conclusion we have used the fact that $f_0(x) = f_0(-x)$). Let us prove (4.9) in all generality.

Theorem 4.8. *For all $f \in \mathcal{S}(\mathbb{R}^n)$, we have* (4.9).

Proof. Assume first that $f(0) = 0$. Then

$$f(x) = \int_0^1 \frac{d}{dt}[f(tx)]\,dt = \sum_j x^j \int_0^1 (\partial_j f)(tx)\,dt = \sum_j x^j g_j(x)$$

for some $g_j \in C^\infty(\mathbb{R}^n, \mathbb{C})$, $j = 1, \dots, n$. Let $\phi \in C_0^\infty(\mathbb{R}^n)$ equal 1 on $\bar{B}_1(0)$. We can then write

$$f(x) = \phi(x) f(x) + [1-\phi(x)] f(x) = \sum_j x^j \phi(x) g_j(x) + \sum_j x^j \frac{x^j[1-\phi(x)]f}{|x|^2}.$$

Note that

$$\phi g_j, \quad \frac{x^j[1-\phi(x)]f}{|x|^2}$$

are both in $\mathcal{S}(\mathbb{R}^n)$. Thus, there are $h_j \in \mathcal{S}(\mathbb{R}^n)$, $j = 1, \dots, n$, such that

$$f = \sum_j x^j h_j.$$

By Fourier transforming this equality and using (4.6), we obtain

$$\hat{f}(\xi) = \sum_j i\partial_{\xi^j} \hat{h}_j.$$

Evaluating the right-hand side of (4.9) at 0, we have

$$(2\pi)^{-n} \int_{\mathbb{R}^n} \hat{f}(\xi)\, d\xi = (2\pi)^{-n} \int_{\mathbb{R}^n} \sum_j i\partial_{\xi^j} \hat{h}_j(\xi)\, d\xi = 0.$$

Since the left-hand side of (4.9) vanishes at 0 by assumption, we conclude that (4.9) holds at $x = 0$ for all functions f such that $f(0) = 0$. Let $f \in \mathcal{S}(\mathbb{R}^n)$ be arbitrary and decompose

$$f = f(0) f_0 + (f - f(0) f_0),$$

where f_0 is defined in (4.10). Since the second term vanishes at zero and since (4.9) holds for f_0, as was noted before the statement of the theorem, we obtain (4.9) for $x = 0$ and arbitrary $f \in \mathcal{S}(\mathbb{R}^n)$. In order to prove the equality for arbitrary x, let $x_0 \in \mathbb{R}^n$ and let $g(x) = f(x + x_0)$. By a change of variables, one can then compute that

$$\hat{g}(\xi) = e^{ix_0\cdot\xi} \hat{f}(\xi).$$

Consequently,

$$f(x_0) = g(0) = (2\pi)^{-n} \int_{\mathbb{R}^n} \hat{g}(\xi)\, d\xi = (2\pi)^{-n} \int_{\mathbb{R}^n} e^{ix_0\cdot\xi} \hat{f}(\xi)\, d\xi,$$

which proves (4.9) in all generality. □

Note that

$$\int_{\mathbb{R}^n} \hat{f} h\, dx = \int_{\mathbb{R}^n} f \hat{h}\, dx, \tag{4.12}$$

since both integrals equal

$$\int_{\mathbb{R}^n} \int_{\mathbb{R}^n} f(x) h(\xi) e^{-ix\cdot\xi}\, d\xi dx,$$

after a change of the order of integration. Let us apply (4.12) to h and f, where

$$h(x) = (2\pi)^{-n} \bar{\hat{g}}(x) = (2\pi)^{-n} \int_{\mathbb{R}^n} e^{ix\cdot\xi} \bar{g}(\xi)\, d\xi.$$

Due to the Fourier inversion formula, we conclude that $\hat{h} = \bar{g}$. By applying (4.12), we obtain

$$(2\pi)^{-n} \int_{\mathbb{R}^n} \hat{f}\,\bar{\hat{g}}\,d\xi = \int_{\mathbb{R}^n} f\,\bar{g}\,dx. \tag{4.13}$$

This identity is referred to as *Parseval's formula*, and it is a very useful tool. Let us for instance apply it with $f = g = \partial^\alpha u$, where $u \in \mathcal{S}(\mathbb{R}^n)$. Since

$$\widehat{\partial^\alpha u} = i^{|\alpha|}\xi^\alpha \hat{u},$$

due to (4.7), we obtain

$$(2\pi)^{-n} \int_{\mathbb{R}^n} \xi^{2\alpha}|\hat{u}(\xi)|^2\,d\xi = \int_{\mathbb{R}^n} |\partial^\alpha u(x)|^2\,dx.$$

In some sense, this equality allows us to relate the degree of differentiability of a function with the rate of decay of its Fourier transform as $|\xi| \to \infty$.

5 Sobolev spaces

For reasons mentioned in the introduction to the previous chapter, Sobolev spaces are of central importance in the proof of local existence of solutions to non-linear wave equations. The present chapter is concerned with establishing the basic properties of such spaces. One way to define them is as the completion of the space of smooth functions with compact support endowed with some suitable norm. However, this perspective is somewhat more abstract than necessary. Another perspective is to first define the concept of a weak derivative: u is said to be k times weakly differentiable on $\mathbb{R}^n$ if, for every multiindex α such that $|\alpha| \leq k$, there is a function u_α such that

$$\int_{\mathbb{R}^n} u \partial^\alpha \phi \, dx = (-1)^{|\alpha|} \int_{\mathbb{R}^n} u_\alpha \phi \, dx$$

for all $\phi \in C_0^\infty(\mathbb{R}^n)$. To be precise, one of course has to specify to which function spaces u and u_α have to belong, cf. Definition 5.2. The question then arises to what extent the weak derivative is well defined. In Section 5.1, we prove that two weak derivatives corresponding to the same multiindex α can at most differ on a set of measure zero. Furthermore, we introduce the concept of a mollifier, which is an essential tool when proving results such as the statement that smooth functions with compact support are dense in the Sobolev spaces we define. The Sobolev space $H^k(\mathbb{R}^n)$ can then be defined as the space of k times weakly differentiable functions, all of whose weak derivatives are in $L^2(\mathbb{R}^n)$. The norm is defined by (4.2) if we interpret $\partial^\alpha u$ as the weak derivatives of u. As a result, we obtain a less abstract definition. However, with such a definition, it is not clear that the space is complete nor is it clear that the space of smooth functions with compact support is dense in it. In Section 5.2, we give the formal definition of some of the Sobolev spaces we shall be using, and we prove that they are complete and that the smooth functions with compact support form a dense subset.

The Sobolev spaces we discussed above consisted of functions with a certain, non-negative, number of weak derivatives in $L^2(\mathbb{R}^n)$. In the proof of existence of solutions to linear symmetric hyperbolic systems, it is however necessary to consider Sobolev spaces corresponding to a negative number of derivatives. There is no obvious classical definition of this concept, but one can prove that the Sobolev spaces $H^k(\mathbb{R}^n)$ mentioned above (for k a non-negative integer) can be characterized as the space of functions u such that their Fourier transform $\hat{u}$ has the property that

$$\int (1 + |\xi|^2)^k |\hat{u}(\xi)|^2 \, d\xi < \infty. \tag{5.1}$$

Furthermore, by taking the square root of the left-hand side, one obtains a norm equivalent to the Sobolev space norm. Note that in the expression appearing on the left-hand side of (5.1), there is no problem with replacing the non-negative integer k with any real number s. Doing so leads to the Sobolev spaces $H_{(s)}(\mathbb{R}^n)$, defined for any real s. For $s < 0$, we interpret these functions as having a negative number of derivatives in $L^2(\mathbb{R}^n)$. Note, however, that how to define these spaces is not immediately obvious; if

$k \geq 0$, then $\hat{u}$ is the Fourier transform of an L^2 function assuming that (5.1) is satisfied, but if $k < 0$, then this need not be the case. As a consequence, $H_{(s)}(\mathbb{R}^n)$ can not be described as a space of functions. However, it is a space of temperate distributions, a concept we define in the beginning of Section 5.3. Following the definition of this concept, we define the spaces $H_{(s)}(\mathbb{R}^n)$ and analyze their basic properties. In particular, we prove estimates that will be of importance when deriving energy estimates for a negative number of derivatives. The chapter ends with a proof of the exact form of the duality we shall be using in the proof of existence of solutions to linear symmetric hyperbolic systems.

5.1 Mollifiers

It will be of interest to approximate functions in e.g. L^p spaces by smooth functions with compact support. We shall begin this section by developing the necessary tools. Let $\phi \in C_0^\infty(\mathbb{R}^n)$ be a non-zero function such that $\phi(x) \geq 0$ for all x. We refer the reader to Section A.6 for a proof of the existence of such functions. By dividing the function by its integral, we can assume that its integral is unity. Assume $u \in L^1_{\text{loc}}(\mathbb{R}^n)$, i.e., assume that $u\colon \mathbb{R}^n \to \mathbb{R}$ is measurable and that it is integrable when restricted to any compact subset of $\mathbb{R}^n$. Then we define, for any $\varepsilon > 0$,

$$(J_\varepsilon u)(x) = \int_{\mathbb{R}^n} \phi_\varepsilon(x-y)u(y)\,dy,$$

where $\phi_\varepsilon(x) = \varepsilon^{-n}\phi(x/\varepsilon)$. Note that $J_\varepsilon u$ is a smooth function and that

$$(J_\varepsilon u)(x) - u(x) = \int_{\mathbb{R}^n} \phi_\varepsilon(y)[u(x-y) - u(x)]\,dy,$$

since the integral of ϕ is one. If u is continuous, we thus see that $J_\varepsilon u - u$ converges to zero uniformly on compact subsets of $\mathbb{R}^n$ as $\varepsilon \to 0$. In particular, if u is continuous with compact support, the convergence is uniform. Note that if $1 < p < \infty$, $u \in L^p(\mathbb{R}^n)$ and $1/p + 1/q = 1$,

$$\begin{aligned} |(J_\varepsilon u)(x)| &\leq \int_{\mathbb{R}^n} \phi_\varepsilon^{1/q}(x-y)\phi_\varepsilon^{1/p}(x-y)|u(y)|\,dy \\ &\leq \left(\int_{\mathbb{R}^n} \phi_\varepsilon(x-y)\,dy\right)^{1/q} \left(\int_{\mathbb{R}^n} \phi_\varepsilon(x-y)|u(y)|^p\,dy\right)^{1/p} \\ &= \left(\int_{\mathbb{R}^n} \phi_\varepsilon(x-y)|u(y)|^p\,dy\right)^{1/p}, \end{aligned} \tag{5.2}$$

where we have used Hölder's inequality, cf. Lemma 6.3. Consequently,

$$\|J_\varepsilon u\|_p \leq \|u\|_p. \tag{5.3}$$

The same is true for $p = 1$ and $p = \infty$, the argument being less involved. By a similar argument,

$$\|J_\varepsilon u - u\|_{L^p} \leq \left(\int_{\mathbb{R}^n} \int_{\mathbb{R}^n} \phi_\varepsilon(y) |u(x-y) - u(x)|^p \, dy dx \right)^{1/p}.$$

If we can prove that $u(\cdot - y)$ converges to u in L^p as $y \to 0$, we can conclude that $J_\varepsilon u \to u$ in L^p. Since continuous functions with compact support are dense in L^p for $1 \leq p < \infty$, cf. Theorem 3.14 of [79], this is however not difficult to prove, assuming that $1 \leq p < \infty$.

We shall need the following technical lemma.

Lemma 5.1. *Let $\Omega \subseteq \mathbb{R}^n$ be open and assume that $u \in L^1_{\mathrm{loc}}(\Omega)$, i.e., that $u : \Omega \to \mathbb{R}$ is measurable and that $u\chi_K$ is integrable for every compact subset $K \subseteq \Omega$. If*

$$\int_\Omega u\phi \, dx = 0$$

for every $\phi \in C_0^\infty(\Omega)$, then $u = 0$ a.e.

Proof. Let A_j denote the subset of Ω on which $|u(x)| \geq 1/j$, let K_l be an increasing sequence of compact subsets of Ω whose union is Ω and let $A_{j,l} = A_j \cap K_l$. If we let

$$v_{j,l}(x) = \frac{u(x)}{|u(x)|} \chi_{A_{j,l}}(x),$$

cf. the terminology of Section A.1, then $J_\varepsilon v_{j,l} \in C_0^\infty(\Omega)$ for ε small enough. In fact there is an $\varepsilon_0 > 0$ and a compact subset $K_{l,0}$ of Ω such that $|J_\varepsilon v_{j,l}| \leq \chi_{K_{l,0}}$ for all $\varepsilon \leq \varepsilon_0$. Thus $|(J_\varepsilon v_{j,l})u| \leq |u|\chi_{K_{l,0}}$, and the right-hand side is integrable. Choose any sequence $0 < \varepsilon_i \leq \varepsilon_0$ converging to zero. By the observations made prior to the statement of the lemma and the fact that $v_{j,l} \in L^p$ for any $1 \leq p < \infty$, we are allowed to conclude that $J_{\varepsilon_i} v_{j,l}$ converges to $v_{j,l}$ with respect to any L^p norm, for $1 \leq p < \infty$. Due to Theorem 3.12 of [79], we can conclude that there is a subsequence ε_{i_k} such that $J_{\varepsilon_{i_k}} v_{j,l}$ converges to $v_{j,l}$ a.e. Since $J_{\varepsilon_{i_k}} v_{j,l} u$ is bounded by an integrable function and since it converges to $|u|\chi_{A_{j,l}}$ a.e. we can apply Lebesgue's dominated convergence theorem in order to conclude that

$$\frac{1}{j}\mu(A_{j,l}) \leq \int_{A_{j,l}} |u(x)| \, dx = \lim_{k \to \infty} \int_\Omega (J_{\varepsilon_{i_k}} v_{j,l})(x) u(x) \, dx.$$

Since the right-hand side is zero by assumption, we conclude that the measure of the union of all the $A_{j,l}$ is zero. Since this union coincides with the set on which u is non-zero, we conclude that $u = 0$ a.e. □

5.2 Weak differentiability, $W^{k,p}$ spaces

Let us define the concept of weak differentiability.

Definition 5.2. A function $u \in L^1_{\mathrm{loc}}(\mathbb{R}^n)$ is said to be k times *weakly differentiable* if for every multiindex α with $|\alpha| \le k$ there is a function $u_\alpha \in L^1_{\mathrm{loc}}(\mathbb{R}^n)$ with the property that

$$\int_{\mathbb{R}^n} u \partial^\alpha \phi \, dx = (-1)^{|\alpha|} \int_{\mathbb{R}^n} u_\alpha \phi \, dx \tag{5.4}$$

for all $\phi \in C_0^\infty(\mathbb{R}^n)$. The functions u_α are referred to as the *weak derivatives* of u.

Remark 5.3. If (5.4) holds with v_α and u_α on the right-hand side, then $u_\alpha = v_\alpha$ a.e., cf. Lemma 5.1. We shall write $\partial^\alpha u$ instead of u_α. Note that one can replace $\mathbb{R}^n$ with any open subset Ω of $\mathbb{R}^n$ in the definition.

A locally integrable complex or vector-valued function is said to be weakly differentiable if its components are. Let us give a first definition of Sobolev spaces.

Definition 5.4. Let $1 \le p < \infty$ and let $\mathcal{W}^{k,p}(\mathbb{R}^n, \mathbb{C}^m)$ denote the set of k times weakly differentiable functions such that all the weak derivatives are in $L^p(\mathbb{R}^n, \mathbb{C}^m)$. Let $W^{k,p}(\mathbb{R}^n, \mathbb{C}^m)$ denote the set of equivalence classes of elements in $\mathcal{W}^{k,p}(\mathbb{R}^n, \mathbb{C}^m)$, two elements being equivalent if the set on which they differ has measure zero. For an element u of $\mathcal{W}^{k,p}(\mathbb{R}^n, \mathbb{C}^m)$, we define

$$\|u\|_{W^{k,p}} = \Big(\sum_{|\alpha| \le k} \int_{\mathbb{R}^n} |\partial^\alpha u|^p \, dx \Big)^{1/p}. \tag{5.5}$$

Remark 5.5. We can define $W^{k,p}(\mathbb{R}^n, \mathbb{R}^m)$ analogously, as well as $W^{k,p}(\Omega, \mathbb{C}^m)$ for any open subset Ω of $\mathbb{R}^n$.

Lemma 5.6. *Let $1 \le p < \infty$ and let $k \ge 0$ be an integer. Then the object $\|\cdot\|_{W^{k,p}}$ defined in* (5.5) *yields a norm on $W^{k,p}(\mathbb{R}^n, \mathbb{C}^m)$ with respect to which $W^{k,p}(\mathbb{R}^n, \mathbb{C}^m)$ is a Banach space.*

Proof. The only non-trivial aspect of proving that $\|\cdot\|_{W^{k,p}}$ is a norm is to prove that

$$\|u + v\|_{W^{k,p}} \le \|u\|_{W^{k,p}} + \|v\|_{W^{k,p}}$$

for $u, v \in W^{k,p}(\mathbb{R}^n, \mathbb{C}^m)$. If $p = 1$, this is clear, so let us assume $1 < p < \infty$. We have

$$\|u+v\|^p_{W^{k,p}} \le \sum_{|\alpha| \le k} \int_{\mathbb{R}^n} |\partial^\alpha u||\partial^\alpha (u+v)|^{p-1} \, dx + \sum_{|\alpha| \le k} \int_{\mathbb{R}^n} |\partial^\alpha v||\partial^\alpha (u+v)|^{p-1} \, dx.$$

Note that if we let $q = p/(p-1)$,

$$\begin{aligned}\sum_{|\alpha| \le k} \int_{\mathbb{R}^n} |\partial^\alpha u||\partial^\alpha (u+v)|^{p-1} \, dx &\le \int_{\mathbb{R}^n} \Big(\sum_{|\alpha| \le k} |\partial^\alpha u|^p \Big)^{1/p} \Big(\sum_{|\alpha| \le k} |\partial^\alpha (u+v)|^p \Big)^{1/q} dx \\ &\le \|u\|_{W^{k,p}} \|u+v\|^{p-1}_{W^{k,p}},\end{aligned}$$

where we have used Hölder's inequality twice, cf. Lemma 6.3; in the first inequality with respect to the counting measure and in the second inequality with respect to the Lebesgue measure. We have a similar inequality for u replaced by v, and together these inequalities yield the desired conclusion. In order to prove that $W^{k,p}(\mathbb{R}^n, \mathbb{C}^m)$ is a Banach space, let us assume u_j is a Cauchy sequence. Then all the weak derivatives of u_j converge in L^p so that for every multiindex α with $|\alpha| \leq k$ there is a $u_\alpha \in L^p(\mathbb{R}^n, \mathbb{C}^m)$ such that $\partial^\alpha u_j \to u_\alpha$ in L^p. Define $u = u_\alpha$ for $\alpha = 0$. Due to the definition of weak derivatives, it is clear that u is weakly differentiable and that u_α are the weak derivatives. □

The special case $p = 2$ yields a Hilbert space.

Definition 5.7. Let $H^k(\mathbb{R}^n, \mathbb{C}^m) = W^{k,2}(\mathbb{R}^n, \mathbb{C}^m)$. For $u, v \in H^k(\mathbb{R}^n, \mathbb{C}^m)$, we define

$$(u, v) = \sum_{|\alpha| \leq k} \int_{\mathbb{R}^n} \partial^\alpha u(x) \cdot \overline{\partial^\alpha v}(x)\, dx. \tag{5.6}$$

Remark 5.8. Note that $H^k(\mathbb{R}^n, \mathbb{C}^m)$ is a complex Hilbert space with inner product given by (5.6).

Lemma 5.9. *The space $C_0^\infty(\mathbb{R}^n, \mathbb{C}^m)$ is dense in $W^{k,p}(\mathbb{R}^n, \mathbb{C}^m)$ for $1 \leq p < \infty$.*

Proof. Let $u \in W^{k,p}(\mathbb{R}^n, \mathbb{C}^m)$, let $\phi \in C_0^\infty(\mathbb{R}^n)$ be such that $\phi(x) = 1$ for $|x| \leq 1$ and let $\phi_l(x) = \phi(x/l)$. That there is a function of this form is proved in Section A.6. Inductively, one can see that $\phi_l u \in W^{k,p}(\mathbb{R}^n, \mathbb{C}^m)$ with

$$\partial^\alpha(\phi_l u) = \sum_{\beta \leq \alpha} \binom{\alpha}{\beta} \partial^{\alpha-\beta} \phi_l \partial^\beta u,$$

cf. Section A.1 for conventions. Note that if $\beta \neq \alpha$, then $\partial^{\alpha-\beta}\phi_l \partial^\beta u$ converges to zero pointwise everywhere and is bounded by a function in L^p. By Lebesgue's dominated convergence theorem it thus converges to zero in L^p. On the other hand $\phi_l \partial^\alpha u$ converges to $\partial^\alpha u$ by a similar argument. Thus $\phi_l u$ converges to u in $W^{k,p}(\mathbb{R}^n, \mathbb{C}^m)$. In other words, we can assume that u has compact support. If u has compact support, then $J_\varepsilon u$ is a smooth function with compact support. By the definition of weak derivatives,

$$\partial^\alpha(J_\varepsilon u)(x) = \int_{\mathbb{R}^n} \phi_\varepsilon(x - y) \partial^\alpha u(y)\, dy,$$

where we use the conventions and assumptions of the first section of the present chapter. In order to prove the lemma, all we need to show is thus that $J_\varepsilon u$ converges to u in L^p. This we already did prior to the statement of Lemma 5.1. □

5.3 Temperate distributions, H^s spaces

Definition 5.7 yields the Sobolev spaces H^k where k is a non-negative integer. We shall however need to use these spaces for k negative. In order to be able to define such spaces, we need to introduce the concept of a temperate distribution.

Definition 5.10. A continuous linear form on $\mathcal{S}(\mathbb{R}^n)$ is called a *temperate distribution*. The set of all temperate distributions is denoted $\mathcal{S}'(\mathbb{R}^n)$.

A linear form u on $\mathcal{S}(\mathbb{R}^n)$ is a map from $\mathcal{S}(\mathbb{R}^n)$ to $\mathbb{C}$ such that

$$u(af + bg) = au(f) + bu(g),$$

for $f, g \in \mathcal{S}(\mathbb{R}^n)$ and $a, b \in \mathbb{C}$. A temperate distribution is a special case of a distribution, which is a continuous linear form on $C_0^\infty(\mathbb{R}^n, \mathbb{C})$ endowed with a particular topology (though one does not necessarily need to define this topology, cf. [49]). We shall however only be interested in temperate distributions here. The reason for introducing the concept of a temperate distribution is the desire to take the Fourier transform.

Definition 5.11. If $u \in \mathcal{S}'(\mathbb{R}^n)$, the Fourier transform $\hat{u}$ of u is defined by

$$\hat{u}(\phi) = u(\hat{\phi})$$

for every $\phi \in \mathcal{S}(\mathbb{R}^n)$. We also define, for every $\phi \in \mathcal{S}(\mathbb{R}^n)$,

$$\partial^\alpha u(\phi) = (-1)^{|\alpha|} u(\partial^\alpha \phi).$$

Since the Fourier transform is a linear and continuous map from $\mathcal{S}(\mathbb{R}^n)$ to itself, we conclude that $\hat{u} \in \mathcal{S}'(\mathbb{R}^n)$. It is also clear that $\partial^\alpha u \in \mathcal{S}'(\mathbb{R}^n)$. Let us consider a special case. If $u : \mathbb{R}^n \to \mathbb{C}$ is a measurable function such that for some real number s, $(1 + |x|^2)^{s/2} u(x)$ defines a function in $L^2(\mathbb{R}^n, \mathbb{C})$, then we can define $U \in \mathcal{S}'(\mathbb{R}^n)$ by

$$U(\phi) = \int_{\mathbb{R}^n} \phi u \, dx. \tag{5.7}$$

In this case, we shall abuse notation and identify U with u. Note that the type of functions we are discussing here are locally integrable, so that if we could represent U in (5.7) by two different functions u and v on the right-hand side, then $u = v$ a.e. due to Lemma 5.1. If $u \in \mathcal{S}(\mathbb{R}^n)$, there are then two interpretations for $\hat{u}$, either as the Fourier transform of the associated temperate distribution or as the standard Fourier transform. That these two different points of view are compatible is ensured by (4.12). If u is a temperate distribution and all the derivatives of order $\leq k$ are in $L^2(\mathbb{R}^n, \mathbb{C})$, then u can be considered to be an element of $H^k(\mathbb{R}^n, \mathbb{C})$ and vice versa.

Let $u \in L^2(\mathbb{R}^n, \mathbb{C})$. Due to Lemma 5.9, there is a sequence $\phi_l \in \mathcal{S}(\mathbb{R}^n)$ such that $\phi_l \to u$ in L^2. Due to (4.13), $\hat{\phi}_l$ is a Cauchy sequence so that it converges to a function $v \in L^2(\mathbb{R}^n, \mathbb{C})$. Furthermore, for $\psi \in \mathcal{S}(\mathbb{R}^n)$,

$$\hat{u}(\psi) = u(\hat{\psi}) = \int_{\mathbb{R}^n} u\hat{\psi} \, dx = \lim_{l\to\infty} \int_{\mathbb{R}^n} \phi_l \hat{\psi} \, dx = \lim_{l\to\infty} \int_{\mathbb{R}^n} \hat{\phi}_l \psi \, dx = \int_{\mathbb{R}^n} v\psi \, dx,$$

where we have used (4.12). In other words, $\hat{u} \in L^2(\mathbb{R}^n, \mathbb{C})$ and $\hat{u} = v$, where v was constructed above. Furthermore, we have (4.13) for elements of $L^2(\mathbb{R}^n, \mathbb{C})$. Finally, let us note that if $\check{\phi}$ denotes the inverse Fourier transform of ϕ for $\phi \in \mathcal{S}(\mathbb{R}^n)$, then we can define the inverse Fourier transform of a temperate distribution u by $\check{u}(\phi) = u(\check{\phi})$. In this way, we see that the Fourier transform defines an invertible map from $\mathcal{S}'(\mathbb{R}^n)$ to itself.

Definition 5.12. Let $u \in \mathcal{S}'(\mathbb{R}^n)$ and let s be a real number. We say that $u \in H_{(s)}(\mathbb{R}^n)$ if $\hat{u}$ is a measurable function such that $\hat{u}(\xi)(1 + |\xi|^2)^{s/2}$ is square integrable. If $u \in H_{(s)}(\mathbb{R}^n)$, we define

$$\|u\|_{(s)} = \left[\frac{1}{(2\pi)^n} \int_{\mathbb{R}^n} |\hat{u}(\xi)|^2 (1 + |\xi|^2)^s \, d\xi\right]^{1/2}.$$

We define $H_{(s)}(\mathbb{R}^n, \mathbb{C}^m)$ by demanding that the components be in $H_{(s)}(\mathbb{R}^n)$.

Note that $H_{(s)}(\mathbb{R}^n, \mathbb{C}^m)$ is a complex Hilbert space with inner product

$$(u, v) = \frac{1}{(2\pi)^n} \int_{\mathbb{R}^n} \hat{u}(\xi) \cdot \bar{\hat{v}}(\xi)(1 + |\xi|^2)^s \, d\xi.$$

Furthermore, $H_{(s)}(\mathbb{R}^n) \subseteq H_{(t)}(\mathbb{R}^n)$ for $s \leq t$, the identity map being a bounded linear map, and $H_{(0)}(\mathbb{R}^n) = L^2(\mathbb{R}^n, \mathbb{C})$. There is a canonical way of relating $H_{(s)}(\mathbb{R}^n)$ for different s.

Definition 5.13. Let $u \in H_{(s)}(\mathbb{R}^n)$ and let t be a real number. Then we define $(1-\Delta)^t u$ to be the temperate distribution whose Fourier transform is given by $(1 + |\xi|^2)^t \hat{u}(\xi)$. Consequently, $(1 - \Delta)^t u$ is in $H_{(s-2t)}(\mathbb{R}^n)$.

Note that, for $t \in \mathbb{R}$,

$$\|(1 - \Delta)^{t/2} u\|_{(s-t)} = \|u\|_{(s)}.$$

Thus $(1-\Delta)^{t/2}$ is a bounded map with a bounded inverse from $H_{(s)}(\mathbb{R}^n)$ to $H_{(s-t)}(\mathbb{R}^n)$. Consequently, $H_{(s)}(\mathbb{R}^n)$ can be considered to be the image of the space $L^2(\mathbb{R}^n, \mathbb{C})$ under the map $(1 - \Delta)^{-s/2}$. By Lemma 5.9, $\mathcal{S}(\mathbb{R}^n)$ is dense in $L^2(\mathbb{R}^n, \mathbb{C})$ and since $(1 - \Delta)^t$ maps $\mathcal{S}(\mathbb{R}^n)$ into itself, we conclude that $\mathcal{S}(\mathbb{R}^n)$ is dense in $H_{(s)}(\mathbb{R}^n)$ for any real s. Finally, let us note that if $u \in \mathcal{S}(\mathbb{R}^n)$ and k is a non-negative integer, then $(1 - \Delta)^k u$ can be interpreted in two different ways. Either we interpret it as above or we interpret it as a differential operator acting on u where Δ is the standard Laplacian. One can check that these two interpretations yield the same result. In fact, for this conclusion to hold it is enough that $u \in H^{2k}(\mathbb{R}^n, \mathbb{C})$.

It is of some interest to note that $H_{(-s)}(\mathbb{R}^n)$ is the dual of $H_{(s)}(\mathbb{R}^n)$ in the sense that given f in the dual of $H_{(s)}(\mathbb{R}^n)$, there is a $\phi \in H_{(-s)}(\mathbb{R}^n)$ such that

$$f(\psi) = \int_{\mathbb{R}^n} \hat{\psi} \bar{\hat{\phi}} \, d\xi$$

for all $\psi \in H_{(s)}(\mathbb{R}^n)$. The reason is that, given f as above, we can define an element g of the dual of $L^2(\mathbb{R}^n, \mathbb{C})$ by

$$g(\psi) = f[(1-\Delta)^{-s/2}\psi].$$

Due to Theorem 6.16 of [79] and (4.13), there is a $\chi \in L^2(\mathbb{R}^n, \mathbb{C})$ such that

$$g(\psi) = \int_{\mathbb{R}^n} \hat{\psi}\bar{\hat{\chi}}\, d\xi$$

for all $\psi \in L^2(\mathbb{R}^n, \mathbb{C})$. Thus, for $\psi \in H_{(s)}(\mathbb{R}^n)$ we obtain

$$f(\psi) = \int_{\mathbb{R}^n} \hat{\psi}(1+|\xi|^2)^{s/2}\bar{\hat{\chi}}\, d\xi.$$

We obtain the desired statement by letting $\phi = (1-\Delta)^{s/2}\chi$.

Let us relate $H_{(k)}(\mathbb{R}^n)$ to $H^k(\mathbb{R}^n, \mathbb{C})$ when k is a non-negative integer. Note that for $\phi, \psi \in \mathcal{S}(\mathbb{R}^n)$,

$$\sum_{|\alpha|\le k} \int_{\mathbb{R}^n} \partial^\alpha\phi\partial^\alpha\bar{\psi}\, dx = \frac{1}{(2\pi)^n}\sum_{|\alpha|\le k} \int_{\mathbb{R}^n} \xi^{2\alpha}\hat{\phi}(\xi)\bar{\hat{\psi}}(\xi)\, d\xi,$$

where we have used (4.7) and (4.13). Since there are constants $c_{i,k} > 0$, $i = 1, 2$ such that

$$c_{1,k}(1+|\xi|^2)^k \le \sum_{|\alpha|\le k} \xi^{2\alpha} \le c_{2,k}(1+|\xi|^2)^k$$

and since $\mathcal{S}(\mathbb{R}^n)$ is dense in $H^k(\mathbb{R}^n, \mathbb{C})$ and $H_{(k)}(\mathbb{R}^n)$, we conclude that if $u \in H^k(\mathbb{R}^n, \mathbb{C})$, then $u \in H_{(k)}(\mathbb{R}^n)$ and vice versa. Furthermore, there are constants $C_{i,k} > 0$, $i = 1, 2$ such that for all $u \in H_{(k)}(\mathbb{R}^n)$,

$$C_{1,k}\|u\|_{(k)} \le \|u\|_{H^k} \le C_{2,k}\|u\|_{(k)},$$

i.e., the norms are equivalent.

The following result will be of interest.

Lemma 5.14. *Let α be a multiindex and $s \in \mathbb{R}$. Then there is a constant C depending on α and s such that for all $f \in \mathcal{S}(\mathbb{R}^n)$,*

$$\|\partial^\alpha f\|_{(s-|\alpha|)} \le C\|f\|_{(s)}. \tag{5.8}$$

Due to the fact that $\mathcal{S}(\mathbb{R}^n)$ is dense in $H_{(s)}(\mathbb{R}^n)$, we conclude that ∂^α can be extended to be a bounded linear operator from $H_{(s)}(\mathbb{R}^n)$ to $H_{(s-|\alpha|)}(\mathbb{R}^n)$.

Remark 5.15. If $s \ge |\alpha|$ and $f \in H_{(s)}(\mathbb{R}^n)$, $\partial^\alpha f$ is the αth weak derivative of f.

Proof. Note that there is a constant C such that

$$|\xi^\alpha|^2(1+|\xi|^2)^{s-|\alpha|} \leq C(1+|\xi|^2)^s.$$

Consequently

$$\|\partial^\alpha f\|_{(s-|\alpha|)}^2 = \frac{1}{(2\pi)^n}\int_{\mathbb{R}^n}(1+|\xi|^2)^{s-|\alpha|}|\xi^\alpha \hat{f}(\xi)|^2\, d\xi \leq C\|f\|_{(s)}^2,$$

and the lemma follows. □

Lemma 5.16. *Assume $u, v \in H_{(s)}(\mathbb{R}^n)$ and that α is a multiindex with $|\alpha| \leq s$. Then u and v are $|\alpha|$ times weakly differentiable, $\partial^\alpha v, \partial^\alpha u \in L^2(\mathbb{R}^n, \mathbb{C})$ and*

$$(u, \partial^\alpha v)_{L^2} = (-1)^{|\alpha|}(\partial^\alpha u, v)_{L^2}, \tag{5.9}$$

where

$$(u, v)_{L^2} = \int_{\mathbb{R}^n} u\bar{v}\, dx.$$

Furthermore, if $u, v \in H_{(s)}(\mathbb{R}^n)$, $s \geq 0$ and $t \leq s$, then $(1-\Delta)^{t/2}u, (1-\Delta)^{t/2}v \in L^2(\mathbb{R}^n, \mathbb{C})$ and

$$((1-\Delta)^{t/2}u, v)_{L^2} = (u, (1-\Delta)^{t/2}v)_{L^2}. \tag{5.10}$$

Proof. One can check the equalities for $u, v \in \mathcal{S}(\mathbb{R}^n)$. Due to the fact that $\mathcal{S}(\mathbb{R}^n)$ is dense in $H_{(s)}(\mathbb{R}^n)$, they then follow in general. □

We shall need the following lemma.

Lemma 5.17. *Let $u \in \mathcal{S}(\mathbb{R}^n)$ and assume $\phi \in C^\infty(\mathbb{R}^n, \mathbb{C})$ is such that all its derivatives are bounded. Then there is a constant C, depending on k and the sup norm of up to $|k|$ derivatives of ϕ, such that*

$$\|\phi u\|_{(k)} \leq C\|u\|_{(k)}.$$

Remark 5.18. Due to the assumptions, $\phi u \in \mathcal{S}(\mathbb{R}^n)$.

Proof. For $k \geq 0$, this is clear due to the equivalence of the norms $\|\cdot\|_{H^k}$ and $\|\cdot\|_{(k)}$. In order to prove that it holds for negative k, let us note that if $u, v \in \mathcal{S}(\mathbb{R}^n)$, then

$$\int_{\mathbb{R}^n} u\bar{v}\, dx = \frac{1}{(2\pi)^n}\int_{\mathbb{R}^n}\hat{u}\bar{\hat{v}}\, d\xi = \frac{1}{(2\pi)^n}\int_{\mathbb{R}^n}(1+|\xi|^2)^{k/2}\hat{u}(1+|\xi|^2)^{-k/2}\bar{\hat{v}}\, d\xi. \tag{5.11}$$

Applying Hölder's inequality to this, we obtain that

$$\sup_{v\in\mathcal{A}_k}\left|\int_{\mathbb{R}^n} u\bar{v}\, dx\right| \leq \|u\|_{(k)},$$

where $\mathcal{A}_k = \{v \in \mathcal{S}(\mathbb{R}^n) : \|v\|_{(-k)} \leq 1\}$. However, we can choose $v \in \mathcal{S}(\mathbb{R}^n)$ to be such that

$$\hat{v}(\xi) = (1+|\xi|^2)^k\hat{u}(\xi)\|u\|_{(k)}^{-1},$$

assuming $\|u\|_{(k)} \neq 0$. Then $\|v\|_{(-k)} = 1$. Inserting this v in (5.11), we obtain

$$\int_{\mathbb{R}^n} u\bar{v}\, dx = \|u\|_{(k)}.$$

For $u \neq 0$, we thus obtain that

$$\sup_{v \in \mathcal{A}_k} \left| \int_{\mathbb{R}^n} u\bar{v}\, dx \right| = \|u\|_{(k)}.$$

That this equality holds for $u = 0$ is quite clear as well. Assume that k is negative, $u, v \in \mathcal{S}(\mathbb{R}^n)$, that ϕ is as in the statement of the lemma and compute

$$\left| \int_{\mathbb{R}^n} \phi u\bar{v}\, dx \right| \leq \|u\|_{(k)} \|\bar{\phi} v\|_{(-k)} \leq C \|u\|_{(k)} \|v\|_{(-k)},$$

where we have used an argument similar to (5.11) in order to obtain the first inequality and the fact that the result holds for k non-negative in order to obtain the second inequality. Taking the supremum over $v \in \mathcal{A}_k$, we obtain the statement of the lemma. □

Corollary 5.19. *Let m and l be non-negative integers, α be a multiindex with $|\alpha| \leq l + m$, $u \in \mathcal{S}(\mathbb{R}^n)$ and $f \in C^\infty(\mathbb{R}^n, \mathbb{C})$, where we assume f and all its derivatives to be bounded. Then*

$$\|f \partial^\alpha u\|_{(-m)} \leq C \|u\|_{(l)},$$

where the constant depends on m, l and a bound of $\partial^\alpha f$ for $|\alpha| \leq m$.

Proof. Combine Lemma 5.14 and 5.17. □

When we prove existence of solutions to non-linear wave equations, it will be of interest to have the following interpolation inequality.

Lemma 5.20. *Let $s_1 < s_2 < s_3$ be real numbers and assume that $u \in H_{(s_3)}(\mathbb{R}^n)$. Then if $a, b \in (0, 1)$ are such that $a + b = 1$ and a is small enough, we have the inequality*

$$\|u\|_{(s_2)} \leq \|u\|_{(s_3)}^b \|u\|_{(s_1)}^a.$$

In fact,

$$\|u\|_{(s_2)} \leq \|u\|_{(s_1)}^{\frac{s_3 - s_2}{s_3 - s_1}} \|u\|_{(s_3)}^{\frac{s_2 - s_1}{s_3 - s_1}}.$$

Proof. Note that if $s = ts_1 + (1 - t)s_3$, then

$$\int_{\mathbb{R}^n} (1 + |\xi|^2)^s |\hat{u}(\xi)|^2 d\xi = \int_{\mathbb{R}^n} [(1 + |\xi|^2)^{s_1} |\hat{u}(\xi)|^2]^t [(1 + |\xi|^2)^{s_3} |\hat{u}(\xi)|^2]^{1-t} d\xi.$$

Applying Hölder's inequality yields the desired estimate. □

The point of the above lemma is that one can apply it to the following situation. Say that u_l is a bounded sequence in $H_{(s_3)}(\mathbb{R}^n)$ and say that u_l is a Cauchy sequence in $H_{(s_1)}(\mathbb{R}^n)$ where $s_1 < s_3$. Then we can use the above inequality to prove that u_l is a Cauchy sequence with respect to any $H_{(s_2)}(\mathbb{R}^n)$ such that $s_1 < s_2 < s_3$. By an application of Fatou's lemma, one can then conclude that the limit u is in fact in $H_{(s_3)}(\mathbb{R}^n)$. Finally, one can prove that u_l converges to u with respect to the weak topology on $H_{(s_3)}(\mathbb{R}^n)$. In fact, let f be an element of the dual of $H_{(s_3)}(\mathbb{R}^n)$. Then there is a $\phi \in H_{(-s_3)}(\mathbb{R}^n)$ such that

$$f(v) = \int_{\mathbb{R}^n} \hat{v}\bar{\hat{\phi}}\, d\xi$$

for all $v \in H_{(s_3)}(\mathbb{R}^n)$. Let $\psi_m \in \mathcal{S}(\mathbb{R}^n)$ converge to ϕ with respect to $H_{(-s_3)}(\mathbb{R}^n)$. Then

$$f(u_l) - f(u) = \int_{\mathbb{R}^n} \hat{u}_l(\bar{\hat{\phi}} - \bar{\hat{\psi}}_m)\, d\xi + \int_{\mathbb{R}^n} (\hat{u}_l - \hat{u})\bar{\hat{\psi}}_m\, d\xi + \int_{\mathbb{R}^n} \hat{u}(\bar{\hat{\psi}}_m - \bar{\hat{\phi}})\, d\xi.$$

Given $\varepsilon > 0$, one can fix m, independent of l, so that the first and the third terms are bounded by $\varepsilon/2$. For this fixed m, one can then let l be large enough that the second term is bounded by $\varepsilon/2$. Thus u_l converges to u weakly.

5.4 Dualities

Note that since $L^2(\mathbb{R}^n, \mathbb{C})$ and $H_{(s)}(\mathbb{R}^n)$ are isomorphic, $H_{(s)}(\mathbb{R}^n)$ is separable. Let

$$u \in L^p\{[0,T], H_{(s)}(\mathbb{R}^n, \mathbb{C}^m)\}, \quad v \in L^q\{[0,T], H_{(-s)}(\mathbb{R}^n, \mathbb{C}^m)\},$$

where $1/p + 1/q = 1$. Then

$$(1-\Delta)^{s/2}u \in L^p\{[0,T], L^2(\mathbb{R}^n, \mathbb{C}^m)\}, \quad (1-\Delta)^{-s/2}v \in L^q\{[0,T], L^2(\mathbb{R}^n, \mathbb{C}^m)\}.$$

We can thus define

$$\langle u, v\rangle = \int_0^T ((1-\Delta)^{s/2}u, (1-\Delta)^{-s/2}v)_{L^2}\, dt, \tag{5.12}$$

since the object inside the integral is measurable and bounded by an integrable function. The definition seems arbitrary in the sense that if

$$u \in L^p\{[0,T], H_{(r)}(\mathbb{R}^n, \mathbb{C}^m)\}$$

with $r > s$, then of course

$$v \in L^q\{[0,T], H_{(-r)}(\mathbb{R}^n, \mathbb{C}^m)\}$$

and so we could have used r instead of s on the right-hand side of (5.12). However, due to (5.10)

$$\begin{aligned}\int_0^T &((1-\Delta)^{r/2}u,(1-\Delta)^{-r/2}v)_{L^2}\,dt \\ &= \int_0^T ((1-\Delta)^{(r-s)/2}(1-\Delta)^{s/2}u,(1-\Delta)^{-r/2}v)_{L^2}\,dt \\ &= \int_0^T ((1-\Delta)^{s/2}u,(1-\Delta)^{-s/2}v)_{L^2}\,dt.\end{aligned}$$

Which s we use in (5.12) is consequently not important, as long as the right-hand side is defined. Finally, let us point out that if

$$u \in L^p\{[0,T],H_{(k)}(\mathbb{R}^n,\mathbb{C}^m)\}, \quad v \in L^q\{[0,T],H_{(k)}(\mathbb{R}^n,\mathbb{C}^m)\},$$

for some non-negative integer k and if α is a multiindex with $|\alpha| \le k$, then

$$\partial^\alpha u \in L^p\{[0,T],H_{(0)}(\mathbb{R}^n,\mathbb{C}^m)\}, \quad \partial^\alpha v \in L^q\{[0,T],H_{(0)}(\mathbb{R}^n,\mathbb{C}^m)\},$$

cf. Lemma 5.14, so that due to (5.9),

$$\langle \partial^\alpha u, v\rangle = (-1)^{|\alpha|}\langle u,\partial^\alpha v\rangle. \tag{5.13}$$

Proposition 5.21. *Let*

$$\begin{aligned}X_{(s)} &= L^1\{[0,T],H_{(s)}(\mathbb{R}^n,\mathbb{C}^m)\}\\ Y_{(-s)} &= L^\infty\{[0,T],H_{(-s)}(\mathbb{R}^n,\mathbb{C}^m)\}.\end{aligned}$$

Given $F \in X_{(s)}^*$ there is a $y \in Y_{(-s)}$ such that

$$F(x) = \langle x,y\rangle$$

for all $x \in X_{(s)}$ and $\|y\|_{Y_{(-s)}} = \|F\|_{X_{(s)}^*}$.

Proof. Given $F \in X_{(s)}^*$, define $G \in X_{(0)}^*$ by

$$G(x_0) = F[(1-\Delta)^{-s/2}x_0].$$

Note that $\|G\|_{X_{(0)}^*} = \|F\|_{X_{(s)}^*}$. Since $X_{(0)} = L^1\{[0,T],L^2(\mathbb{R}^n,\mathbb{C}^m)\}$, we can use Proposition 3.6 in order to obtain a $y_0 \in Y_{(0)}$ such that

$$G(x_0) = \langle x_0,y_0\rangle, \quad \|y_0\|_{Y_{(0)}} = \|F\|_{X_{(s)}^*}$$

for all $x_0 \in X_{(0)}$. Let us define $y = (1-\Delta)^{s/2}y_0$. Then, for $x \in X_{(s)}$, we obtain

$$F(x) = G[(1-\Delta)^{s/2}x] = \langle (1-\Delta)^{s/2}x,(1-\Delta)^{-s/2}y\rangle = \langle x,y\rangle.$$

Since $y \in Y_{(-s)}$ and $\|y\|_{Y_{(-s)}} = \|y\|_{Y_{(0)}}$, the proposition follows. □

6 Sobolev embedding

In Section 6.1, we begin the chapter by reminding the reader of some basic inequalities such as Hölder's inequality. We then proceed, in Section 6.2, to establish Sobolev embedding, the importance of which was pointed out in the introduction to Chapter 4. However, the main purpose of the present chapter is to prove the following inequality: let $\phi_1, \dots, \phi_l \in C_0^\infty(\mathbb{R}^n)$ and assume $\alpha_1, \dots, \alpha_l$ to be multiindices such that $\sum |\alpha_i| = k$. Then

$$\|\partial^{\alpha_1}\phi_1 \dots \partial^{\alpha_l}\phi_l\|_2 \leq C \sum_{i=1}^{l} \|D^k \phi_i\|_2 \prod_{j \neq i} \|\phi_j\|_\infty, \tag{6.1}$$

where

$$\|D^k f\|_p = \Big(\sum_{|\alpha|=k} \int_{\mathbb{R}^n} |(\partial^\alpha f)(x)|^p \, dx \Big)^{1/p}.$$

Considering (6.1), we see that there is a special case: if all the α_i except one are zero, then we trivially get the estimate. Due to (6.1), we see that, in practice, this is the only case we need to consider. At this stage, the estimate (6.1) may seem a bit technical, but it is of central importance both in proofs of local existence, as well as in proofs of global existence of solutions to non-linear wave equations. Let us give an example as a motivation for its importance (though it should be remarked that it is not necessary to read the remaining part of the present section in order to be able to understand the rest of the chapter, nor the rest of these notes).

Consider the equation

$$-\partial_t^2 u + \Delta u = F(u), \tag{6.2}$$
$$u(0, \cdot) = u_0, \tag{6.3}$$
$$\partial_t u(0, \cdot) = u_1 \tag{6.4}$$

on $\mathbb{R}^{n+1}$, where Δ is the ordinary Laplacian on $\mathbb{R}^n$ and F is a smooth function such that $F(0) = 0$. Let us state a standard local existence result concerning solutions to this equation.

Theorem 6.1. *If $u_0 \in H^{k+1}(\mathbb{R}^n)$ and $u_1 \in H^k(\mathbb{R}^n)$ for $k > n/2$, then there is a $T > 0$, depending only on $\|u_0\|_{H^{k+1}}$ and $\|u_1\|_{H^k}$, and a unique continuously differentiable solution u to* (6.2)–(6.4) *such that*

$$u \in C([-T, T], H^{k+1}(\mathbb{R}^n)) \cap C^1([-T, T], H^k(\mathbb{R}^n)). \tag{6.5}$$

Remark 6.2. In Chapter 9 we shall prove much more general results than this, though we shall require somewhat more differentiability of the initial data.

It is important to note that this result does not yield local existence of smooth solutions given smooth initial data; the existence time T depends on $\|u_0\|_{H^{k+1}}$ and

$\|u_1\|_{H^k}$, and in principle, these quantities could tend to infinity as $k \to \infty$. A related problem is that of finding a continuation criterion, i.e., to find an expression of the data u and $\partial_t u$ at time t such that if this expression remains bounded on $[0, T)$, then the solution can be continued beyond T. In order to resolve these issues, let us consider a solution to (6.2)–(6.4) with regularity as in (6.5). Define

$$E_k = \frac{1}{2} \sum_{|\alpha| \leq k} \int_{\mathbb{R}^n} [(\partial^\alpha \partial_t u)^2 + |\nabla \partial^\alpha u|^2 + (\partial^\alpha u)^2] \, dx.$$

Due to the degree of regularity we assume, we are formally not allowed to differentiate E_k with respect to time. However, it is possible to carry out more technical arguments that in practice amount to the same thing. Below, we shall therefore ignore this issue. Differentiating with respect to time, integrating by parts and using the equation and Hölder's inequality, we obtain

$$\begin{aligned} \frac{dE_k}{dt} &= \sum_{|\alpha| \leq k} \int_{\mathbb{R}^n} [\partial^\alpha (\partial_t^2 u - \Delta u) \partial^\alpha \partial_t u + \partial^\alpha u \partial^\alpha \partial_t u] \, dx \\ &\leq C \|F(u)\|_{H^k} E_k^{1/2} + E_k \end{aligned} \tag{6.6}$$

for some constant C. The essential step thus consists of estimating $F(u)$ in H^k. Note that $\partial^\alpha F(u)$ can, up to numerical factors, be written as a sum of terms of the form

$$F^{(j)}(u) \partial^{\alpha_1} u \ldots \partial^{\alpha_j} u,$$

where $\alpha_1 + \cdots + \alpha_j = \alpha$ (note also that $F(u) = g(u)u$ for some smooth function g due to the fact that $F(0) = 0$). Let us, for the moment, assume $u(t, \cdot)$ to have a uniform bound in $L^\infty(\mathbb{R}^n)$ on $t \in [0, T)$. Then we can extract the factors $F^{(j)}(u)$ in L^∞, and what remains to be estimated is

$$\|\partial^{\alpha_1} u \ldots \partial^{\alpha_j} u\|_{L^2}.$$

However, this is exactly the type of expression one can estimate with the help of (6.1). In fact, it is bounded by

$$C(\|u\|_\infty) \|u\|_{H^k}.$$

Combining this estimate with (6.6), we obtain

$$\frac{dE_k}{dt} \leq C(\|u\|_\infty) E_k.$$

As a consequence, if we assume $\|u(t, \cdot)\|_\infty$ to be uniformly bounded on an interval $[0, T)$, then $\|u(t, \cdot)\|_{H^{k+1}}$ and $\|\partial_t u(t, \cdot)\|_{H^k}$ remain uniformly bounded on the same interval. Due to the local existence theorem, Theorem 6.1, it then follows that the solution can be continued beyond T. As an immediate consequence of these observations (and uniqueness), we obtain local existence of smooth solutions given smooth, compactly supported initial data. Furthermore, either the solution is global to the future, or $\|u(t,\cdot)\|_\infty$ becomes unbounded within a finite time.

It is possible to prove a local existence theorem relying only on Sobolev embedding. However, using such methods, it would with all probability be necessary to demand a higher degree of regularity of the initial data than is required in order to apply Theorem 6.1. One can also prove that there are continuation criteria using only Sobolev embedding. Nevertheless, to the best of our knowledge, it would then be necessary to demand more than the boundedness of u in L^∞.

Finally, let us give an example of how (6.1) (via the above continuation criterion) can be used to prove global existence of solutions. Consider the equation

$$u_{tt} - u_{xx} + V'(u) = 0 \tag{6.7}$$

on $\mathbb{R}^{1+1}$, where $V \in C^\infty(\mathbb{R})$ is such that $V(\xi) \geq 0$ for all $\xi \in \mathbb{R}$ and $V(0) = V'(0) = 0$. Given initial data in $C_0^\infty(\mathbb{R})$ to this equation, we, by the above, get a smooth local solution. Furthermore, we can define

$$E = \frac{1}{2}\int_{\mathbb{R}} [u_t^2 + u_x^2 + 2V(u)]\, dx.$$

Note that, due to the equation, E is preserved. Furthermore,

$$H = \frac{1}{2}\int_{\mathbb{R}} u^2\, dx$$

has the property that

$$\left|\frac{dH}{dt}\right| \leq \int_{\mathbb{R}} |uu_t|\, dx \leq 2H^{1/2}E^{1/2}.$$

Since E is conserved, this implies that $H^{1/2}$ cannot grow faster than linearly. To conclude,

$$\int_{\mathbb{R}} (u^2 + u_x^2)\, dx$$

cannot become unbounded in a finite time. Combining this observation with Sobolev embedding, we conclude that $\|u\|_\infty$ cannot become unbounded in finite time. Due to the continuation criterion mentioned earlier, it is thus clear that all solutions to (6.7), where V satisfies the conditions specified above, are global in time if the initial data are smooth and of compact support (though it should be mentioned that the requirement that the support be compact can be removed; as long as the initial data are smooth, global existence follows).

6.1 Basic inequalities

Let us prove that if p and q are positive real numbers such that

$$\frac{1}{p} + \frac{1}{q} = 1,$$

then *Young's inequality*,

$$ab \le \frac{a^p}{p} + \frac{b^q}{q}, \tag{6.8}$$

holds for all non-negative a, b. If either a or b are zero, then the inequality holds trivially. If we let $t = a/b^{q-1}$, then (6.8) is equivalent to

$$t \le \frac{1}{q} + \frac{t^p}{p}.$$

However, the function

$$\frac{t^{-1}}{q} + \frac{t^{p-1}}{p}$$

tends to infinity as $t \to 0+$ and as $t \to \infty$. Furthermore it has a unique minimum for $t = 1$. This proves (6.8). In fact, Young's inequality can be generalized in the following way. Assume $p_1, \dots, p_k$ are positive numbers such that

$$\frac{1}{p_1} + \dots + \frac{1}{p_k} = 1. \tag{6.9}$$

Then if $a_1, \dots, a_k$ are non-negative numbers, we have

$$a_1 \dots a_k \le \frac{a_1^{p_1}}{p_1} + \dots + \frac{a_k^{p_k}}{p_k}. \tag{6.10}$$

We know that (6.10) holds for $k = 2$. Assume it holds for some $k \ge 2$, and let us prove that it holds for $k + 1$. Given $p_1, \dots, p_{k+1}$ satisfying (6.9) with k replaced by $k + 1$, let us define $r_i = p_i$ for $i = 1, \dots, k - 1$ and

$$r_k = \frac{p_k p_{k+1}}{p_k + p_{k+1}}.$$

Then $r_1, \dots, r_k$ are positive numbers satisfying a condition of the form (6.9). Consequently

$$a_1 \dots a_{k+1} \le \frac{a_1^{p_1}}{p_1} + \dots + \frac{a_{k-1}^{p_{k-1}}}{p_{k-1}} + \frac{(a_k a_{k+1})^{r_k}}{r_k}.$$

However, if we let $p = p_k/r_k$ and $q = p_{k+1}/r_k$, we can apply (6.8) in order to obtain

$$(a_k a_{k+1})^{r_k} \le \frac{a_k^{r_k p}}{p} + \frac{a_{k+1}^{r_k q}}{q} = r_k\left(\frac{a_k^{p_k}}{p_k} + \frac{a_{k+1}^{p_{k+1}}}{p_{k+1}}\right).$$

This completes the induction. As a consequence, we obtain the following result.

Lemma 6.3. *Let $p_1, \dots, p_k$ be positive numbers such that* (6.9) *holds. Assume $u_i \in L^{p_i}(\mathbb{R}^n)$ for $i = 1, \dots, k$. Then $u_1 \dots u_k \in L^1(\mathbb{R}^n)$ and*

$$\int_{\mathbb{R}^n} |u_1 \dots u_k|\, dx \le \|u_1\|_{p_1} \dots \|u_k\|_{p_k}. \tag{6.11}$$

Remark 6.4. We can of course allow some of the p_i to equal ∞. The special case $k = 2$ is Hölder's inequality.

Proof. Note that if $\|u_i\|_{p_i} = 0$ for some i, then the right and the left-hand side of (6.11) equal zero. We can consequently assume $\|u_i\|_{p_i} > 0$. Define $v_i = u_i/\|u_i\|_{p_i}$. Due to (6.10) we have

$$\int_{\mathbb{R}^n} |v_1 \dots v_k|\, dx \le \int_{\mathbb{R}^n} \left(\frac{|v_1|^{p_1}}{p_1} + \dots + \frac{|v_k|^{p_k}}{p_k} \right) dx = 1,$$

where we have used (6.9) and the fact that $\|v_i\|_{p_i} = 1$. Multiplying this inequality with $\|u_1\|_{p_1} \dots \|u_k\|_{p_k}$, we obtain the desired result. □

6.2 Sobolev embedding

We shall need the following Sobolev inequality.

Theorem 6.5. *Let k be a non-negative integer and assume that $s > k + n/2$. Then there is a constant C, depending on k, n and s such that for all $f \in \mathcal{S}(\mathbb{R}^n)$,*

$$\|f\|_{C_b^k(\mathbb{R}^n,\mathbb{C})} \le C \|f\|_{(s)}. \tag{6.12}$$

Remark 6.6. The function space $C_b^k(\mathbb{R}^n, \mathbb{C})$ consists of all the C^k functions whose derivatives up to order k are bounded. The norm is the sum of the suprema of the derivatives. There are more general Sobolev inequalities for $W^{k,p}(\mathbb{R}^n, \mathbb{C}^m)$ spaces, see e.g. [44]. We shall, however, not need them here. For Sobolev embedding results in more general domains than $\mathbb{R}^n$, we refer the reader to [1].

Proof. Let us first consider the case $k = 0$. Due to (4.9), we have

$$\begin{aligned} |f(x)| &\le (2\pi)^{-n} \int_{\mathbb{R}^n} |\hat{f}(\xi)|\, d\xi = (2\pi)^{-n} \int_{\mathbb{R}^n} (1 + |\xi|^2)^{-s/2} (1 + |\xi|^2)^{s/2} |\hat{f}(\xi)|\, d\xi \\ &\le (2\pi)^{-n} \Big(\int_{\mathbb{R}^n} (1 + |\xi|^2)^{-s}\, d\xi \Big)^{1/2} \Big(\int_{\mathbb{R}^n} (1 + |\xi|^2)^s |\hat{f}(\xi)|^2\, d\xi \Big)^{1/2} \\ &= (2\pi)^{-n/2} \Big(\int_{\mathbb{R}^n} (1 + |\xi|^2)^{-s}\, d\xi \Big)^{1/2} \|f\|_{(s)}. \end{aligned}$$

Since $(1 + |\xi|^2)^{-s}$ is integrable for $s > n/2$, we obtain the desired result. Let α be any multiindex. Then if $s - |\alpha| > n/2$, we obtain

$$\|\partial^\alpha f\|_{C_b(\mathbb{R},\mathbb{C})} \le C \|\partial^\alpha f\|_{(s-|\alpha|)} \le C \|f\|_{(s)},$$

where we have used the inequality (5.8). Adding these inequalities for all α such that $|\alpha| \le k$, we obtain the statement of the theorem. □

Note that (6.12) allows us to think of elements of $H_{(s)}(\mathbb{R}^n)$ as elements of $C_b^k(\mathbb{R}^n, \mathbb{C})$ for $s > k + n/2$. The reason is that if $\phi_l \to u$ in $H_{(s)}(\mathbb{R}^n)$, $\phi_l \in \mathcal{S}(\mathbb{R}^n)$, then the inequality shows that ϕ_l is a Cauchy sequence in $C_b^k(\mathbb{R}^n, \mathbb{C})$. Furthermore, there is a subsequence of ϕ_l that converges to u a.e. due to Theorem 3.12 of [79]. Consequently u equals a function in $C_b^k(\mathbb{R}^n, \mathbb{C})$ a.e. so that we can think of u as being in this space. Since $H_{(k)}(\mathbb{R}^n)$ coincides with $H^k(\mathbb{R}^n, \mathbb{C})$ when k is a non-negative integer, and since the norms are equivalent, we conclude that we have inequalities similar to (6.12) where the norm on the right-hand side is replaced by the H^l norm for $l > n/2 + k$.

It will be of relevance to know the following.

Lemma 6.7. *Let $\Omega \subset \mathbb{R}^n$ be open and assume that $u \in L^2_{\text{loc}}(\Omega)$ is l times weakly differentiable and that the weak derivatives are in $L^2_{\text{loc}}(\Omega)$. If $l > k + n/2$, then $u \in C^k(\Omega)$.*

Proof. Let $\phi \in C_0^\infty(\Omega)$. Then $\phi u \in H^l(\mathbb{R}^n)$. Consequently ϕu is a C^k function. Since, for any compact subset $K \subseteq \Omega$, there is a $\phi \in C_0^\infty(\Omega)$ such that $\phi(x) = 1$ for $x \in K$, cf. Proposition A.12, we obtain the conclusion of the lemma. □

6.3 Gagliardo–Nirenberg inequalities

In the present section we shall prove some inequalities due to Gagliardo, Nirenberg and Moser. The presentation follows that of [83] quite closely, though we shall consider a somewhat more general situation. The reader interested in the original references is referred to [42], [64], [63]. Let Y be a real vector space with norm $|\cdot|_Y$ arising from an inner product $\langle \cdot, \cdot \rangle$. We shall be interested in $C_0^\infty(\mathbb{R}^n, Y)$; those unfamiliar with the concept of the derivative of a function with values in an infinite dimensional vector space are referred to Section A.7, but let us point out that it, for the purposes of these notes, is completely sufficient to only consider the case $Y = \mathbb{R}$ with the standard inner product. Define, for $f \in C_0^\infty(\mathbb{R}^n, Y)$,

$$\|f\|_p = \left(\int_{\mathbb{R}^n} |f(x)|_Y^p \, dx \right)^{1/p}, \quad \|f\|_\infty = \sup_{x \in \mathbb{R}^n} |f(x)|_Y$$

for $1 \le p < \infty$. We shall also use the notation

$$\|D^l f\|_p = \Big(\sum_{|\alpha|=l} \int_{\mathbb{R}^n} |(\partial^\alpha f)(x)|_Y^p \, dx \Big)^{1/p}, \quad \|D^l f\|_\infty = \sup_{x \in \mathbb{R}^n} \sum_{|\alpha|=l} |(\partial^\alpha f)(x)|_Y.$$

Lemma 6.8. *Let $1 \le j \le n$, let $\kappa, r \in \mathbb{R}$ be such that $1 \le r \le \kappa$ and let Y be as above. Then there is a constant C such that for all $f \in C_0^\infty(\mathbb{R}^n, Y)$,*

$$\|\partial_j f\|_{2\kappa/r}^2 \le C \|f\|_{2\kappa/(r-1)} \|\partial_j^2 f\|_{2\kappa/(r+1)}. \tag{6.13}$$

Remark 6.9. When $r = 1$, $2\kappa/(r-1)$ should be interpreted as ∞. It will be of interest to keep in mind that the constant only depends on an upper bound on κ.

Proof. Let $2 \le q \in \mathbb{R}$ and consider ϕ_j, defined by

$$\phi_j(x) = \langle f(x), \partial_j f(x)\rangle \langle \partial_j f(x), \partial_j f(x)\rangle^{\frac{q-2}{2}},$$

where the last factor should be interpreted as 1 if $q = 2$. This function clearly has compact support. We wish to prove that it is continuously differentiable. Due to observations made in Section A.7, the expressions $\langle f(x), \partial_j f(x)\rangle$ and $\langle \partial_j f(x), \partial_j f(x)\rangle$ define smooth functions with compact support. If $q = 2$, ϕ_j is thus smooth, so let us assume $q > 2$. If ξ is such that $(\partial_j f)(\xi) \neq 0$, then ϕ_j is smooth in a neighbourhood of ξ, so consider a ξ such that $(\partial_j f)(\xi) = 0$. Let $\psi_j(x) = \langle \partial_j f(x), \partial_j f(x)\rangle$. Then ψ_j is smooth and $\psi_j(\xi) = \partial_k \psi_j(\xi) = 0$ for all $1 \le k \le n$. Thus

$$\psi_j(x) = O(|x - \xi|^2).$$

Since

$$|\phi_j(x)| \le |f(x)|_Y [\psi_j(x)]^{1/2} [\psi_j(x)]^{\frac{q-2}{2}} = |f(x)|_Y [\psi_j(x)]^{\frac{q-1}{2}} = O(|x-\xi|^{q-1}),$$

we conclude that ϕ_j is differentiable at ξ and that the derivative is zero. If $(\partial_j f)(x) \neq 0$, we can differentiate ϕ_j with respect to the kth variable in order to obtain

$$\begin{aligned}(\partial_k \phi_j)(x) = {} & \langle (\partial_k f)(x), (\partial_j f)(x)\rangle [\psi_j(x)]^{\frac{q-2}{2}} \\ & + \langle f(x), (\partial_k \partial_j f)(x)\rangle [\psi_j(x)]^{\frac{q-2}{2}} \\ & + (q-2)\langle f(x), (\partial_j f)(x)\rangle \langle (\partial_j f)(x), (\partial_k \partial_j f)(x)\rangle [\psi_j(x)]^{\frac{q-4}{2}}.\end{aligned} \tag{6.14}$$

Note that if $q > 2$, then $(\partial_j f)(x_l) \neq 0$ and $x_l \to \xi$ with $(\partial_j f)(\xi) = 0$, then $(\partial_k \phi_j)(x_l) \to 0$. In other words, ϕ_j is continuously differentiable. Integrating (6.14) over $\mathbb{R}^n$ for $k = j$ yields

$$\int_{\mathbb{R}^n} |(\partial_j f)(x)|_Y^q \, dx \le (q-1) \int_{\mathbb{R}^n} |f(x)|_Y |(\partial_j^2 f)(x)|_Y |(\partial_j f)(x)|_Y^{q-2} \, dx. \tag{6.15}$$

For $q = 2$, we obtain the same result if we interpret $|(\partial_j f)(x)|_Y^{q-2}$ as 1. In other words,

$$\int_{\mathbb{R}^n} |(\partial_j f)(x)|_Y^2 \, dx \le \int_{\mathbb{R}^n} |f(x)|_Y |(\partial_j^2 f)(x)|_Y \, dx \le \|f\|_{2\kappa/(r-1)} \|\partial_j^2 f\|_{2\kappa/(r+1)},$$

assuming that $\kappa = r \ge 1$, where we used Hölder's inequality in the last step. We thus get the desired inequality in case $\kappa = r \ge 1$. In case $1 \le r < \kappa$, let

$$q = \frac{2\kappa}{r}, \quad q_1 = \frac{2\kappa}{r-1}, \quad q_2 = \frac{2\kappa}{r+1}, \quad q_3 = \frac{q}{q-2},$$

Then $1/q_1 + 1/q_2 + 1/q_3 = 1$, so that we can apply Hölder's inequality to (6.15) in order to obtain

$$\int_{\mathbb{R}^n} |(\partial_j f)(x)|_Y^q \, dx \le (q-1) \|f\|_{2\kappa/(r-1)} \|\partial_j^2 f\|_{2\kappa/(r+1)} \|\partial_j f\|_q^{q-2}.$$

This implies the desired estimate. □

Lemma 6.10. *Let Y be as above and let $1 \leq j, l, i \in \mathbb{Z}$ and $\kappa, r \in \mathbb{R}$ be such that $j \leq r \leq \kappa + 1 - i$ and $l \geq j$. Then there is a constant C such that for all $\phi \in C_0^\infty(\mathbb{R}^n, Y)$,*

$$\|D^l\phi\|_{2\kappa/r} \leq C[\|D^{l-j}\phi\|_{2\kappa/(r-j)} + \|D^{l+i}\phi\|_{2\kappa/(r+i)}]. \tag{6.16}$$

Remark 6.11. Note that $2\kappa/(r-j)$ should be interpreted as ∞ when $r = j$. Again, it will be of importance to keep in mind that the constant only depends on n and an upper bound on κ and $l + i$.

Proof. By an application of (6.13), we have

$$\|D^l\phi\|_{2\kappa/r}^2 \leq C\|D^{l-1}\phi\|_{2\kappa/(r-1)}\|D^{l+1}\phi\|_{2\kappa/(r+1)},$$

assuming $l \geq 1$ and $1 \leq r \leq \kappa$. Note that the constant only depends on n and an upper bound on l and κ. Since

$$ab \leq \frac{1}{2}(\varepsilon a + \varepsilon^{-1}b)^2$$

for all non-negative a, b and $\varepsilon > 0$, we obtain

$$\|D^l\phi\|_{2\kappa/r} \leq C[\varepsilon\|D^{l-1}\phi\|_{2\kappa/(r-1)} + \varepsilon^{-1}\|D^{l+1}\phi\|_{2\kappa/(r+1)}].$$

This yields (6.16) in the case that $i = j = 1$. Let us assume that we have

$$\|D^l\phi\|_{2\kappa/r} \leq C[\varepsilon\|D^{l-j}\phi\|_{2\kappa/(r-j)} + C(\varepsilon)\|D^{l+i}\phi\|_{2\kappa/(r+i)}] \tag{6.17}$$

for arbitrary r, κ, j, l, i satisfying the conditions of the lemma, plus the condition that $j, i \leq \iota$. We already know this to be true for $\iota = 1$. Assume it to be true for ι and let us prove it for $\iota + 1$. First, let us prove that we can increase j to $j + 1$. Assume the conditions of the lemma are satisfied with j replaced by $j + 1$ and that $1 \leq i, j \leq \iota$. By the inductive hypothesis, applied to $r' = r - j$, $\kappa' = \kappa$, $l' = l - j$, $i' = j$ and $j' = 1$, we have

$$\|D^{l-j}\phi\|_{2\kappa/(r-j)} \leq C[\varepsilon_1\|D^{l-j-1}\phi\|_{2\kappa/(r-j-1)} + C(\varepsilon_1)\|D^l\phi\|_{2\kappa/r}].$$

Inserting this inequality into (6.17), fixing ε_1 and assuming ε to be small enough, we see that

$$\|D^l\phi\|_{2\kappa/r}$$

appears on the right-hand side, but with a factor which can be assumed to be smaller than $1/2$. Consequently, we can move it over to the left-hand side in order to obtain (6.17) with j replaced by $j + 1$. Thus (6.17) holds for all r, κ, j, l, i satisfying the conditions of the lemma and $i \leq \iota$, $j \leq \iota + 1$. Assume now that the conditions of the lemma are satisfied with i replaced by $i + 1$ and that $1 \leq i \leq \iota$ and $j \leq \iota + 1$. Due to the induction hypothesis, we can apply (6.17) with $r' = r + i$, $\kappa' = \kappa$, $j' = i$, $l' = l + i$ and $i' = 1$. Combining the resulting estimate with (6.17) and an argument similar to the one presented above, we obtain the induction hypothesis with ι replaced by $\iota + 1$. □

Note that letting $j = l$ and $r + i = \kappa$ yields

$$\|D^l \phi\|_{2\kappa/r} \le C[\|\phi\|_{2\kappa/(r-l)} + \|D^{l+\kappa-r}\phi\|_2] \tag{6.18}$$

for all $\phi \in C_0^\infty(\mathbb{R}^n, Y)$, a positive integer l and real κ, r such that $l \le r$ and $\kappa - r$ is a positive integer.

Lemma 6.12. *Let Y be as above, let l, μ and i be non-negative integers such that $l \le \max\{\mu, i\}$ and let $q, \varrho, \rho \in [1, \infty]$. Define*

$$\alpha = \frac{n}{q} - \frac{n}{\varrho} + \mu - l, \quad \beta = -\frac{n}{q} + \frac{n}{\rho} - i + l \tag{6.19}$$

and assume that neither of these quantities vanish. If there are constants $0 < C_1$, $C_2 \in \mathbb{R}$ such that the inequality

$$\|D^l \phi\|_q \le C_1 \|D^\mu \phi\|_\varrho + C_2 \|D^i \phi\|_\rho$$

holds for all $\phi \in C_0^\infty(\mathbb{R}^n, Y)$, then α and β have the same sign and, for all $\phi \in C_0^\infty(\mathbb{R}^n, Y)$,

$$\|D^l \phi\|_q \le (C_1 + C_2)\|D^\mu \phi\|_\varrho^{\beta/(\alpha+\beta)} \|D^i \phi\|_\rho^{\alpha/(\alpha+\beta)}.$$

Remark 6.13. If $q = \infty$, then n/q should be interpreted as 0 and similarly for n/ϱ and n/ρ.

Proof. Let us write the assumed inequality $Q \le C_1 R + C_2 P$. Replacing $\phi(x)$ by $\phi(sx)$, where $0 < s \in \mathbb{R}$, we get

$$s^{l-n/q} Q \le C_1 s^{\mu - n/\varrho} R + C_2 s^{i-n/\rho} P.$$

This implies $Q \le C_1 s^\alpha R + C_2 s^{-\beta} P$. If α and β had different signs, we could let s tend to zero or ∞ in order to conclude that $Q = 0$. In other words that $\|D^l \phi\|_q = 0$ for all $\phi \in C_0^\infty(\mathbb{R}^n, Y)$. Since we know that this is not true, we conclude that α and β have different signs. If P or R is zero, we conclude that ϕ is zero so that the inequality holds. Otherwise, we choose $s = (P/R)^{1/(\alpha+\beta)}$. This yields the desired inequality. □

As a corollary, we get the following estimates.

Corollary 6.14. *Let Y be as above, $1 \le l \in \mathbb{Z}$ and $\kappa, r \in \mathbb{R}$ be such that $l \le r$ and $\kappa - r$ is a positive integer. Then there is a constant C such that for all $\phi \in C_0^\infty(\mathbb{R}^n, Y)$,*

$$\|D^l \phi\|_{2\kappa/r} \le C \|\phi\|_{2\kappa/(r-l)}^{(\kappa-r)/(\kappa+l-r)} \|D^{\kappa+l-r}\phi\|_2^{l/(\kappa+l-r)}. \tag{6.20}$$

Proof. The idea is to apply Lemma 6.12 to (6.18). However, this is not always possible. If (and only if) $n = 2\kappa$, then $\alpha = \beta = 0$, where α and β are given in (6.19). In order to prove that the inequality holds in this case as well, we note that the constant appearing in

(6.18) only depends on an upper bound on n, on κ and on $l+\kappa-r$. Assuming $\kappa = 2n$, let $\kappa_\varepsilon = \kappa + \varepsilon$ and $r_\varepsilon = r + \varepsilon$ for $\varepsilon \in (0,1)$. The corollary is applicable for $l, \kappa_\varepsilon, r_\varepsilon$, and the constant can be chosen to be independent of ε. Since $\kappa_\varepsilon - r_\varepsilon = \kappa - r$, all we need to prove is that

$$\lim_{t\to t_0} \|\phi\|_t = \|\phi\|_{t_0}$$

when ϕ is a smooth function with compact support and $1 < t_0 \leq \infty$. Since $|\phi(x)|_Y$ defines a continuous function with compact support, it is enough to prove that $\|\psi\|_t \to \|\psi\|_{t_0}$, where ψ is a continuous real-valued function with compact support. Let us first consider the case $1 < t_0 < \infty$. Then $|\psi|^t$ converges to $|\psi|^{t_0}$ everywhere and is bounded by an integrable function. By Lebesgue's dominated convergence theorem, $\|\psi\|_t^t \to \|\psi\|_{t_0}^{t_0}$. From this one can deduce that $\|\psi\|_t \to \|\psi\|_{t_0}$. Consider the case $t_0 = \infty$. Let K be a compact set such that $\psi = 0$ outside of K. Then, for $1 < t < \infty$,

$$\|\psi\|_t \leq \|\psi\|_\infty [\mu(K)]^{1/t}.$$

As a consequence,

$$\limsup_{t\to\infty} \|\psi\|_t \leq \|\psi\|_\infty.$$

Let, for any real number $\alpha > 0$, A_α be the set of $x \in \mathbb{R}^n$ for which $|\psi(x)| \geq \alpha$. Then

$$\alpha[\mu(A_\alpha)]^{1/t} \leq \|\psi\|_t.$$

If $\mu(A_\alpha) > 0$, we conclude that

$$\alpha \leq \liminf_{t\to\infty} \|\psi\|_t.$$

By combining the above observations, we conclude that $\lim_{t\to\infty} \|\psi\|_t = \|\psi\|_\infty$. The lemma follows. □

Corollary 6.15. *Let Y be as above and k, l be a positive integers such that $k \geq l+1$. Then there is a constant C such that for all $\phi \in C_0^\infty(\mathbb{R}^n, Y)$,*

$$\|D^l\phi\|_{2k/l} \leq C\|\phi\|_\infty^{1-l/k}\|D^k\phi\|_2^{l/k}. \tag{6.21}$$

This inequality implies the result we have been heading for.

Lemma 6.16. *Let $\phi_1, \dots, \phi_l \in C_0^\infty(\mathbb{R}^n)$ and assume that $\alpha_1, \dots, \alpha_l$ are multiindices such that $\sum |\alpha_i| = k$. Then*

$$\|\partial^{\alpha_1}\phi_1 \dots \partial^{\alpha_l}\phi_l\|_2 \leq C\sum_{i=1}^{l} \|D^k\phi_i\|_2 \prod_{j\neq i} \|\phi_j\|_\infty. \tag{6.22}$$

Proof. Let us define $k_i = |\alpha_i|$ and $p_i = k/k_i$. Then $1/p_1 + \dots + 1/p_l = 1$, so that we can use (6.11) in order to obtain

$$\|\partial^{\alpha_1}\phi_1 \dots \partial^{\alpha_l}\phi_l\|_2 \leq \|\partial^{\alpha_1}\phi_1\|_{2k/k_1} \cdots \|\partial^{\alpha_l}\phi_l\|_{2k/k_l}.$$

If only one $k_i \neq 0$, the inequality is true, so let us assume this is not the case. Then $k_i \leq k-1$, so that we can apply (6.21) with $Y = \mathbb{R}$. We obtain

$$\|\partial^{\alpha_1}\phi_1 \dots \partial^{\alpha_l}\phi_l\|_2 \leq C\|\phi_1\|_\infty^{1-k_1/k}\|D^k\phi_1\|_2^{k_1/k} \dots \|\phi_l\|_\infty^{1-k_l/k}\|D^k\phi_l\|_2^{k_l/k}. \tag{6.23}$$

Note that since $1 - k_i/k = \sum_{j\neq i} k_j/k$, we can arrange the factors in l groups of the form

$$\Big(\|D^k\phi_i\|_2 \prod_{j\neq i}\|\phi_j\|_\infty\Big)^{k_i/k}.$$

We are allowed to use (6.10) in order to obtain the result. □

Let us remark that if $\phi_1, \dots, \phi_l \in H^k(\mathbb{R}^n)$, are such that $\|\phi_i\|_\infty < \infty$ for $i = 1, \dots, l$, then (6.22) holds. The reason is that there is a sequence $\phi_{i,m} \in C_0^\infty(\mathbb{R}^n)$ converging to ϕ_i. Furthermore, we can assume that $\|\phi_{i,m}\|_\infty \leq \|\phi_i\|_\infty$, cf. Chapter 5, the proof of Lemma 5.9 and (5.3) for $p = \infty$. Finally, we can choose subsequences of $\phi_{i,m}$ so that $\partial^{\alpha_i}\phi_{i,m}$ converges to $\partial^{\alpha_i}\phi_i$ almost everywhere, due to Theorem 3.12 of [79]. The result then follows by Fatou's lemma; see 1.28 of [79].

Lemma 6.17. *Let $F \in C^\infty(\mathbb{R}^{n+1})$ be such that $F(x,0) = 0$ for all $x \in \mathbb{R}^n$ and such that for every non-negative integer j and multiindex α, there is a continuous, increasing function $f_{\alpha,j}$ such that*

$$|(\partial^\alpha \partial_\xi^j F)(x,\xi)| \leq f_{\alpha,j}(|\xi|)$$

for all $(x,\xi) \in \mathbb{R}^{n+1}$. If $u \in H^k(\mathbb{R}^n) \cap L^\infty(\mathbb{R}^n)$, then $F(\cdot\,,u)$ is k times weakly differentiable, and the weak derivatives are given by the expressions one would have obtained if u were smooth, the only difference being that all derivatives that occur should be interpreted as the weak derivatives of u. In fact $F(\cdot\,,u) \in H^k(\mathbb{R}^n)$ and

$$\|F(\cdot\,,u)\|_{H^k} \leq C(\|u\|_\infty)\|u\|_{H^k},$$

where the constant C increases as $\|u\|_\infty$ increases.

Remark 6.18. The exact statement of the lemma is not important. What we wish to point out here is that there is a family of statements one can make given the tools we have. In particular, we have the same statement if u is $\mathbb{R}^N$-valued.

Proof. Note that

$$\begin{aligned} F[x,u(x)] - F[x,v(x)] &= \int_0^1 \partial_t\{F[x,tu(x)+(1-t)v(x)]\}\,dt \\ &= \int_0^1 \partial_u F[x,tu(x)+(1-t)v(x)]\,dt \cdot [u(x)-v(x)]. \end{aligned} \tag{6.24}$$

By applying (6.22) to this equality with $v = 0$, we obtain the desired estimate for $u \in C_0^\infty(\mathbb{R}^n)$. Note that we first differentiate (6.24), take out all derivatives of F with respect to u and x in the sup norm, and then apply (6.22) to what remains. In order to prove the statement in general, we let $\phi_l \in C_0^\infty(\mathbb{R}^n)$ converge to u with respect to $\|\cdot\|_{H^k}$ and be such that $\|\phi_l\|_\infty \le \|u\|_\infty$. By choosing a suitable subsequence, cf. Theorem 3.12 of [79], we can assume that $\partial^\alpha \phi_l$ converges to $\partial^\alpha u$ pointwise a.e. for all multiindices with $|\alpha| \le k$. Applying (6.23) to (6.24) with $u = \phi_l$ and $v = \phi_m$, we see that $F(\cdot, \phi_l)$ is a Cauchy sequence in H^k. However, by our choice of sequence, $\partial^\alpha[F(\cdot, \phi_l)]$ converges pointwise a.e. to $\partial^\alpha[F(\cdot, u)]$, which we define to be the expression obtained by assuming u to be smooth, writing down the ordinary formula, and replacing all derivatives of u by weak derivatives. We conclude that $F(\cdot, u)$ is k times weakly differentiable, the weak derivatives being given by the expressions mentioned in the statement of the lemma. Furthermore, the desired estimate holds. □

7 Symmetric hyperbolic systems

The main purpose of the present chapter is to prove the existence of solutions to linear symmetric hyperbolic systems. The proof is based on a combination of energy estimates, the Hahn–Banach theorem and uniqueness. Proving the standard energy estimates is not very complicated, but what is needed is estimates for a negative number of derivatives. This leads to some technical complications, and we present the necessary details in Section 7.2. When deriving energy estimates, the immediate conclusion is often that the object one wishes to estimate is bounded in terms of itself. It seems unclear how to extract any meaningful information from such an estimate, but the type of inequalities that appear have the property that a result called Grönwall's lemma can be applied. We prove this inequality in Section 7.1, and we shall have reason to apply it, not only in this chapter, but also in the proof of local existence of solutions to nonlinear wave equations. In order to prove that, given suitable assumptions concerning the initial data and the coefficients of the equation, there are smooth solutions, it is necessary to prove that

- for any degree of regularity, there is a solution with the corresponding degree of regularity,
- any two solutions of a sufficiently high degree of regularity have to coincide.

The latter statement is simply that of uniqueness, and we establish it in Section 7.3. Needless to say, uniqueness is of interest in its own right. The uniqueness statement we establish in Section 7.3 is of a rather primitive form and is based on Stokes' theorem. In Chapter 12 we shall derive more satisfactory results in the context of linear wave equations on a Lorentz manifold background. Finally, in Section 7.4, we establish existence of solutions. Some of the steps of the proof are quite technical in nature and are therefore presented in the Appendix.

7.1 Grönwall's lemma

Before turning to the main topic of the section, let us mention a lemma which will be important in what follows.

Lemma 7.1 (Grönwall's lemma). *Let $T_0 \in \mathbb{R}$, $T > T_0$ and f, k and G be non-negative functions on $[T_0, T]$ such that $f \in L^\infty([T_0, T])$, $k \in L^1([T_0, T])$ and G is non-decreasing. Assuming*

$$f(t) \leq G(t) + \int_{T_0}^{t} k(s) f(s)\, ds, \tag{7.1}$$

for all $t \in [T_0, T]$, we obtain

$$f(t) \leq G(t) \exp\left(\int_{T_0}^{t} k(s)\, ds\right)$$

for all $t \in [T_0, T]$.

Remark 7.2. The weaker assumption that $k, kf \in L^1([T_0, T])$ immediately implies $f \in L^\infty([T_0, T])$ due to (7.1), assuming that G is bounded. The above lemma is often referred to as Grönwall's lemma.

Proof. Since G is non-decreasing, and since it is clear that it is enough to prove the statement for $t = T$, we can assume $G = G(T)$, i.e., that it is a constant. Let us extend k and f to the entire real line by requiring that they be zero outside the interval $[T_0, T]$. Let

$$F(t) = G + \int_{T_0}^{t} k(s) f(s)\, ds.$$

Due to Theorem 7.11 of [79], F is differentiable almost everywhere and $F' = kf$. Furthermore, it is clear that F satisfies the conditions of Theorem 7.18 (c) of [79]. Thus F is absolutely continuous (AC) on $[T_0, T]$ in the sense of Definition 7.17 in [79]. For similar reasons $\int_{T_0}^{t} k(s)\, ds$ is AC on $[T_0, T]$. As a consequence, one can check that

$$g(t) = F(t) \exp\left(-\int_{T_0}^{t} k(s)\, ds\right)$$

is AC. We already know this function to be differentiable a.e., and the derivative is

$$g' = kf \exp\left(-\int_{T_0}^{t} k(s)\, ds\right) - kF \exp\left(-\int_{T_0}^{t} k(s)\, ds\right) \leq 0,$$

where we have used (7.1). Since $g(T_0) \leq G$, we would like to say that $g(t) \leq G$. This does however require some justification, which is supplied by Theorem 7.20 of [79]. The lemma follows. □

7.2 The basic energy inequality

Consider an equation of the form

$$A^\mu \partial_\mu u + Bu = f \tag{7.2}$$

$$u(0, \cdot) = u_0. \tag{7.3}$$

Here u is an $\mathbb{R}^N$-valued function on $\Omega \subseteq \mathbb{R}^{n+1}$, A^μ, $\mu = 0, \dots, n$, and B are smooth $N \times N$ real matrix-valued functions on Ω all of whose derivatives are bounded, f is a smooth $\mathbb{R}^N$-valued function on Ω and we use the Einstein summation convention and the convention concerning coordinates described in Section A.1. Furthermore u_0 is a smooth function on $\mathbb{R}^n$. Finally, we demand that A^μ, $\mu = 0, \dots, n$, be symmetric and that A^0 be positive definite with a uniform positive lower bound, i.e., that there is a constant $c_0 > 0$ such that $A \geq c_0$. We shall use the notation Lu for the left-hand side of (7.2).

Let us assume we have a smooth solution u to (7.2)–(7.3) on $S_T = [0, T] \times \mathbb{R}^n$, a set we shall refer to as the *slab*, and let us assume that u and $\partial_t u$ satisfy uniform

Schwartz bounds in the sense that for every pair of multiindices α and β, there is a constant $C_{\alpha,\beta}$ such that

$$|x^\alpha|\,[|\partial^\beta u| + |\partial^\beta \partial_0 u|](t,x) \le C_{\alpha,\beta}$$

on the slab. Note that as a consequence of our assumptions, f necessarily satisfies uniform Schwartz bounds.

Let us derive the basic energy inequality in detail. Let

$$E = \frac{1}{2}\int_{\mathbb{R}^n} u^t A^0 u \, dx.$$

Differentiating and using the symmetry of A^0, we conclude that

$$\partial_t E = \int_{\mathbb{R}^n} \left[\frac{1}{2} u^t (\partial_t A^0) u + u^t A^0 \partial_t u\right] dx.$$

To deal with the second term, we use the equation in order to obtain

$$\int_{\mathbb{R}^n} u^t A^0 \partial_t u \, dx = \int_{\mathbb{R}^n} [-u^t A^i \partial_i u - u^t B u + u^t f] \, dx.$$

Note that the first term can be rewritten

$$\begin{aligned} -\int_{\mathbb{R}^n} u^t A^i \partial_i u \, dx &= -\frac{1}{2}\int_{\mathbb{R}^n} \left[\partial_i (u^t A^i u) - u^t (\partial_i A^i) u\right] dx \\ &= \frac{1}{2}\int_{\mathbb{R}^n} u^t (\partial_i A^i) u \, dx \end{aligned}$$

due to the symmetry of A^i. Adding up the above, we obtain

$$\partial_t E = \int_{\mathbb{R}^n} u^t \left(\frac{1}{2}\partial_0 A^0 + \frac{1}{2}\partial_j A^j - B\right) u \, dx + \int_{\mathbb{R}^n} u^t f \, dx.$$

Note that the first term on the right-hand side can be bounded by CE for some constant C due to the fact that B and the derivatives of A have an upper bound and A^0, considered as a positive definite matrix, has a positive lower bound. In the end, we thus conclude that

$$\partial_t E \le CE + CE^{1/2} \| f(t,\cdot)\|_2. \tag{7.4}$$

In this inequality and below, we shall use C to denote a constant, the value of which might change from line to line. Let $\varepsilon > 0$ and $E_\varepsilon = E + \varepsilon$ (this definition is motivated by the desire to divide by $E^{1/2}$). Then the same inequality holds for E_ε and we can divide by $\sqrt{E_\varepsilon}$ in order to obtain

$$\partial_t E_\varepsilon^{1/2} \le C E_\varepsilon^{1/2} + C\| f(t,\cdot)\|_2.$$

Integrating, we obtain, for $t \ge 0$,

$$E_\varepsilon^{1/2}(t) \le E_\varepsilon^{1/2}(0) + C\int_0^t \| f(s,\cdot)\|_2 \, ds + C\int_0^t E_\varepsilon^{1/2}(s)\, ds.$$

At this stage, we can apply Grönwall's lemma, Lemma 7.1, with G given by the first two terms on the right-hand side and $k(t) = C$. We conclude that

$$E_\varepsilon(t) \leq \left(E_\varepsilon^{1/2}(0) + C\int_0^t \|f(s,\cdot)\|_2\,ds\right)e^{Ct}.$$

Letting $\varepsilon \to 0+$, we conclude that

$$E(t) \leq \left(E^{1/2}(0) + C\int_0^t \|f(s,\cdot)\|_2\,ds\right)e^{Ct}.$$

Note that this inequality immediately gives uniqueness of solutions to (7.2)–(7.3) given the presupposed conditions; consider the energy of the difference. In the end, we shall need to have estimates for energies involving derivatives of the solution. In fact, we shall need estimates for a negative number of derivatives, and to obtain such estimates is the goal of this section.

Lemma 7.3. *Consider a solution to* (7.2)–(7.3) *under the assumptions stated above. Define*

$$E_k[u] = \frac{1}{2}\sum_{|\alpha|\leq k}\int_{\mathbb{R}^n}(\partial^\alpha u)^t A^0 \partial^\alpha u\,dx.$$

Then

$$\partial_t E_k \leq CE_k + CE_k^{1/2}\|f\|_{H^k}, \tag{7.5}$$

where the constants depend on the bounds on A^μ and B. In particular, they depend on c_0.

Proof. Note that $E_0 = E$. Due to (7.4), we thus have (7.5) for $k = 0$. In order to get the inequality in all generality, observe that

$$L\partial^\alpha u = \partial^\alpha f + [L,\partial^\alpha]u.$$

Thus, using (7.5) for $k = 0$, we obtain

$$\partial_t E[\partial^\alpha u] \leq CE[\partial^\alpha u] + CE^{1/2}[\partial^\alpha u]\|\partial^\alpha f + [L,\partial^\alpha]u\|_2.$$

Note that for $|\alpha| \leq k$,

$$\|[L,\partial^\alpha]u\|_2 \leq C\|\partial_0 u\|_{H^{k-1}} + CE_k^{1/2}.$$

However, by using the equation, we see that the first term on the right can be estimated by

$$\|\partial_0 u\|_{H^{k-1}} \leq CE_k^{1/2} + C\|f\|_{H^{k-1}}.$$

Adding up these observations, we conclude that (7.5) holds. □

Corollary 7.4. *Consider a solution to* (7.2)–(7.3) *under the assumptions stated above. Then, for* $t \in [0, T]$,

$$E_k^{1/2}(t) \le C\Big[E_k^{1/2}(0) + \int_0^t \|f(s,\cdot)\|_{H^k}\,ds\Big]. \tag{7.6}$$

Here, the constant depends on k*, the bounds on* A^μ *and* B *and on* T.

Proof. Let $\mathcal{E}_k = e^{-Ct}E_k + \varepsilon$, where C is the first constant appearing on the right-hand side of (7.5) and $\varepsilon > 0$. Then

$$\partial_t \mathcal{E}_k \le C e^{-Ct/2}\mathcal{E}_k^{1/2}\|f\|_{H^k}.$$

Since we allow the constant on the right-hand side of (7.6) to depend on T, we can estimate $e^{-Ct/2}$ by a constant. Dividing the inequality by $\mathcal{E}_k^{1/2}$, which is allowed, since $\mathcal{E}_k \ge \varepsilon > 0$, and integrating, we obtain

$$\mathcal{E}_k^{1/2}(t) \le \mathcal{E}_k^{1/2}(0) + C\int_0^t \|f(s,\cdot)\|_{H^k}\,ds.$$

Letting ε tend to zero and noting that the difference between E_k and $\mathcal{E}_k$ with $\varepsilon = 0$ is only a constant factor, since we allow the constants to depend on T, we get the desired conclusion. □

Note that we can speak of solutions to (7.2) with values in $\mathbb{C}^N$; such a solution can be considered as two $\mathbb{R}^N$-valued solutions with different f's. If u is $\mathbb{C}^N$-valued, we define

$$E_k[u] = E_k[\mathrm{Re}\, u] + E_k[\mathrm{Im}\, u],$$

and with this notation, we get the same estimates for $\mathbb{C}^N$-valued solutions as for $\mathbb{R}^N$-valued ones.

In order to be able to prove existence, we shall need the following inequality.

Lemma 7.5. *Assume* u *is a solution of* (7.2) *under the assumptions made in the beginning of the section and let* k *be any integer. Then, for* $t \in [0, T]$,

$$\|u(t,\cdot)\|_{(k)} \le C\Big[\|u(0,\cdot)\|_{(k)} + \int_0^t \|f(s,\cdot)\|_{(k)}\,ds\Big]. \tag{7.7}$$

Here, the constant depends on k*, the bounds on* A^μ *and* B *and on* T.

Proof. Due to (7.6), (7.7) holds for $k \ge 0$, so let us assume k is a negative integer. Define

$$U(t,\cdot) = (1-\Delta)^k u(t,\cdot)$$

for all $t \in [0, T]$. Note that due to our assumptions U satisfies uniform Schwartz bounds. Furthermore,

$$\begin{aligned}\|u(t,\cdot)\|_{(k)} = \|U(t,\cdot)\|_{(-k)} &\leq CE_{-k}^{1/2}[U](t) \\ &\leq C\Big[E_{-k}^{1/2}[U](0) + \int_0^t \|LU(s,\cdot)\|_{(-k)}\,ds\Big] \\ &\leq C\Big[\|u(0,\cdot)\|_{(k)} + \int_0^t \|LU(s,\cdot)\|_{(-k)}\,ds\Big],\end{aligned}$$

where we have used (7.6) in the second to last step. What remains to be estimated is the last term on the right-hand side. We have

$$f = Lu = (1-\Delta)^{-k}LU + [L, (1-\Delta)^{-k}]U.$$

This equality yields that

$$\|LU(t,\cdot)\|_{(-k)} \leq \|f(t,\cdot)\|_{(k)} + \|[L,(1-\Delta)^{-k}]U(t,\cdot)\|_{(k)}.$$

By Corollary 5.19, the last term can be estimated by

$$C\left[\|U(t,\cdot)\|_{(-k)} + \|\partial_t U(t,\cdot)\|_{(-k-1)}\right].$$

In order to be able to estimate the last term of this expression, let us define

$$L_0 u = (A^0)^{-1}Lu.$$

Then

$$(A^0)^{-1}f = L_0 u = (1-\Delta)^{-k}L_0 U + [L_0, (1-\Delta)^{-k}]U.$$

This equality can be used to obtain an expression for $(1-\Delta)^{-k}\partial_0 U$. We wish to estimate this expression in $H_{(k-1)}$. Note that

$$\|(1-\Delta)^{-k}(L_0 - \partial_t)U(t,\cdot)\|_{(k-1)} \leq C\,\|U(t,\cdot)\|_{(-k)}.$$

The term $[L_0, (1-\Delta)^{-k}]U$ does not contain any time derivatives and in fact satisfies a better estimate in $H_{(k-1)}$. To conclude, we obtain

$$\|\partial_t U(t,\cdot)\|_{(-k-1)} \leq C[\|U(t,\cdot)\|_{(-k)} + \|f(t,\cdot)\|_{(k-1)}].$$

Note that we used Lemma 5.17 in order to estimate $(A^0)^{-1}f$ in $H_{(k-1)}$. Adding up the above observations, we obtain

$$\|u(t,\cdot)\|_{(k)} \leq C\Big[\|u(0,\cdot)\|_{(k)} + \int_0^t (\|u(s,\cdot)\|_{(k)} + \|f(s,\cdot)\|_{(k)})\,ds\Big].$$

An application of Grönwall's lemma, Lemma 7.1, yields the desired result. □

Corollary 7.6. *Assume that u is a solution of (7.2) under the assumptions made in the beginning of the section and let k be any integer. Then, for $t \in [0, T]$,*

$$\|u(t,\cdot)\|_{(k)} \leq C\left[\|u(T,\cdot)\|_{(k)} + \int_t^T \|f(s,\cdot)\|_{(k)}\,ds\right]. \tag{7.8}$$

Here, the constant depends on k, the bounds on A^μ and B and on T.

Proof. Define

$$\begin{aligned}(\tilde{L}u)(t,x) &= -A^0(T-t,x)(\partial_t u)(t,x) + A^j(T-t,x)(\partial_j u)(t,x)\\ &\quad + B(T-t,x)u(t,x),\\ v(t,x) &= u(T-t,x).\end{aligned}$$

Then $-\tilde{L}$ is an operator of the same type as L, so that

$$\|v(t,\cdot)\|_{(k)} \leq C\left[\|v(0,\cdot)\|_{(k)} + \int_0^t \|(\tilde{L}v)(s,\cdot)\|_{(k)}\,ds\right]$$

for all integers k and all $t \in [0, T]$ due to (7.7). Since $(\tilde{L}v)(t,x) = (Lu)(T-t,x)$, this inequality can be reformulated to (7.8). □

7.3 Uniqueness

Before we state uniqueness, let us introduce notation for a set that will come up repeatedly in the statement. Given $x_0 \in \mathbb{R}^n$, $r > 0$, $s_0 > 0$ and $T_1 < 0 < T_2$, define

$$C_{x_0,r,s_0,T_1,T_2} = \{(t,x) \in [T_1,T_2] \times \mathbb{R}^n : |t| < r/s_0,\ x \in B_{r-s_0|t|}(x_0)\}. \tag{7.9}$$

Proposition 7.7. *Assume A^μ and B to be maps from $\mathbb{R}^{n+1}$ to the set of real-valued $N \times N$ matrices, with A^μ symmetric and C^1 and B in C^0. Assume furthermore that for any compact interval $[T_1, T_2]$, A^0 is positive definite on $[T_1, T_2] \times \mathbb{R}^n$ with a constant positive lower bound and that the matrices A^μ are all bounded on the same set. Concerning the function $f : \mathbb{R}^{n+1} \to \mathbb{R}^N$, we assume that it is continuous. Let us assume we have two C^1-solutions u_1 and u_2 to (7.2)–(7.3), defined on $(a,b) \times \mathbb{R}^n$, where $a < 0$ and $b > 0$, and corresponding to initial data u_{01} and u_{02}. Let $[T_1, T_2]$ be a compact subinterval of (a,b) with $T_1 \leq 0$ and $T_2 \geq 0$. Then there is an $s_0 > 0$, depending on the lower bound on A^0 and the upper bound on A^i in $[T_1, T_2]$ such that if $u_{01}(x) = u_{02}(x)$ for $x \in B_r(x_0)$, then $u_1(t,x) = u_2(t,x)$ for $(t,x) \in C = C_{x_0,r,s_0,T_1,T_2}$. Analogously, if u is a C^1 solution to (7.2)–(7.3) on $[T_1,T_2] \times \mathbb{R}^n$, $u_0(x) = 0$ for $x \in B_r(x_0)$ and $f(t,x) = 0$ for $(t,x) \in C$, then $u(t,x) = 0$ for $(t,x) \in C$.*

Remark 7.8. One special conclusion of the proposition is of course that if the initial data coincide, then the solutions coincide where they are defined. Furthermore, if u_0 has compact support and for any $[T_1, T_2]$, there is a compact set K such that $f(t,x) = 0$ for $t \in [T_1, T_2]$ if $x \notin K$, then there is a compact set K_1 such that $u(t,x) = 0$ for $t \in [T_1, T_2]$ if $x \notin K_1$.

Proof. By considering $u_1 - u_2$, we see that the first statement is a consequence of the second. Define

$$\mathcal{D} = C_{x_0,r,s_0,0,T_2}.$$

Let us use the notation $\partial_0 = \partial_t$ and consider

$$\partial_\alpha[e^{-kt}u^t A^\alpha u] = e^{-kt}u^t[-kA^0 + (\partial_\alpha A^\alpha) - 2B]u + 2e^{-kt}u^t f,$$

where k is some constant. Let us integrate this equality over $\mathcal{D}$. The integral of the left-hand side can be reformulated to an integral over the boundary using Stokes' theorem, cf. (10.3) in the case of the ordinary Euclidean metric on $\mathbb{R}^{n+1}$. There are two or three terms. The integral over the part of $\partial\mathcal{D}$ with $t = 0$ is zero. The outward normal on the remaining surfaces, say n_α, can be made to be such that $n_\alpha A^\alpha$ is a positive definite matrix by choosing s_0 to be large enough (note that this can be done independently of the region and only depending on a lower bound on A^0 and an upper bound on A^i on the set $[T_1, T_2] \times \mathbb{R}^n$). Fix the corresponding s_0 once and for all. Then the boundary integrals are all non-negative. By assumption, f is zero in $\mathcal{D}$. The integral of the right-hand side can be assumed to be smaller than

$$-c_0 \int_{\mathcal{D}} e^{-kt}u^t u \, dx$$

for some positive constant c_0 by choosing k large enough (this is possible since A^0 has a positive uniform lower bound considered as a positive definite matrix, and the other matrices involved have a uniform upper bound, since the region is bounded and the relevant quantities are continuous). We thus have an equality in which the left-hand side is non-negative and the right-hand side is non-positive. Both sides must thus equal zero, and we conclude that u has to equal zero in $\mathcal{D}$. The argument for negative times follows by reversing time. □

7.4 Existence

We are finally in a position to prove existence of solutions to (7.2).

Theorem 7.9. *Let $S_T = [0, T] \times \mathbb{R}^n$, where $T > 0$. Consider the initial value problem (7.2)–(7.3), where $u_0 \in C_0^\infty(\mathbb{R}^n, \mathbb{R}^N)$, $f \in C_0^\infty(\mathbb{R}^{n+1}, \mathbb{R}^N)$, A^μ and B are smooth functions from $\mathbb{R}^{n+1}$ to the set of real-valued $N \times N$-matrices, all of whose derivatives are bounded, with A^μ symmetric and A^0 positive definite with a uniform positive lower bound. Then there is a unique $u \in C^\infty([0, T) \times \mathbb{R}^n, \mathbb{R}^N)$ solving (7.2)–(7.3) and a compact set $K \subset \mathbb{R}^n$ such that $u(t, x) = 0$ for $t \in [0, T]$ and $x \notin K$.*

Proof. Given L as above, let us define L^* by

$$\begin{aligned} L^*u &= -A^0\partial_t u - A^j\partial_j u + (-\partial_t A^0 - \partial_j A^j + B^t)u \\ &= -\partial_t(A^0 u) - \partial_j(A^j u) + B^t u. \end{aligned}$$

Then $-L^*$ is an operator of the same type as L, so that (7.8) holds with L replaced by L^*. For every $\phi \in C_0^\infty(\mathbb{R}^{n+1}, \mathbb{C}^N)$ such that $\phi(t,x) = 0$ for all $t \geq T$, we thus get

$$\|\phi(t,\cdot)\|_{(-k)} \leq C \int_t^T \|(L^*\phi)(s,\cdot)\|_{(-k)}\, ds \tag{7.10}$$

for all $t \in [0,T]$. Note that as a consequence, if $\phi, \psi \in C_0^\infty(\mathbb{R}^{n+1}, \mathbb{C}^N)$ both vanish for $t \geq T$ and $L^*\phi = L^*\psi$, then $\phi(t,\cdot) = \psi(t,\cdot)$ for all $t \in [0,T]$. Given a ϕ as above and an $f \in L^1\{[0,T], H_{(k)}(\mathbb{R}^n, \mathbb{C}^N)\}$, define

$$F(L^*\phi) = \langle f, \phi\rangle = \int_0^T (\phi(t), f(t))_{L^2}\, dt,$$

where the last equality is a definition of $\langle \cdot\,, \cdot\rangle$. Strictly speaking, the definition goes along the same lines of the discussion preceding Proposition 5.21, but due to the regularity of ϕ, the right-hand side makes sense as it stands, assuming that $k \geq 0$. By the above observations, the definition of F makes sense. By (7.10), we have

$$|F(L^*\phi)| \leq C \int_0^T \|(L^*\phi)(t,\cdot)\|_{(-k)}\, dt.$$

Note that $L^*\phi$ can be considered as an element of

$$X = L^1\{[0,T], H_{(-k)}(\mathbb{R}^n, \mathbb{C}^N)\}.$$

Let M be the subspace of X spanned by $L^*\phi$ for ϕ as above. Then F as defined above is a bounded linear functional on M. By the Hahn–Banach theorem, cf. Theorem 5.16 of [79], F can be extended to a bounded linear functional on X which we, by abuse of notation, shall also denote F. Furthermore, the norm of the extension coincides with the norm of the functional restricted to M. By Proposition 5.21, we conclude that there is a $u \in L^\infty\{[0,T], H_{(k)}(\mathbb{R}^n, \mathbb{C}^N)\}$ such that, for all ϕ of the above type,

$$\int_0^T (\phi(t), f(t))_{L^2}\, dt = \int_0^T (L^*\phi(t), u(t))_{L^2}\, dt.$$

Let us first assume $f \in C_0^\infty(\mathbb{R}^{n+1}, \mathbb{R}^N)$ is such that $f(t,\cdot) = 0$ for all $t \leq 0$. We can then extend u to be an element of $L^\infty\{(-\infty, T], H_{(k)}(\mathbb{R}^n, \mathbb{C}^N)\}$ by demanding that it be zero for $t < 0$, so that

$$\int_{-\infty}^T (\phi(t), f(t))_{L^2}\, dt = \int_{-\infty}^T (L^*\phi(t), u(t))_{L^2}\, dt$$

for all $\phi \in C_0^\infty(\mathbb{R}^{n+1}, \mathbb{C}^N)$ such that $\phi(t,\cdot) = 0$ for all $t \geq T$. Due to Lemma A.5, there is a $U \in L^2_{\text{loc}}[(-\infty, T) \times \mathbb{R}^n, \mathbb{C}^N]$ which is k times weakly differentiable with respect to x. Furthermore,

$$\int_{-\infty}^T \int_{\mathbb{R}^n} \phi \cdot \bar{f}\, dx dt = \int_{-\infty}^T \int_{\mathbb{R}^n} L^*\phi \cdot \bar{U}\, dx dt \tag{7.11}$$

for all $\phi \in C_0^\infty[(-\infty, T) \times \mathbb{R}^n, \mathbb{C}^N)$. Let us make the inductive assumption that if $j+|\alpha| \leq k$ and $j \leq l \leq k-1$, then there is a function $U_{j,\alpha} \in L^2_{\text{loc}}[(-\infty, T)\times\mathbb{R}^n, \mathbb{C}^N]$ such that

$$\int_{-\infty}^T \int_{\mathbb{R}^n} \phi \cdot \bar{U}_{j,\alpha}\, dx dt = (-1)^{j+|\alpha|} \int_{-\infty}^T \int_{\mathbb{R}^n} \partial_t^j \partial^\alpha \phi \cdot \bar{U}\, dx dt$$

for all $\phi \in C_0^\infty[(-\infty, T) \times \mathbb{R}^n, \mathbb{C}^N)$. We shall also write $\partial_t^j \partial^\alpha U = U_{j,\alpha}$. If $l = 0$, the inductive statement is true. We can reformulate (7.11) to

$$\int_{-\infty}^T \int_{\mathbb{R}^n} \psi \cdot \bar{g}\, dx dt = -\int_{-\infty}^T \int_{\mathbb{R}^n} \partial_t \psi \cdot \bar{U}\, dx dt, \tag{7.12}$$

where $\psi = A^0 \phi$ and

$$g = (A^0)^{-1}[f - A^j \partial_j U - BU].$$

Note that any $\psi \in C_0^\infty[(-\infty, T) \times \mathbb{R}^n, \mathbb{C}^N)$ can be written as $\psi = A^0 \phi$ where ϕ is in the same class as ψ. Furthermore, for any multiindex α and non-negative integer j with $|\alpha| + j \leq k - 1$ and $j \leq l$, the weak derivative $\partial^\alpha \partial_t^j g$ exists and is in $L^2_{\text{loc}}[(-\infty, T) \times \mathbb{R}^n, \mathbb{C}^N]$. For any α with $|\alpha| \leq k - l - 1$, we can thus replace ψ in (7.12) with $\partial_t^l \partial^\alpha \psi$ and reformulate the left-hand side to be the integral of the scalar product of ψ and a function which is in $L^2_{\text{loc}}[(-\infty, T) \times \mathbb{R}^n, \mathbb{C}^N]$. Thus the inductive assumption holds with l replaced by $l + 1$ until we reach the conclusion that U is k times weakly differentiable with respect to x and t in $(-\infty, T) \times \mathbb{R}^n$. For k large enough, we conclude that U is continuously differentiable due to Lemma 6.7. Then (7.11) implies that $LU = f$. Furthermore, $U = 0$ for $t \leq 0$. We would like to conclude that U is smooth, but the above procedure gives a U for each k, and we are not allowed to assume apriori that the different U's coincide. Using Proposition 7.7, we are, however, allowed to assume that the solutions coincide as long as they are C^1. As a consequence, we see that all the solutions constructed above for different k's (large enough) coincide. Consequently the solution is smooth.

Let us consider the case that $f \in C_0^\infty(\mathbb{R}^{n+1}, \mathbb{R}^N)$ but that f does not necessarily vanish for $t \leq 0$. Let $\eta \in C_0^\infty(\mathbb{R}, \mathbb{R})$ be such that $0 \leq \eta \leq 1$, $\eta(t) = 0$ for all $t \leq 0$ and $\eta(t) = 1$ for all $t \geq 1$. Define

$$f_\varepsilon(t, x) = \eta(t/\varepsilon) f(t, x).$$

For every $\varepsilon > 0$, we get a smooth solution u_ε to the equation $Lu_\varepsilon = f_\varepsilon$ such that $u_\varepsilon(t, \cdot) = 0$ for all $t \leq 0$. By Proposition 7.7, there is a compact set K such that $u_\varepsilon(t, x) = 0$ for $x \notin K$ and $t \leq T$ and all $\varepsilon > 0$. Due to (7.7), we conclude that

$$\|(u_{\varepsilon_1} - u_{\varepsilon_2})(t, \cdot)\|_{(k)} \leq C \int_0^t |\eta(s/\varepsilon_1) - \eta(s/\varepsilon_2)| \|f(s, \cdot)\|_{(k)}\, ds.$$

Thus for any $t \in [0, T]$, $u_\varepsilon(t, \cdot)$ converges in any H^k-norm, and thus with respect to any C^k norm, as $\varepsilon \to 0+$. Using the equation, we get convergence of any number of time derivatives as well, assuming we stay away from $t = 0$. In this way, we obtain a smooth function u on $(0, T) \times \mathbb{R}^n$ which solves the equation. We would like to extend the definition of this function to $t = 0$. How to define the function at $t = 0$ is clear; $u(0, \cdot) = 0$, and the higher time derivatives are given recursively by the equation. The question is then if $\partial_t u$ converges to what it should converge to as $t \to 0+$ etc. Due to (7.7), we have

$$\|u_\varepsilon(t, \cdot)\|_{(k)} \le C \int_0^t \|f_\varepsilon(s, \cdot)\|_{(k)}\, ds \le C \int_0^t \|f(s, \cdot)\|_{(k)}\, ds.$$

Since the right-hand side is independent of ε, this inequality must hold for u as well. Choosing k large enough that the H^k-norm dominates the C^0-norm, we conclude that $u(t, \cdot)$ converges to zero in C^0 as $t \to 0+$ (in fact it converges to zero in any C^k norm). Inserting this information into the equation, we conclude that $\partial_t u$ converges to what it ought to converge to in any C^k norm. Recursively, the same holds for all the higher time derivatives. The equation (7.2)–(7.3) thus has a smooth solution on $[0, T) \times \mathbb{R}^n$ assuming $f \in C_0^\infty(\mathbb{R}^{n+1}, \mathbb{R}^N)$ and that $u_0 = 0$. To get the same statement with $u_0 \neq 0$, we simply consider the equation for $u - u_0\psi$, where $\psi \in C_0^\infty(\mathbb{R})$ is such that $\psi(t) = 1$ for $t \in [-1, T+1]$. Thus we get a smooth solution if we, in addition to the assumptions made concerning f, also assume that $u_0 \in C_0^\infty(\mathbb{R}^n, \mathbb{R}^N)$. □

Corollary 7.10. *Let $u_0 \in C^\infty(\mathbb{R}^n, \mathbb{R}^N)$, $f \in C^\infty(\mathbb{R}^{n+1}, \mathbb{R}^N)$, and A^μ and B be smooth functions from $\mathbb{R}^{n+1}$ to the set of real-valued $N \times N$-matrices. Assume that the matrices A^μ are symmetric and that for any compact interval $[T_1, T_2]$, there are constants $a_0, b_0 > 0$ such that $A^0 \ge a_0$ and $\|A^\mu\| \le b_0$, $\mu = 0, \dots, n$, on $[T_1, T_2] \times \mathbb{R}^n$. Then there is a unique solution $u \in C^\infty(\mathbb{R}^{n+1}, \mathbb{R}^N)$ to the initial value problem* (7.2)–(7.3).

Remark 7.11. One can of course also prove existence results in lower regularity.

Proof. The uniqueness part of the statement follows from Proposition 7.7. In order to prove existence, let us construct the solution on $[0, T) \times \mathbb{R}^n$ for a given $T \in (0, \infty)$. Let s_0 be as in the statement of Proposition 7.7, given $[T_1, T_2] = [0, T]$. Let $r \ge Ts_0 + 1$, and define

$$C_r = \{(t, x) \in [0, T] \times \mathbb{R}^n : x \in B_{r - s_0 t}(0)\}.$$

Let $\psi_r \in C_0^\infty(\mathbb{R}^{n+1})$ be such that $\psi_r(t, x) = 1$ on C_{2r+2s_0T} and $0 \le \psi_r \le 1$, and let $\phi_r \in C_0^\infty(\mathbb{R}^n)$ be such that $\phi_r(x) = 1$ on $B_r(0)$ and $\phi_r(x) = 0$ for $x \notin B_{2r}(0)$. Define

$$A_r^0 = \psi_r A^0 + (1 - \psi_r) A^0(0, 0), \quad A_r^i = \psi_r A^i, \quad B_r = \psi_r B, \quad u_{0r} = \phi_r u_0,$$

and $f_r(t, x) = \psi_r(t, x)\phi_r(x) f(t, x)$. Due to Theorem 7.9, the equation

$$A_r^\mu \partial_\mu u + B_r u = f_r, \tag{7.13}$$

$$u(0, \cdot) = u_{0r} \tag{7.14}$$

has a smooth solution; let us call it u_r. The s_0 mentioned above only depends on a lower bound on A^0 and an upper bound on A^i. A^0_r satisfies the same lower bound as A^0 and the A^i_r satisfy a better upper bound. Consequently we can apply Proposition 7.7 with the same s_0 as for A^μ to the solution of (7.13)–(7.14). In particular, we conclude that $u_r(t,x) = 0$ for $t \in [0,T]$ and $x \notin B_{2r+s_0t}(0)$. Consequently, whenever $u_r(t,x) \neq 0$ and $t \in [0,T]$, we have $A^\mu_r(t,x) = A^\mu(t,x)$ and $B_r(t,x) = B(t,x)$. Consequently, u_r is a solution to the equation

$$\begin{aligned} A^\mu \partial_\mu u + Bu &= f_r, \\ u(0,\,\cdot\,) &= u_{0r} \end{aligned}$$

on $[0,T) \times \mathbb{R}^n$. In the region C_r, u_r is of course a solution to the original equation. Let us consider two solutions u_{r_i}, $i = 1,2$ of the above type, where $r_1 < r_2$. By uniqueness, $u_{r_1} = u_{r_2}$ on C_{r_1}. If $(t,x) \in [0,T) \times \mathbb{R}^n$, let r be such that (t,x) is in the interior of C_r and define $u(t,x) = u_r(t,x)$. By the above observation, the choice of r does not matter, and we get a smooth solution to the equation on $[0,T) \times \mathbb{R}^n$. Since T was arbitrary, we can define the solution for all future times due to a similar uniqueness argument. The argument in the opposite time direction follows by reversing time. □

8 Linear wave equations

Let us begin this chapter by giving a detailed derivation of an energy estimate for a linear wave equation. The reason for the importance of the energy estimates is that they form the basis for existence and uniqueness proofs. They will not play a central role in the present chapter, since we shall here rely on the results derived in the symmetric hyperbolic setting, but they will in the arguments used to prove local existence of solutions to non-linear wave equations.

Consider an equation of the form

$$g^{\mu\nu}\partial_\mu\partial_\nu u + a^\mu\partial_\mu u + bu = f, \tag{8.1}$$

where $g^{\mu\nu}$ are the components of a smooth $(n+1)\times(n+1)$ real symmetric matrix-valued function on $\mathbb{R}^{n+1}$ which is such that $g^{00} < 0$ and g^{ij}, $i, j = 1, \dots, n$, are the components of a positive definite matrix. Assume furthermore a^μ and b to be smooth $N\times N$ real matrix-valued functions and f to be a smooth $\mathbb{R}^N$-valued function on $\mathbb{R}^{n+1}$. Finally, assume we have a smooth $\mathbb{R}^N$-valued solution u to this equation on $\mathbb{R}^{n+1}$ with the property that for a compact interval $I \subseteq \mathbb{R}$, there is a compact set $K_I \subset \mathbb{R}^n$ such that $u(t,x) = 0$ if $t \in I$ and $x \notin K_I$. Under these circumstances, the energy defined by

$$\mathcal{E} = \frac{1}{2}\int_{\mathbb{R}^n}[-g^{00}|u_t|^2 + g^{ij}\partial_i u\cdot\partial_j u + |u|^2]\,dx \tag{8.2}$$

makes sense.

Lemma 8.1. *Given the situation described above, assume $g^{\mu\nu}$, its first derivatives, b and a^μ all have uniform bounds. Assume furthermore that g^{00} has a uniform negative upper bound and that the positive definite matrix with components g^{ij}, $i, j = 1, \dots, n$ has a uniform positive lower bound. Then there is a constant C, depending on the bounds, such that for $t \geq 0$, the energy defined by* (8.2) *satisfies the inequality*

$$\mathcal{E}^{1/2}(t) \leq \left(\mathcal{E}^{1/2}(0) + C\int_0^t \|f(s,\cdot)\|_2\,ds\right)e^{Ct}. \tag{8.3}$$

Remark 8.2. The result is only intended to illustrate the type of arguments one can carry out; it is of course possible to obtain the same conclusions under weaker conditions. Note that one consequence of this inequality is that if we have two solutions to the equation (8.1) satisfying the conditions stated above, and if the initial data of these solutions coincide for $t = 0$, then they have to coincide for all $t \geq 0$; consider the inequality obtained for the difference of the solutions.

Proof. Differentiating with respect to time, we obtain

$$\begin{aligned}\partial_t\mathcal{E} = \int_{\mathbb{R}^n}\Big[&-\frac{1}{2}(\partial_t g^{00})|u_t|^2 - g^{00}u_t\cdot u_{tt}\\ &+\frac{1}{2}(\partial_t g^{ij})\partial_i u\cdot\partial_j u + g^{ij}\partial_i u\cdot\partial_j\partial_t u + u\cdot u_t\Big]\,dx.\end{aligned} \tag{8.4}$$

The first, the third and the fifth terms in the integrand on the right-hand side are harmless since they can be bounded in terms of the energy. In order for the resulting constants to be uniform, we do, however, need to have a uniform negative upper bound on g^{00} and a uniform positive lower bound on the positive definite matrix with components g^{ij}, $i, j = 1, \dots, n$. We then obtain that

$$\int_{\mathbb{R}^n} \Big[-\frac{1}{2}(\partial_t g^{00})|u_t|^2 + \frac{1}{2}(\partial_t g^{ij})\partial_i u \cdot \partial_j u + u \cdot u_t \Big] dx \leq C\mathcal{E} \tag{8.5}$$

for some constant C (in what follows C will change from line to line without comment in order that the notation not become to cumbersome). In considering the remaining terms of (8.4), note that

$$\begin{aligned}
&\int_{\mathbb{R}^n} g^{ij}\partial_i u \cdot \partial_j \partial_t u \, dx \\
&\quad = \int_{\mathbb{R}^n} [\partial_j(g^{ij}\partial_i u \cdot \partial_t u) - (\partial_j g^{ij})\partial_i u \cdot \partial_t u - g^{ij}\partial_i \partial_j u \cdot \partial_t u]\, dx \\
&\quad = -\int_{\mathbb{R}^n} [(\partial_j g^{ij})\partial_i u \cdot \partial_t u + g^{ij}\partial_i \partial_j u \cdot \partial_t u]\, dx.
\end{aligned}$$

The first term in the integrand on the extreme right-hand side of this equation can be estimated as in (8.5). Adding up the above observations, we thus conclude that

$$\partial_t \mathcal{E} \leq C\mathcal{E} - \int_{\mathbb{R}^n} (g^{00}u_{tt} + g^{ij}\partial_i \partial_j u) \cdot u_t \, dx.$$

In the end we wish to use the equation, but considering the above expression, it is clear that there is one term missing: $-2g^{0i}\partial_i \partial_t u \cdot u_t$. However,

$$\int_{\mathbb{R}^n} 2g^{0i}\partial_i \partial_t u \cdot u_t \, dx = \int_{\mathbb{R}^n} g^{0i}\partial_i(|u_t|^2)\, dx = -\int_{\mathbb{R}^n} (\partial_i g^{0i})|u_t|^2 \, dx.$$

Since the right-hand side of this equation can be estimated by $C\mathcal{E}$, we conclude that

$$\partial_t \mathcal{E} \leq C\mathcal{E} - \int_{\mathbb{R}^n} g^{\mu\nu}\partial_\mu \partial_\nu u \cdot u_t \, dx.$$

At this stage we are of course allowed to use the equation. Assuming a^μ and b to be bounded, we conclude that

$$\partial_t \mathcal{E} \leq C\mathcal{E} + C\mathcal{E}^{1/2}\|f(t,\cdot)\|_2.$$

Let $\varepsilon > 0$ and define $\mathcal{E}_\varepsilon = \mathcal{E} + \varepsilon$ (the motivation for this is a technicality; we wish to divide by $\sqrt{\mathcal{E}_\varepsilon}$). Then we get the same estimate for $\mathcal{E}_\varepsilon$ and we can divide by $\sqrt{\mathcal{E}_\varepsilon}$ in order to obtain

$$\partial_t \mathcal{E}_\varepsilon^{1/2} \leq C\mathcal{E}_\varepsilon^{1/2} + C\|f(t,\cdot)\|_2.$$

Integrating, we obtain, for $t \geq 0$,

$$\mathcal{E}_\varepsilon^{1/2}(t) \leq \mathcal{E}_\varepsilon^{1/2}(0) + C \int_0^t \|f(s,\cdot)\|_2\, ds + C \int_0^t \mathcal{E}_\varepsilon^{1/2}(s)\, ds.$$

Applying Grönwall's lemma, Lemma 7.1, we conclude that

$$\mathcal{E}_\varepsilon^{1/2}(t) \leq \Big(\mathcal{E}_\varepsilon^{1/2}(0) + C \int_0^t \|f(s,\cdot)\|_2\, ds\Big) e^{Ct}.$$

At this stage, we can let $\varepsilon \to 0+$ in order to obtain (8.3). □

In the above argument, we made certain assumptions concerning $g^{\mu\nu}$. However, in the end it will be natural to consider $g^{\mu\nu}$ to be the components of the inverse of a Lorentz metric. In the next section, we write down conditions on the components of the Lorentz metric that ensure that the components of the inverse have these properties.

8.1 Linear algebra

Let us introduce some notation. Let g be a symmetric $(n+1)\times(n+1)$-dimensional real-valued matrix with components $g_{\mu\nu}$, $\mu,\nu = 0,\dots,n$. We shall denote the $n\times n$ matrix with components g_{ij}, $i,j = 1,\dots,n$, by $g_\flat$ and if g is invertible, we shall denote the components of the inverse by $g^{\mu\nu}$, $\mu,\nu = 0,\dots,n$ and the $n\times n$ matrix with components g^{ij}, $i,j = 1,\dots,n$, by $g^\sharp$. We shall use $v[g]$ to denote the n-vector with components g_{0i}, $i = 1,\dots,n$ and for any symmetric and positive definite $n\times n$-matrix ξ and any n-vector v, we shall write

$$|v|_\xi = \Big(\sum_{i,j=1}^n \xi_{ij} v^i v^j\Big)^{1/2}.$$

We shall also use the notation $|v| = |v|_\delta$, where δ_{ij} is the Kronecker delta. Furthermore, if A is an $n\times n$ real-valued matrix (not necessarily symmetric), we shall denote the $(n+1)\times(n+1)$-dimensional matrix with 00-component 1, $0i$ and $i0$-components 0 and ij components given by A_{ij} by M_A. Finally, if ρ is a symmetric, real-valued $(n+1)\times(n+1)$-matrix with one negative eigenvalue and n positive ones, we shall say that it is a *Lorentz matrix*.

Lemma 8.3. *Let ρ be a symmetric $(n+1)\times(n+1)$ real-valued matrix. Assume that $\rho_{00} < 0$ and that $\rho_\flat$ is positive definite. Then ρ is a Lorentz matrix.*

Proof. Let A be an orthogonal $n\times n$-matrix diagonalizing $\rho_\flat$ and $h = M_A^t \rho M_A$. Then $h_\flat = A^t \rho_\flat A$ is diagonal, with diagonal elements $\lambda_i > 0$, $i = 1,\dots,n$. Furthermore $h_{00} = \rho_{00} < 0$ and the eigenvalues of ρ and h coincide. If we compute the determinant of $h - \lambda \mathrm{Id}$, we obtain

$$p(\lambda) = \Big(\rho_{00} - \lambda - \frac{h_{01}^2}{\lambda_1 - \lambda} - \dots - \frac{h_{0n}^2}{\lambda_n - \lambda}\Big)(\lambda_1 - \lambda)\dots(\lambda_n - \lambda).$$

Let us define

$$f(\lambda) = \rho_{00} - \lambda - \frac{h_{01}^2}{\lambda_1 - \lambda} - \cdots - \frac{h_{0n}^2}{\lambda_n - \lambda}.$$

If we differentiate this function, we obtain

$$f'(\lambda) = -1 - \frac{h_{01}^2}{(\lambda_1 - \lambda)^2} - \cdots - \frac{h_{0n}^2}{(\lambda_n - \lambda)^2}.$$

Note that $\lambda_1, \dots, \lambda_n$ are all positive. Let us denote the smallest of the λ_i by $\lambda_{\min}$. For λ belonging to the interval $(-\infty, \lambda_{\min})$, we obtain the conclusion that $f'(\lambda) < 0$. Furthermore, $f(-\infty) = \infty$ and $f(0) < 0$. Thus there is a unique negative value of λ, say λ_0, for which $f(\lambda) = 0$. This is clearly an eigenvalue of ρ. Since it is easy to see that $p'(\lambda_0) \neq 0$, we see that λ_0 is a root with multiplicity one to the polynomial equation $p(\lambda) = 0$. There is in other words only one eigenvalue in the interval $(-\infty, \lambda_{\min})$. Since ρ is a symmetric matrix, it only has real eigenvalues, so the remaining n eigenvalues have to be positive. □

In what follows, it will be convenient to introduce some terminology for the type of matrices discussed in the above lemma.

Definition 8.4. A *canonical Lorentz matrix* is a symmetric $(n+1)\times(n+1)$-dimensional real-valued matrix g with components $g_{\mu\nu}$, $\mu, \nu = 0, \dots, n$ such that $g_{00} < 0$ and $g_\flat > 0$. Let $\mathfrak{C}_n$ denote the set of $(n+1) \times (n+1)$-dimensional canonical Lorentz matrices. If a_1, a_2 and a_3 are positive real numbers and $\boldsymbol{a} = (a_1, a_2, a_3)$, we define the $\mathfrak{C}_{n,\boldsymbol{a}}$ to be the subset of $\mathfrak{C}_n$ consisting of matrices g satisfying

$$g_{00} \le -a_1, \quad g_\flat \ge a_2, \quad \sum_{\mu,\nu=0}^{n} |g_{\mu\nu}| \le a_3.$$

We shall sometimes write $\boldsymbol{a} > 0$ to indicate that all the components of $\boldsymbol{a}$ are strictly positive.

Lemma 8.5. *Let $g \in \mathfrak{C}_n$. Then g is a Lorentz matrix,*

$$g^{00} = \frac{1}{g_{00} - d^2}, \tag{8.6}$$

where $d = |v[g]|_{g_\flat^{-1}}$, $g^\sharp$ is positive definite, with

$$\frac{g_{00}}{g_{00} - d^2} |w|^2_{g_\flat^{-1}} \le |w|^2_{g^\sharp} \le |w|^2_{g_\flat^{-1}} \tag{8.7}$$

for any $w \in \mathbb{R}^n$ and

$$v[g^{-1}] = \frac{1}{d^2 - g_{00}} g_\flat^{-1} v[g]. \tag{8.8}$$

Note that g^{00} is negative since g_{00} is negative and that there is an upper bound on this quantity depending only on g_{00} and d. In particular, $g^{-1} \in \mathfrak{C}_n$.

Proof. That g is a Lorentz matrix follows from Lemma 8.3. Let A be the square root of $g_\flat^{-1}$, i.e., the positive definite, symmetric matrix with the property that $A^2 = g_\flat^{-1}$. Then $A^t g_\flat A = \mathrm{Id}$. Consider $h = M_A^t g M_A$. Then $h_{00} = g_{00}$, $h_\flat = \mathrm{Id}$ and $v[h] = A^t v[g]$. Let B be an orthogonal matrix such that $B^t A^t v[g] = |A^t v[g]| e_1$, where $e_1 = (1, 0, \dots, 0)^t$. Note that

$$d = |A^t v[g]| = |v[g]|_{g_\flat^{-1}}.$$

Consider $\rho = M_B^t M_A^t g M_A M_B$. Then $\rho_{00} = g_{00}$, $\rho_\flat = \mathrm{Id}$ and $v[\rho] = d e_1$. Note that the inverse of the 2×2-matrix with components $\rho_{\mu\nu}$, $\mu, \nu = 0, 1$ is given by

$$\frac{1}{g_{00} - d^2} \begin{pmatrix} 1 & -d \\ -d & g_{00} \end{pmatrix}$$

Since

$$g^{-1} = M_A M_B \rho^{-1} M_B^t M_A^t,$$

and the matrices M_A and M_B preserve the 00-component of a matrix, we obtain (8.6). Furthermore $g^\sharp = A B \rho^\sharp B^t A^t$. We are interested in the supremum and infimum, for $w \neq 0$, of

$$\frac{|w|^2_{g^\sharp}}{|w|^2_{g_\flat^{-1}}} = \frac{(g^\sharp w, w)}{(Aw, Aw)} = \frac{(\rho^\sharp B^t A^t w, B^t A^t w)}{(B^t A^t w, B^t A^t w)},$$

where we have used the fact that B is orthogonal and A is symmetric. Since $\rho^\sharp$ is diagonal with 11-component equal to $g_{00}/(g_{00} - d^2) \leq 1$ and the ii-components equal 1 for $i > 1$, we obtain (8.7). Since

$$v[\rho^{-1}] = -\frac{d}{g_{00} - d^2} e_1, \quad d B e_1 = A^t v[g],$$

we obtain

$$v[g^{-1}] = A B v[\rho^{-1}] = -\frac{1}{g_{00} - d^2} A^2 v[g] = -\frac{1}{g_{00} - d^2} g_\flat^{-1} v[g],$$

which implies (8.8). Note that one could also have obtained this equality by applying $g_\flat^{-1}$ to $g^{0i} g_{ij} + g^{00} g_{0j} = 0$ and using the fact that (8.6) holds. □

8.2 Existence of solutions to linear wave equations

Let g_I, $I = 1, \dots, N$ be smooth functions from $\mathbb{R}^{n+1}$ into $\mathcal{C}_n$, cf. Definition 8.4, with components $g_{I\mu\nu}$, $\mu, \nu = 0, \dots, n$. Assume that for any compact interval $[T_1, T_2]$, there are constants $a_i > 0$, $i = 1, 2, 3$, such that $g_I(t, x) \in \mathcal{C}_{n,\boldsymbol{a}}$ (where $\boldsymbol{a} = (a_1, a_2, a_3)$) for all $(t, x) \in [T_1, T_2] \times \mathbb{R}^n$ and $I = 1, \dots, N$. Note that g_I is a Lorentz matrix-valued function for $I = 1, \dots, N$. We shall denote the components of the inverse by $g_I^{\mu\nu}$. Assume that $b_I^{J\alpha}, c_J^I, f^I \in C^\infty(\mathbb{R}^{n+1})$, where $I, J = 1, \dots, N$

and $\alpha = 0, \dots, n$, and that $u_0^I, u_1^I \in C^\infty(\mathbb{R}^n)$. We are interested in the initial value problem

$$g_I^{\mu\nu}\partial_\mu\partial_\nu u^I + b_J^{I\alpha}\partial_\alpha u^J + c_J^I u^J = f^I, \tag{8.9}$$

$$u^I(0,\cdot) = u_0^I, \tag{8.10}$$

$$\partial_t u^I(0,\cdot) = u_1^I, \tag{8.11}$$

where I and J range from 1 to N, μ, ν and α range from 0 to n and we sum over all indices except I. Due to Lemma 8.5, $g_I^\sharp$ is positive definite and $g_I^{00} < 0$. Furthermore, for any compact interval $[T_1, T_2]$, there are constants $b_i > 0$, $i = 1, 2, 3$, such that $g_I{}^{-1}(t, x) \in \mathcal{C}_{n,\boldsymbol{b}}$ (where $\boldsymbol{b} = (b_1, b_2, b_3)$) for all $(t, x) \in [T_1, T_2] \times \mathbb{R}^n$ and $I = 1, \dots, N$. Consequently, when convenient, we shall divide (8.9) by $-g_I^{00}$, so that we can assume $g_I^{00} = -1$. Let

$$v^I = (\partial_1 u^I, \dots, \partial_n u^I, \partial_0 u^I, u^I)^t. \tag{8.12}$$

We shall denote the components of this vector by $v_1^I, \dots, v_{n+2}^I$. Define the symmetric $(n+2) \times (n+2)$-matrices A^{I0} and A^{Ik} by

$$A_{ij}^{I0} = g_I^{ij}, \quad A_{n+1n+1}^{I0} = A_{n+2n+2}^{I0} = 1, \quad A_{in+1}^{Ik} = g_I^{ik},$$

$$A_{n+1i}^{Ik} = g_I^{ik}, \quad A_{n+1n+1}^{Ik} = 2g_I^{0k},$$

where $i, j, k = 1, \dots, n$ and $I = 1, \dots, N$ and we demand that the remaining components be zero. Let furthermore

$$d_{Jn+1i}^I = -b_J^{Ii}, \quad d_{Jn+1n+1}^I = -b_J^{I0}, \quad d_{Jn+1n+2}^I = -c_J^I,$$

$$d_{Jn+2n+1}^I = -\delta_J^I, \quad h_{n+1}^I = -f^I,$$

where $i = 1, \dots, n$, $I, J = 1, \dots, N$, and the remaining components be zero. If u satisfies (8.9) and v is given by (8.12), then

$$A^{I0}\partial_0 v^I - A^{Ik}\partial_k v^I + \sum_J d_J^I v^J = h^I, \tag{8.13}$$

where we sum over k but not over I. Here d_J^I is a smooth function with values in the set of $(n+2) \times (n+2)$-matrices and h^I is a smooth $\mathbb{R}^{n+2}$-valued function. If we put all the v^I into one vector v, we see that this equation is symmetric hyperbolic. By the above observations, we also see that for any compact interval $[T_1, T_2]$, there is a constant $a_0 > 0$ such that A^0, which is obtained by putting all the A^{I0} into one matrix, satisfies $A^0 \geq a_0$ on $[T_1, T_2] \times \mathbb{R}^n$. Furthermore, the A^k are bounded on the same set.

Assume we have a smooth solution to (8.9)–(8.11). If we define v^I by (8.12), we obtain a solution to (8.13) such that the initial data have the property that

$$\partial_i v_{n+2}^I(0,\cdot) = v_i^I(0,\cdot) \tag{8.14}$$

for $i = 1, \dots, n$. Assume we have a smooth solution to (8.13) such that the initial data satisfy (8.14). Due to (8.13),

$$\partial_0 v_j^I = \partial_j v_{n+1}^I, \quad \partial_0 v_{n+2}^I = v_{n+1}^I. \tag{8.15}$$

Define $u^I = v_{n+2}^I$. Then, due to (8.14), $\partial_i u^I(0,\cdot) = v_i^I(0,\cdot)$. Thus, using (8.15),

$$\begin{aligned}\partial_i u^I(t,x) &= v_i^I(0,x) + \int_0^t \partial_i \partial_0 v_{n+2}^I(\tau,x)\,d\tau \\ &= v_i^I(0,x) + \int_0^t \partial_0 v_i^I(\tau,x)\,d\tau = v_i^I(t,x).\end{aligned}$$

This observation, together with (8.15), yields the conclusion that v^I is of the form (8.12). Thus u^I is a solution of (8.9)–(8.11) with $u^I(0,x) = v_{n+2}^I(0,x)$ and $\partial_t u^I(0,x) = v_{n+1}^I(0,x)$.

Due to Proposition 7.7, Remark 7.8 and Corollary 7.10, we obtain the following theorem.

Theorem 8.6. *Let g_I, $I = 1, \dots, N$ be smooth functions from $\mathbb{R}^{n+1}$ into $\mathcal{C}_n$, with components $g_{I\mu\nu}$, $\mu, \nu = 0, \dots, n$. Assume that for any compact interval $[T_1, T_2]$, there are constants $a_i > 0$, $i = 1, 2, 3$, such that $g_I(t,x) \in \mathcal{C}_{n,\mathbf{a}}$, where $\mathbf{a} = (a_1, a_2, a_3)$, for all $(t,x) \in [T_1, T_2] \times \mathbb{R}^n$ and $I = 1, \dots, N$. Denote the components of the inverse by $g_I^{\mu\nu}$. Assume that $b_I^{J\alpha}$, c_J^I, $f^I \in C^\infty(\mathbb{R}^{n+1})$, where $I, J = 1, \dots, N$ and $\alpha = 0, \dots, n$, and that $u_0^I, u_1^I \in C^\infty(\mathbb{R}^n)$. Then (8.9)–(8.11) has a unique solution $u \in C^\infty(\mathbb{R}^{n+1}, \mathbb{R}^N)$. Furthermore, if the initial data have compact support and $f^I(t,x) = 0$ for $t \in [T_1, T_2]$ and $x \notin K_1$, where $T_1 < 0 < T_2$ and K_1 is compact, then there is a compact set K such that $u(t,x) = 0$ for $t \in [T_1, T_2]$ if $x \notin K$.*

It is also of interest to have a uniqueness statement in greater generality.

Proposition 8.7. *Let g_I, $I = 1, \dots, N$ be C^1 functions from $\mathbb{R}^{n+1}$ into $\mathcal{C}_n$, with components $g_{I\mu\nu}$, $\mu, \nu = 0, \dots, n$. Assume that for any compact interval $[T_1, T_2]$, there are constants $a_i > 0$, $i = 1, 2, 3$, such that $g_I(t,x) \in \mathcal{C}_{n,\mathbf{a}}$, where $\mathbf{a} = (a_1, a_2, a_3)$, for all $(t,x) \in [T_1, T_2] \times \mathbb{R}^n$ and $I = 1, \dots, N$. Denote the components of the inverse by $g_I^{\mu\nu}$. Assume that $b_I^{J\alpha}$, c_J^I and f^I, where $I, J = 1, \dots, N$ and $\alpha = 0, \dots, n$, are continuous functions. Then, given a compact interval $[T_1, T_2]$ with $T_1 \le 0$ and $T_2 \ge 0$, there is an $s_0 > 0$ such that the following holds: if u is a C^2 solution to (8.9), the initial data vanish inside $B_r(x_0)$, and $f^I(t,x) = 0$ for $(t,x) \in C_{x_0,r,s_0,T_1,T_2} = C$, defined in (7.9), then $u(t,x) = 0$ in C.*

Remark 8.8. One consequence of the above statement is of course that we have uniqueness of solutions.

Proof. The statement is an immediate consequence of Proposition 7.7. □

9 Local existence, non-linear wave equations

The argument presented in the present chapter, proving local existence of solutions to non-linear wave equations, is based on the construction of a convergent sequence of solutions to a family of linear wave equations. In order to prove convergence, it is necessary to require that the metric and non-linearity satisfy certain technical conditions. The main purpose of Section 9.1 is to specify these requirements and to introduce the associated terminology. The proof of the fact that the sequence is convergent can conveniently be divided into several steps, cf. the proof of Proposition 9.12, see also [59]. The first step consists of proving that the sequence is bounded with respect to the "strong" norm, associated with the space

$$C^0\{[T_0, T_0 + T], H^{k+1}(\mathbb{R}^n, \mathbb{R}^N)\} \cap C^1\{[T_0, T_0 + T], H^k(\mathbb{R}^n, \mathbb{R}^N)\}, \tag{9.1}$$

assuming that the data are in the space

$$H^{k+1}(\mathbb{R}^n, \mathbb{R}^N) \times H^k(\mathbb{R}^n, \mathbb{R}^N)$$

for some large enough k ($k > n/2 + 1$). The main tool in taking this step is Lemma 9.9 of Section 9.2, the result of which is an energy estimate quite similar to the ones carried out in the linear setting, though it is necessary to keep track of more information. In the second step, one proves that the sequence converges in a "weak" norm, associated with the space

$$C^0\{[T_0, T_0 + T], H^1(\mathbb{R}^n, \mathbb{R}^N)\} \cap C^1\{[T_0, T_0 + T], L^2(\mathbb{R}^n, \mathbb{R}^N)\},$$

and Lemma 9.11 of Section 9.2 is the basic tool in achieving this goal. By interpolation, cf. Lemma 5.20, one then obtains convergence in the spaces

$$C^0\{[T_0, T_0 + T], H_{(s+1)}(\mathbb{R}^n, \mathbb{R}^N)\} \cap C^1\{[T_0, T_0 + T], H_{(s)}(\mathbb{R}^n, \mathbb{R}^N)\}$$

for any $0 \le s < k$. Furthermore, one can prove weak continuity. In order to prove strong continuity, some additional technical arguments are necessary.

The basic existence result states that there are solutions with regularity of the form (9.1), with an existence time T depending on k. Due to the dependence of T on k, one does not immediately obtain local existence of smooth solutions. In Section 9.4, we provide a continuation criterion, demonstrating, among other things, that the maximal time interval on which the solution exists is independent of k. As a corollary, there are local smooth solutions.

The chapter ends with a proof of Cauchy stability.

9.1 Terminology

Definition 9.1. Let $1 \le n, N \in \mathbb{Z}$ and let k be such that either $0 \le k \in \mathbb{Z}$ or $k = \infty$. Let g be a C^k function from $\mathbb{R}^{nN+2N+n+1}$ to $\mathcal{C}_n$. If

- for every multiindex $\alpha = (\alpha_1, \dots, \alpha_{nN+2N+n+1})$ such that $|\alpha| < k+1$, and compact interval $I = [T_1, T_2]$, there is a continuous, increasing function $h_{I,\alpha}\colon \mathbb{R} \to \mathbb{R}$ such that

$$|(\partial^\alpha g_{\mu\nu})(t, x, \xi)| \leq h_{I,\alpha}(|\xi|) \tag{9.2}$$

for all $\mu, \nu = 0, \dots, n$, $t \in I$, $x \in \mathbb{R}^n$ and $\xi \in \mathbb{R}^{nN+2N}$, and

- for every compact interval $I = [T_1, T_2]$, there are $a_i > 0$, $i = 1, 2, 3$, such that $g(t, x, \xi) \in \mathcal{C}_{n,\boldsymbol{a}}$ for every $(t, x, \xi) \in I \times \mathbb{R}^{nN+2N+n}$, where $\boldsymbol{a} = (a_1, a_2, a_3)$,

we shall say that g is a C^k *N, n-admissible metric.*

Remark 9.2. By $g_{\mu\nu}$ we mean the components of the matrix-valued function. Recall that the notation $\mathcal{C}_n$ and $\mathcal{C}_{n,\boldsymbol{a}}$ was introduced in Definition 8.4.

Assume g is a C^0 N, n-admissible metric. Then g is a Lorentz matrix valued function due to Lemma 8.3 and we shall denote the components of the inverse by $g^{\mu\nu}$. As a consequence of our assumptions and Lemma 8.5, there are, for every compact interval $I = [T_1, T_2]$, constants $b_i > 0$, $i = 1, 2, 3$ such that $g^{-1}(t, x, \xi) \in \mathcal{C}_{n,\boldsymbol{b}}$ for all $(t, x, \xi) \in I \times \mathbb{R}^{nN+2N+n}$, where $\boldsymbol{b} = (b_1, b_2, b_3)$. Given a differentiable function $u\colon \Omega \to \mathbb{R}^N$ for some $\Omega \subseteq \mathbb{R}^{n+1}$, we define $g[u]$ to be the function on Ω given by

$$g[u](t, x) = g\{t, x, u(t, x), \partial_0 u(t, x), \dots, \partial_n u(t, x)\}.$$

Before specifying the class of non-linearities we wish to consider, let us introduce the following concept.

Definition 9.3. Let $1 \leq m, n \in \mathbb{Z}$. A function $h\colon \mathbb{R}^{n+1} \to \mathbb{R}^m$ is said to be of *locally x-compact support* if for any compact interval $[T_1, T_2]$ there is a compact set $K \subset \mathbb{R}^n$ such that $h(t, x) = 0$ if $t \in [T_1, T_2]$ and $x \notin K$.

We shall use this concept frequently below. The reason for its importance is due to the fact that a smooth function of locally x-compact support $u\colon \mathbb{R}^{n+1} \to \mathbb{R}^m$ can be viewed as an element of $C^l[\mathbb{R}, H^k(\mathbb{R}^n, \mathbb{R}^m)]$ for any l, k (those readers unfamiliar with the concept of differentiability of functions from one Banach space to another are referred to, e.g., [49], p. 7, cf. also Section A.7). Note that this is no longer true if we consider smooth functions with the property that for every fixed t, $u(t, \cdot)$ has compact support. A simple counterexample is obtained by taking $\phi \in C_0^\infty(\mathbb{R}^n)$ which is not identically zero and defining $u(t, x) = \phi(x^1 - 1/t, x^2, \dots, x^n)$ for $t > 0$ and $u(t, x) = 0$ for $t \leq 0$. Then u is smooth for $t > 0$, for $t < 0$ and for each point of the form $(0, x)$, there is a neighbourhood of that point such that $u(t, y) = 0$ for all (t, y) in the neighbourhood (since ϕ has compact support). Thus u is smooth. For each fixed t, $u(t, \cdot)$ has compact support, for $t \leq 0$, $\|u(t, \cdot)\|_2 = 0$ and for $t > 0$, $\|u(t, \cdot)\|_2 = c_0 > 0$.

Definition 9.4. Let $1 \leq n, N \in \mathbb{Z}$ and let k be such that either $0 \leq k \in \mathbb{Z}$ or $k = \infty$. Let $f \in C^k(\mathbb{R}^{nN+2N+n+1}, \mathbb{R}^N)$. If

- for every multiindex $\alpha = (\alpha_1, \dots, \alpha_{nN+2N+n+1})$ such that $|\alpha| < k+1$ and compact interval $I = [T_1, T_2]$, there is a continuous, increasing function $h_{I,\alpha}\colon \mathbb{R} \to \mathbb{R}$ such that

$$|(\partial^\alpha f)(t, x, \xi)| \leq h_{I,\alpha}(|\xi|) \tag{9.3}$$

  for all $t \in I$, $x \in \mathbb{R}^n$ and $\xi \in \mathbb{R}^{nN+2N}$, and

- the function f_b, defined by $f_b(t, x) = f(t, x, 0)$, has locally x-compact support,

we shall say that f is a C^k *N, n-admissible non-linearity.*

We shall use similar conventions concerning f as we do concerning g, in particular we shall write $f[u]$, the meaning being analogous to the case of g.

In the arguments proving local existence of solutions, constants and functions depending on the metric and non-linearity will appear that have special properties. It is convenient to introduce some terminology for them.

Definition 9.5. Let $1 \leq N, n \in \mathbb{Z}$. A map κ which to every C^∞ N, n-admissible metric g, C^∞ N, n-admissible non-linearity f and compact interval $I = [T_1, T_2]$ associates a continuous function $\kappa_I[g, f]$ from some $\mathbb{R}^m$ to the non-negative real numbers such that $\kappa_{I_1}[g, f] \leq \kappa_{I_2}[g, f]$ if $I_1 \subseteq I_2$ will be referred to as an *N, n-admissible majorizer.* A map C which to every C^∞ N, n-admissible metric g, C^∞ N, n-admissible non-linearity f and compact interval $I = [T_1, T_2]$ associates a real number $C_I[g, f] > 0$ such that $C_{I_1}[g, f] \leq C_{I_2}[g, f]$ if $I_1 \subseteq I_2$ will be referred to as an *N, n-admissible constant.*

Remark 9.6. What g and f are will below usually be clear from the context, and so we shall, for the sake of brevity, omit the argument $[g, f]$ and write κ_I and C_I. Furthermore, it will not be of any interest to keep track of the different constants and functions, and we shall therefore use the same notation, κ_I and C_I, for all N, n-admissible majorizers and constants; what they actually are might change from line to line but the notation will remain the same.

In this section we are interested in the initial value problem

$$g^{\mu\nu}\partial_\mu\partial_\nu u = f, \tag{9.4}$$

$$u(T_0, \cdot) = U_0, \tag{9.5}$$

$$\partial_t u(T_0, \cdot) = U_1, \tag{9.6}$$

where $T_0 \in \mathbb{R}$ and we write g instead of $g[u]$ and f instead of $f[u]$. Let us remark that all the assumptions made hold if we divide the equation by $-g^{00}$. It is thus not a restriction to assume that $g^{00} = -1$, and we shall make this assumption whenever it is convenient.

Lemma 9.7. *Let $1 \leq N, n \in \mathbb{Z}$, let g be a C^1 N, n-admissible metric and f be a C^1 N, n-admissible non-linearity. Consider two C^2 solutions u and v to* (9.4) *on*

$[T_0, T) \times \mathbb{R}^n$, where we do not assume that the initial data have compact support. If the initial data coincide at $t = T_0$, then $u = v$ on $[T_0, T) \times \mathbb{R}^n$. The statement in the opposite time direction is similar.

Remark 9.8. It is possible to make local in space uniqueness statements. The condition that f be such that f_b have locally x-compact support is not necessary.

Proof. Let us write $g_v = g[v]$ and similarly for the other objects involved in the equations. Then

$$g_u^{\mu\nu}\partial_\mu\partial_\nu(u - v) = (g_v^{\mu\nu} - g_u^{\mu\nu})\partial_\mu\partial_\nu v + f_u - f_v. \tag{9.7}$$

Due to equalities of the type

$$h(u) - h(v) = \int_0^1 h'(tu + (1-t)v)\,dt \cdot (u - v), \tag{9.8}$$

there are continuous functions f_1 and f_2^μ such that

$$f_u - f_v = f_1 \cdot (u - v) + f_2^\mu \cdot \partial_\mu(u - v),$$

and similarly for $g_v^{\mu\nu} - g_u^{\mu\nu}$. Inserting this information into (9.7), we see that Proposition 8.7 is applicable. The lemma follows. □

9.2 Preliminaries

Below, we shall use the notation

$$\begin{aligned} M_k[v](t) &= \|v(t,\cdot)\|_{H^{k+1}} + \|\partial_t v(t,\cdot)\|_{H^k}, \\ m[v](t) &= \sum_{|\alpha|+j\le 2} \sup_{x\in\mathbb{R}^n} |\partial^\alpha \partial_t^j v(t,x)|, \end{aligned}$$

given that the right-hand sides are defined. We shall also write $M[v] = M_0[v]$. The following lemma will, among other things, be used to prove that the elements of the sequence we set up in order to prove existence of a local solution are uniformly bounded on a suitable time interval.

Lemma 9.9. *Let $1 \le N, n \in \mathbb{Z}$. Then there are N, n-admissible majorizers κ_1, κ_2 and N, n-admissible constants C_1, C_2, C_3 such that the following holds. Let g be a C^∞ N, n-admissible metric and f be a C^∞ N, n-admissible non-linearity. Let $U_0, U_1 \in C_0^\infty(\mathbb{R}^n, \mathbb{R}^N)$ and assume $v \in C^\infty(\mathbb{R}^{n+1}, \mathbb{R}^N)$ is of locally x-compact support. Let $T_0 \le T_1 \in \mathbb{R}$, $g_v = g[v]$, $f_v = f[v]$ and u be the solution of*

$$g_v^{\mu\nu}\partial_\mu\partial_\nu u = f_v, \tag{9.9}$$

$$u(T_0,\cdot) = U_0, \tag{9.10}$$

$$\partial_t u(T_0,\cdot) = U_1. \tag{9.11}$$

Let $I = [T_0, T_1]$. *If* $t \in I$, *then*

$$\begin{aligned} M_k[u](t) \leq C_{1,I} M_k[u](T_0) \\ + \int_{T_0}^{t} [C_{2,I} + \kappa_{1,I}(m[v])\{(1 + m[u])M_k[v] + M_k[u]\}]\, ds. \end{aligned} \tag{9.12}$$

If $f_b = 0$, *the constant term inside the integral can be omitted. Finally, if*

$$E_k[v,u] = \frac{1}{2} \sum_{|\alpha|\leq k} \int_{\mathbb{R}^n} [-g_v^{00}|\partial^\alpha \partial_t u|^2 + g_v^{ij}\partial^\alpha \partial_i u \cdot \partial^\alpha \partial_j u + |\partial^\alpha u|^2]\, dx, \tag{9.13}$$

the following inequality holds:

$$\partial_t E_k[v,u] \leq C_{3,I} + \kappa_{2,I}(m[u], m[v])(M_k^2[v] + E_k[v,u]). \tag{9.14}$$

Remark 9.10. By Theorem 8.6, there is a unique solution $u \in C^\infty(\mathbb{R}^{n+1}, \mathbb{R}^N)$ to (9.9)–(9.11). Furthermore, the solution is of locally x-compact support. In (9.13), $g_v^{00} = -1$, but we write down the energy as above to point out that there is no problem in defining it when this is not satisfied.

Proof. Note that due to the assumptions we have made on g, there is an N, n-admissible constant C such that for $t \in I = [T_0, T_1]$,

$$\frac{1}{C_I} E_k^{1/2}[v,w](t) \leq M_k[w](t) \leq C_I E_k^{1/2}[v,w](t), \tag{9.15}$$

assuming w is such that the objects involved in the inequality are defined. From now on we shall write E_k instead of $E_k[v,u]$ and $E = E_0$. Compute

$$\begin{aligned} \partial_t E = \int_{\mathbb{R}^n} \Big[& - g_v^{\mu\nu}\partial_\mu \partial_\nu u \cdot \partial_t u - \partial_i(g_v^{0i})|\partial_t u|^2 - \frac{1}{2}(\partial_t g_v^{00})|\partial_t u|^2 \\ & - (\partial_i g_v^{ij})\partial_j u \cdot \partial_t u + \frac{1}{2}(\partial_t g_v^{ij})\partial_j u \cdot \partial_i u + u \cdot \partial_t u \Big]\, dx, \end{aligned} \tag{9.16}$$

where we have used the fact that due to partial integration,

$$\int_{\mathbb{R}^n} g_v^{ij}\partial_i u \cdot \partial_t \partial_j u\, dx = -\int_{\mathbb{R}^n} \partial_j(g_v^{ij}\partial_i u) \cdot \partial_t u\, dx,$$

$$\int_{\mathbb{R}^n} 2g_v^{0i}\partial_t \partial_i u \cdot \partial_t u\, dx = -\int_{\mathbb{R}^n} \partial_i(g_v^{0i})|\partial_t u|^2\, dx.$$

In (9.16), $\partial_t g_v^{00} = 0$, but we write it out for the same reasons mentioned above. Note that $\partial_i g_v^{0i}$ is bounded, the bound depending on $m[v]$ and the interval I, and similarly for $\partial_t g_v^{00}$, etc. Consequently,

$$\partial_t E \leq \kappa_I(m[v])E + C\|f_v(t,\cdot)\|_2 E^{1/2}, \tag{9.17}$$

where the second constant is numerical and κ is an N, n-admissible majorizer. In order to get a similar estimate for E_k, let us note that $\partial^\alpha u$ satisfies the equation

$$g_v^{\mu\nu}\partial_\mu\partial_\nu\partial^\alpha u = \partial^\alpha f_v + [g_v^{\mu\nu}\partial_\mu\partial_\nu, \partial^\alpha]u.$$

Combining this observation with (9.17), we conclude that

$$\partial_t E_k \leq \kappa_I(m[v])E_k + C\|f_v\|_{H^k}E_k^{1/2} + C\sum_{|\alpha|\leq k}\|[g_v^{\mu\nu}\partial_\mu\partial_\nu, \partial^\alpha]u\|_2 E_k^{1/2},$$

where the H^k-norm is only with respect to the x-variable. Note that, up to constant factors,

$$[g_v^{\mu\nu}\partial_\mu\partial_\nu, \partial^\alpha]u$$

can be written as a sum of terms of the form

$$(\partial^\beta\partial_i g_v^{\mu\nu})\partial^\gamma\partial_\mu\partial_\nu u, \tag{9.18}$$

where $|\beta| + |\gamma| = |\alpha| - 1$. Since $g_v^{00} = -1$ and we differentiate $g_v^{\mu\nu}$ with respect to x^i, we can assume that at most one of μ, ν equals zero. Let us rewrite

$$(\partial^\beta\partial_i g_v^{\mu\nu})\partial^\gamma\partial_\mu\partial_\nu u = [\partial^\beta\partial_i(g_v^{\mu\nu} - g_0^{\mu\nu})]\partial^\gamma\partial_\mu\partial_\nu u + (\partial^\beta\partial_i g_0^{\mu\nu})\partial^\gamma\partial_\mu\partial_\nu u,$$

where $g_0 = g[0]$. Note that the sup norm of the first factor in the second term over $[T_0, T] \times \mathbb{R}^n$ only depends on the interval I. When estimating the second term in L^2, we can thus take out the first factor in the sup norm. To the first term, we can apply (6.22), where $\phi_1 = \partial_i(g_v^{\mu\nu} - g_0^{\mu\nu})$ and $\phi_2 = \partial_\mu\partial_\nu u$, and then Lemma 6.17. To conclude,

$$\|(\partial^\beta\partial_i g_v^{\mu\nu})\partial^\gamma\partial_\mu\partial_\nu u\|_2 \leq \kappa_I(m[v])(M_k[u] + m[u]M_k[v]).$$

Thus

$$\|[g_v^{\mu\nu}\partial_\mu\partial_\nu, \partial^\alpha]u\|_2 \leq \kappa_I(m[v])(M_k[u] + m[u]M_k[v]).$$

What remains to be estimated is f_v in H^k. However, Lemma 6.17 yields

$$\|f_v\|_{H^k} \leq C_I + \kappa_I(m[v])M_k[v],$$

where the constant term comes from f_b. Adding up, we obtain

$$\partial_t E_k \leq \kappa_I(m[v])E_k + \{C_I + \kappa_I(m[v])(M_k[v] + M_k[u] + m[u]M_k[v])\}E_k^{1/2}. \tag{9.19}$$

Combining this inequality with Young's inequality and (9.15), we obtain (9.14). Letting $\hat{E}_k = E_k + \varepsilon$, we get

$$\partial_t \hat{E}_k^{1/2} \leq C_I + \kappa_I(m[v])(M_k[v] + m[u]M_k[v] + M_k[u]),$$

which yields (9.12) after integrating and letting $\varepsilon \to 0+$. Note that we have used (9.15). □

The following lemma will be used to prove that the sequence we set up to prove the existence of a local solution converges for a low degree of regularity.

Lemma 9.11. *Let $1 \leq N, n \in \mathbb{Z}$. Then there are N, n-admissible majorizers κ_3, κ_4 and an N, n-admissible constant C_4 such that the following holds. Let g be a C^∞ N, n-admissible metric and f be a C^∞ N, n-admissible non-linearity. Let $T_0 \leq T_1 \in \mathbb{R}$, $U_{0,i}, U_{1,i} \in C_0^\infty(\mathbb{R}^n, \mathbb{R}^N)$ and let $v_i \in C^\infty(\mathbb{R}^{n+1}, \mathbb{R}^N)$ be functions of locally x-compact support, where $i = 1, 2$. Let $g_i = g[v_i]$, $f_i = f[v_i]$ and u_i be solutions to*

$$
\begin{aligned}
g_i^{\mu\nu}\partial_\mu\partial_\nu u_i &= f_i, \\
u_i(T_0, \cdot) &= U_{0,i}, \\
\partial_t u_i(T_0, \cdot) &= U_{1,i}.
\end{aligned}
$$

If $u = u_2 - u_1$ and $v = v_2 - v_1$, then, for $t \in I = [T_0, T_1]$,

$$
\begin{aligned}
M[u](t) \leq C_{4,I} \exp&\left(\int_{T_0}^t \kappa_{4,I}(m[v_2])\, ds\right) \\
&\cdot \left(M[u](T_0) + \int_{T_0}^t \kappa_{3,I}(m[u_1], m[v_1], m[v_2]) M[v]\, ds\right)
\end{aligned}
$$

where we use the notation $M[u] = M_0[u]$.

Proof. Note that

$$
g_2^{\mu\nu}\partial_\mu\partial_\nu u = (g_1^{\mu\nu} - g_2^{\mu\nu})\partial_\mu\partial_\nu u_1 + f_2 - f_1.
$$

Let

$$
\mathcal{E} = \frac{1}{2}\int_{\mathbb{R}^n} [-g_2^{00}|\partial_t u|^2 + g_2^{ij}\partial_i u \cdot \partial_j u + |u|^2]\, dx.
$$

Using equalities of the form (9.8), we have

$$
\partial_t \mathcal{E} \leq \kappa_I(m[v_2])\mathcal{E} + \kappa_I(m[u_1], m[v_1], m[v_2]) M[v]\mathcal{E}^{1/2}.
$$

This implies that

$$
\mathcal{E}^{1/2}(t) \leq \mathcal{E}^{1/2}(T_0) + \int_{T_0}^t \{\kappa_I(m[v_2])\mathcal{E}^{1/2} + \kappa_I(m[u_1], m[v_1], m[v_2]) M[v]\}\, ds.
$$

Due to Grönwall's lemma, Lemma 7.1, we get

$$
\begin{aligned}
&\mathcal{E}^{1/2}(t) \\
&\leq \left(\mathcal{E}^{1/2}(T_0) + \int_{T_0}^t \kappa_I(m[u_1], m[v_1], m[v_2]) M[v]\, ds\right) \exp\left(\int_{T_0}^t \kappa_I(m[v_2])\, ds\right).
\end{aligned}
$$

The lemma follows. □

9.3 Local existence

Proposition 9.12. *Let $1 \leq N, n \in \mathbb{Z}$. Then there is an N, n-admissible majorizer κ and an N, n-admissible constant C such that the following holds. Let g be a C^∞ N, n-admissible metric and f be a C^∞ N, n-admissible non-linearity. Let $k > n/2 + 1$, $U_0 \in H^{k+1}(\mathbb{R}^n, \mathbb{R}^N)$ and $U_1 \in H^k(\mathbb{R}^n, \mathbb{R}^N)$. Given a compact interval $I = [T_1, T_2]$, there is a $T > 0$, depending on I and continuously on the H^{k+1}-norm of U_0 and on the H^k-norm of U_1, such that if $T_0 \in [T_1, T_2]$, then there is a unique solution $u \in C_b^2([T_0, T_0 + T] \times \mathbb{R}^n, \mathbb{R}^N)$ to (9.4)–(9.6). Furthermore,*

$$u \in C\{[T_0, T_0+T], H^{k+1}(\mathbb{R}^n, \mathbb{R}^N)\}, \ \partial_t u \in C\{[T_0, T_0+T], H^k(\mathbb{R}^n, \mathbb{R}^N)\}, \tag{9.20}$$

and, for $t \in [T_0, T_0 + T]$,

$$\mathcal{E}_k(t) \leq [\mathcal{E}_k(T_0) + C_I \cdot (t - T_0)] \exp\left(\int_{T_0}^t \kappa_I(m[u])\, d\tau\right), \tag{9.21}$$

where $\mathcal{E}_k(t) = E_k[u, u](t)$ and $E_k[v, u](t)$ is defined in (9.13). Finally, one can assume the bound on $\mathcal{E}_k(t)$ for $t \in [T_0, T_0 + T]$ to depend only on I, on an upper bound on the H^{k+1}-norm of U_0 and on an upper bound on the H^k-norm of U_1.

Remark 9.13. The above result does not immediately imply the existence of local smooth solutions given smooth initial data since the existence time could depend on the degree of regularity of the initial data. Existence in the opposite time direction follows by time reversal. Recall that the notation C_b^k was introduced in Remark 6.6. There are many presentations of similar results in the literature, cf., e.g., [50] and [59].

Proof. Let $U_{0,l}, U_{1,l} \in C_0^\infty(\mathbb{R}^n, \mathbb{R}^N)$ be sequences which converge to U_0 and U_1 in H^{k+1} and H^k respectively. Let us assume

$$C_0 = \|U_0\|_{H^{k+1}} + \|U_1\|_{H^k}, \quad \|U_{0,l}\|_{H^{k+1}} + \|U_{1,l}\|_{H^k} \leq C_0 + 1,$$

where the second inequality can be obtained by making a suitable choice of sequence. Define w_0 by $w_0(t, x) = U_{0,0}(x)$. Given that w_l has been defined and has x-compact support, let $g_{l+1} = g[w_l]$ and $f_{l+1} = f[w_l]$. Define w_{l+1} as the solution to

$$g_{l+1}^{\mu\nu} \partial_\mu \partial_\nu w_{l+1} = f_{l+1}, \tag{9.22}$$

$$w_{l+1}(T_0, \cdot) = U_{0,l+1}, \tag{9.23}$$

$$\partial_t w_{l+1}(T_0, \cdot) = U_{1,l+1}. \tag{9.24}$$

Due to Theorem 8.6, there is a unique $w_{l+1} \in C^\infty(\mathbb{R}^{n+1}, \mathbb{R}^N)$ solving (9.22)–(9.24) which is furthermore of locally x-compact support.

Boundedness of the sequence. The first step is to prove that the sequence is bounded in the right space. Let us make the inductive assumption that

$$M_k[w_{l-1}](t) \leq \mathcal{M} \tag{9.25}$$

for $T_0 \leq t \leq T_0 + T$ and some constant $\mathcal{M}$. For $l = 1$, this assumption is clearly valid for any T, with $\mathcal{M}$ depending only on C_0. Assume (9.25) holds for l and $l+1$. Then Sobolev embedding and the equation yield

$$m[w_l](t) \leq \kappa_I(\mathcal{M})\{1 + M_k[w_l](t)\} \leq \kappa_I(\mathcal{M})$$

on $[T_0, T_0+T]$, assuming $T \leq 1$, where the first inequality holds assuming only (9.25) for l (strictly speaking, in order to get this estimate we need to control f and g on $I_1 = [T_1, T_2+1]$, and therefore one could consider the notation κ_{I_1} more natural, but note that the definition of an admissible majorizer does not require us to use such a notation). In this inequality, the 1 appearing in the middle is due to the fact that we have used the equation in order to get an estimate for $\partial_t^2 w_l$. Note that when we use the equation, the only control we need to have concerning w_{l-1} is the sup norm of the derivatives of order 0 and 1, which is given by (9.25). Sup norm control of the second derivatives of w_l is given by (9.25) with l replaced by $l+1$, assuming that at least one of the derivatives is spatial. Note that $m[w_0]$ is bounded, so that we do not need to know anything about w_{-1} in the inductive step from $l = 1$ to $l = 2$, cf. below.

Assume either that (9.25) holds for $l = 1$ or for l and $l-1$. Then $m[w_{l-1}] \leq \kappa_I(\mathcal{M})$ and

$$m[w_l](t) \leq \kappa_I(\mathcal{M})\{1 + M_k[w_l](t)\}.$$

Inserting this information into (9.12), we obtain

$$M_k[w_l](t) \leq C_I\left(M_k[w_l](T_0) + \kappa_I(\mathcal{M})\int_{T_0}^t (1 + M_k[w_l])\,ds\right).$$

By Grönwall's lemma, Lemma 7.1, we get

$$M_k[w_l](t) \leq C_I[M_k[w_l](T_0) + \kappa_I(\mathcal{M})(t-T_0)]\exp[(t-T_0)\kappa_I(\mathcal{M})].$$

Choose $\mathcal{M} = 4C_I(C_0+1)$. Then $\mathcal{M} \geq 4C_I M_k[w_l](T_0)$ for all l. By demanding that T be small enough, depending only on $\mathcal{M}$ and I, i.e. on C_0 and I, we can complete the inductive step. Thus (9.25) holds for all l.

Convergence in the low norm. Let us prove that the sequence converges in

$$C^0\{[T_0, T_0+T], H^1(\mathbb{R}^n, \mathbb{R}^N)\} \cap C^1\{[T_0, T_0+T], L^2(\mathbb{R}^n, \mathbb{R}^N)\}.$$

One should of course check that the iterates are in the appropriate space. Let us apply Lemma 9.11 with $v_2 = w_l$, $v_1 = w_{l-1}$, $u_2 = w_{l+1}$ and $u_1 = w_l$. If we let

$$a_l = \sup_{T_0 \leq t \leq T_0+T} M[w_{l+1} - w_l],$$

assume that T is small enough, depending only on C_0 and I, and assume that the initial data sequence has been chosen so that

$$2C_{4,I} M[w_{l+1} - w_l](T_0) \leq 2^{-l},$$

then

$$a_l \le 2^{-l} + \frac{1}{2} a_{l-1}.$$

Consequently,

$$a_l \le \frac{l-1}{2^l} + \frac{1}{2^{l-1}} a_1.$$

In particular, a_l is summable, and we conclude that we have convergence in the desired space.

Convergence in higher norms. Assuming $0 < s < k$, we can use Lemma 5.20 in order to conclude that there are $a_s, b_s \in (0, 1)$ with $a_s + b_s = 1$ such that

$$\|w_l(t,\,\cdot) - w_m(t,\,\cdot)\|_{(s+1)} \le \|w_l(t,\,\cdot) - w_m(t,\,\cdot)\|_{(k+1)}^{b_s} \|w_l(t,\,\cdot) - w_m(t,\,\cdot)\|_2^{a_s}.$$

Since the first factor is bounded for $T_0 \le t \le T_0 + T$ and the second factor converges to zero, we conclude that w_l is a Cauchy sequence in

$$C^0\{[T_0, T_0 + T], H_{(s+1)}(\mathbb{R}^n, \mathbb{R}^N)\}.$$

Similarly, we conclude that w_l is a Cauchy sequence in

$$C^1\{[T_0, T_0 + T], H_{(s)}(\mathbb{R}^n, \mathbb{R}^N)\}.$$

Since we are assuming $k > n/2 + 1$, we can use the above, Sobolev embedding and the equation in order to conclude that w_l converges in $C_b^2([T_0, T_0 + T] \times \mathbb{R}^n, \mathbb{R}^N)$. In particular, we get a C^2 solution to the equation. Let u denote the solution. We need to prove that $u(t,\,\cdot)$ is in H^{k+1} and that $\partial_t u(t,\,\cdot)$ is in H^k. For any $0 < s < k$, we have $u(t,\,\cdot) \in H_{(s+1)}$ and $\partial_t u(t,\,\cdot) \in H_{(s)}$. Furthermore, for $0 < s < k$,

$$\|u(t,\,\cdot)\|_{(s+1)} + \|\partial_t u(t,\,\cdot)\|_{(s)} = \lim_{l\to\infty} [\|w_l(t,\,\cdot)\|_{(s+1)} + \|\partial_t w_l(t,\,\cdot)\|_{(s)}] \le C\mathcal{M}.$$

Since the right-hand side does not depend on s we can use Lebesgue's monotone convergence theorem to conclude that $u(t,\cdot) \in H^{k+1}$ and $\partial_t u(t,\cdot) \in H^k$. Furthermore

$$\|u(t,\,\cdot)\|_{H^{k+1}} + \|\partial_t u(t,\,\cdot)\|_{H^k} \le C\mathcal{M}.$$

Weak continuity. We would like to say that (9.20) holds. However, it is necessary to divide the proof into two steps. First, let us prove that the solution is weakly continuous. Let f be in the dual of $H_{(k+1)}(\mathbb{R}^n, \mathbb{R}^N)$. Then there is a $\phi \in H_{(-k-1)}(\mathbb{R}^n, \mathbb{R}^N)$ such that

$$f(w) = \int_{\mathbb{R}^n} \hat{w}(\xi) \cdot \hat{\phi}(\xi)\, d\xi$$

for all $w \in H_{(k+1)}(\mathbb{R}^n, \mathbb{R}^N)$. Consequently, if ϕ_j is a sequence of Schwartz functions converging to ϕ in $H_{(-k-1)}$, we obtain that

$$|f[u(t,\cdot)] - f[w_l(t,\cdot)]| \le C\mathcal{M}\|\phi - \phi_j\|_{(-k-1)} + \left| \int_{\mathbb{R}^n} [\hat{u}(t,\xi) - \hat{w}_l(t,\xi)] \cdot \hat{\phi}_j(\xi)\, d\xi \right|,$$

where we abuse notation and write $\hat{u}(t, \xi)$ when we mean the Fourier transform of $u(t, \cdot)$ evaluated at ξ. Letting j be large enough that the first term on the right-hand side is less than or equal to $\varepsilon/2$, and then choosing l large enough, depending on j, so that the second term is less than $\varepsilon/2$, we conclude that the right-hand side is less than ε. We conclude that $f[w_l(t, \cdot)]$ converges uniformly to $f[u(t, \cdot)]$. In other words, the latter function is continuous. The argument for $\partial_t u$ is similar, and we conclude that u and $\partial_t u$ are weakly continuous.

Bound. Note that we can apply Lemma 9.9 with v, U_0 and U_1 replaced by w_l, $U_{0,l+1}$ and $U_{1,l+1}$ respectively. The corresponding solution is then w_{l+1}. Applying (9.14), we obtain, for $t \in [T_0, T_0 + T]$,

$$E_k[w_l, w_{l+1}](t) \leq E_k[w_l, w_{l+1}](T_0) + \int_{T_0}^{t} [C_I + \kappa_I(m[w_l], m[w_{l+1}])(M_k^2[w_l] + E_k[w_l, w_{l+1}])]\, dt. \tag{9.26}$$

Note that $m[w_l] \to m[u]$, since w_l converges to u with respect to the C^2-norm. Note also that for $t \in [T_0, T_0 + T]$,

$$\lim_{l\to\infty} [E_k[w_l, w_{l+1}](t) - E_k[u, w_{l+1}](t)] = 0,$$

since w_l converges to u in C^2 and since $M_k[w_l]$ is bounded. Furthermore, we have $M_k^2[w_l](t) \leq C_I E_k[u, w_l](t)$. Finally, $E_k[w_l, w_{l+1}](T_0)$ converges to $E_k[u, u](T_0)$ due to the above observations and the choice of $U_{0,l}$, $U_{1,l}$. Applying the above observations to (9.26) and using Fatou's lemma, Lemma 1.28, p. 23 of [79], together with an argument similar to the proof of Lebesgue's dominated convergence theorem on p. 27 of [79], we obtain

$$\mathfrak{E}_k(t) \leq \mathcal{E}_k(T_0) + \int_{T_0}^{t} [C_I + \kappa_I(m[u])\mathfrak{E}_k(\tau)]\, d\tau,$$

where $\mathcal{E}_k(t) = E_k[u, u](t)$ and

$$\mathfrak{E}_k(t) = \limsup_{l\to\infty} E_k[u, w_l](t).$$

By Grönwall's lemma, we obtain that

$$\mathfrak{E}_k(t) \leq [\mathcal{E}_k(T_0) + C_I \cdot (t - T_0)] \exp\left(\int_{T_0}^{t} \kappa_I(m[u])\, d\tau\right).$$

Note that since we have weak convergence of w_l to u in H^{k+1} and of $\partial_t w_l$ to $\partial_t u$ in H^k, we have

$$\begin{aligned}\mathcal{E}_k &= \lim_{l\to\infty} \frac{1}{2} \sum_{|\alpha|\leq k} \int_{\mathbb{R}^n} [(\partial^\alpha \partial_t u) \cdot (\partial^\alpha \partial_t w_l) + g_u^{ij} (\partial^\alpha \partial_i u) \cdot (\partial^\alpha \partial_j w_l) + \partial^\alpha u \cdot \partial^\alpha w_l]\, dx \\ &\leq \mathcal{E}_k^{1/2} \limsup_{l\to\infty} E_k^{1/2}[u, w_l],\end{aligned}$$

so that $\mathcal{E}_k(t) \leq \mathfrak{E}_k(t)$. Thus (9.21) holds.

Strong continuity. Let us prove that

$$\lim_{t\to T_0+}(\|u(t,\cdot)-u(T_0,\cdot)\|_{H^{k+1}}+\|\partial_t u(t,\cdot)-\partial_t u(T_0,\cdot)\|_{H^k})=0, \tag{9.27}$$

i.e., that u and $\partial_t u$ are right continuous at T_0. To this end, let $h^{ij}(x)=g^{ij}[u](T_0,x)$ and define an inner product on $H^{k+1}(\mathbb{R}^n,\mathbb{R}^N)\times H^k(\mathbb{R}^n,\mathbb{R}^N)$ by

$$\begin{aligned}&\langle(v_1,v_2),(w_1,w_2)\rangle\\&=\frac{1}{2}\sum_{|\alpha|\le k}\int_{\mathbb{R}^n}[(\partial^\alpha v_2)\cdot(\partial^\alpha w_2)+h^{ij}(\partial^\alpha\partial_i v_1)\cdot(\partial^\alpha\partial_j w_1)+(\partial^\alpha v_1)\cdot(\partial^\alpha w_1)]\,dx.\end{aligned}$$

Compute

$$\begin{aligned}&\langle(u(t,\cdot)-U_0,\partial_t u(t,\cdot)-U_1),(u(t,\cdot)-U_0,\partial_t u(t,\cdot)-U_1)\rangle\\&\quad=\langle(u(t,\cdot),\partial_t u(t,\cdot)),(u(t,\cdot),\partial_t u(t,\cdot))\rangle\\&\qquad-2\langle(u(t,\cdot),\partial_t u(t,\cdot)),(U_0,U_1)\rangle+\langle(U_0,U_1),(U_0,U_1)\rangle.\end{aligned} \tag{9.28}$$

Note that the last term on the right-hand side is $\mathcal{E}_k(T_0)$. The second term on the right-hand side converges to $-2\mathcal{E}_k(T_0)$ due to the weak continuity properties of u. Due to (9.21),

$$\limsup_{t\to T_0+}\mathcal{E}_k(t)\le\mathcal{E}_k(T_0).$$

Furthermore, due to the continuity and boundedness properties of u,

$$\lim_{t\to T_0+}[\langle(u(t,\cdot),\partial_t u(t,\cdot)),(u(t,\cdot),\partial_t u(t,\cdot))\rangle-\mathcal{E}_k(t)]=0.$$

Combining these observations with (9.28), we conclude that

$$\limsup_{t\to T_0+}\langle(u(t,\cdot)-U_0,\partial_t u(t,\cdot)-U_1),(u(t,\cdot)-U_0,\partial_t u(t,\cdot)-U_1)\rangle=0,$$

so that (9.27) follows. Note that at each $t\in[T_0,T_0+T]$, we can set up an iteration that converges in a neighbourhood of t and use the above argument and uniqueness to prove that u and $\partial_t u$, considered as functions from $[T_0,T_0+T]$ to H^{k+1} and H^k respectively, are right continuous at t. By time reversal, one obtains left continuity and thus continuity. □

9.4 Continuation criterion, smooth solutions

Lemma 9.14. *Let $1\le N,n\in\mathbb{Z}$, g be a C^∞ N,n-admissible metric, f be a C^∞ N,n-admissible non-linearity and $T_0\in\mathbb{R}$. Assume that for some $T\in(0,\infty)$, $u\in C^2([T_0,T_0+T]\times\mathbb{R}^n,\mathbb{R}^N)$ is a solution to* (9.4)–(9.6), *where the initial data satisfy $U_0\in H^{k+1}(\mathbb{R}^n,\mathbb{R}^N)$ and $U_1\in H^k(\mathbb{R}^n,\mathbb{R}^N)$ for some $k>n/2+1$. Then u satisfies*

(9.20) *and* (9.21)*, where, in the latter estimate,* κ *is an* N, n*-admissible majorizer,* C *is an* N, n*-admissible constant and* $I = [T_0, T_0 + T]$*. Let* T_k *be the supremum of times* T *such that there is a* C^2 *solution on* $[T_0, T_0 + T] \times \mathbb{R}^n$ *satisfying* (9.20)*. Then either* $T_k = \infty$ *or*

$$\lim_{t \to T_k -} \sup_{T_0 \le \tau \le t} m[u](\tau) = \infty. \tag{9.29}$$

Remark 9.15. Note in particular that T_k is independent of k. There is a similar statement in the opposite time direction.

Proof. Assume we have a C^2 solution with initial data in $H^{k+1} \times H^k$. Consider (9.21). Note that we know that this estimate holds on the time interval of interest in Proposition 9.12, i.e. on the interval where the iteration converges, but we do not know that it holds on the interval of interest in the current lemma. Let $\mathcal{A}$ be the set of $T_* \in [T_0, T_0 + T]$ such that (9.20) and (9.21) hold on $[T_0, T_*]$. Due to Proposition 9.12, we know that $\mathcal{A}$ contains an open neighbourhood of T_0, so that it is non-empty. We need to prove that $\mathcal{A}$ is open and closed. To prove openness, assume $T_* \in \mathcal{A}$. Then Proposition 9.12 yields a solution on $[T_*, T_* + \varepsilon]$ for some $\varepsilon > 0$ and we obtain (9.20) on $[T_0, T_* + \varepsilon]$. To prove (9.21) on $[T_0, T_* + \varepsilon]$, note that it holds on $[T_0, T_*]$ by assumption and that on $[T_*, T_* + \varepsilon]$ we have, by Proposition 9.12,

$$\mathcal{E}_k(t) \le [\mathcal{E}_k(T_*) + C_I(t - T_*)]\left(\int_{T_*}^t \kappa_I(m[u])\, d\tau\right).$$

Combining this observation with (9.21) with t replaced by T_*, we get

$$\begin{aligned}\mathcal{E}_k(t) &\le \left[[\mathcal{E}_k(T_0) + C_I(T_* - T_0)] \exp\left(\int_{T_0}^{T_*} \kappa_I(m[u])\, d\tau\right) + C_I(t - T_*)\right] \\ &\qquad \exp\left(\int_{T_*}^t \kappa_I(m[u])\, d\tau\right) \\ &\le [\mathcal{E}_k(T_0) + C_I(t - T_0)] \exp\left(\int_{T_0}^t \kappa_I(m[u])\, d\tau\right).\end{aligned}$$

To complete the proof, we need to demonstrate that $\mathcal{A}$ is closed. Assume T_* is in the closure of $\mathcal{A}$. Due to (9.21), which holds on $[T_0, T_{**}]$ for any $T_{**} < T_*$, we conclude that there is a uniform bound on $u(t, \cdot)$ and $\partial_t u(t, \cdot)$ in H^{k+1} and H^k respectively for $t \in [T_0, T_*)$. Since the existence time in Proposition 9.12 only depends on the size of the initial data and the compact interval in which the data are specified we get a solution beyond T_*, and the conclusions (9.20) and (9.21) follow; the first by the proposition and the second by continuity.

To prove the characterization of T_k given, assume $T_k < \infty$ and that (9.29) does not hold. Then $m[u]$ is bounded on $[T_0, T_k)$ so that (9.21) implies that $\mathcal{E}_k$ has a universal bound on $[T_0, T_k)$. By the local existence result we can then extend the solution beyond T_k, contradicting the definition of T_k. □

Corollary 9.16. *Let $1 \le N, n \in \mathbb{Z}$, g be a C^∞ N, n-admissible metric and f be a C^∞ N, n-admissible non-linearity. Let $U_0, U_1 \in C_0^\infty(\mathbb{R}^n, \mathbb{R}^N)$ and $T_0 \in \mathbb{R}$. Then there are $T_1 < T_0 < T_2$ and a unique solution $u \in C^\infty[(T_1, T_2) \times \mathbb{R}^n, \mathbb{R}^N]$ to* (9.4)–(9.6)*. The solution is of locally x-compact support and either $T_2 = \infty$ or*

$$\lim_{\tau \to T_2 -} \sup_{T_0 \le t \le \tau} \sum_{|\alpha| + j \le 2} \sup_{x \in \mathbb{R}^n} |\partial^\alpha \partial_t^j u(t, x)| = \infty.$$

The statement concerning T_1 is similar.

9.5 Stability

When studying Einstein's equations, it will be of interest to know that if we are given a background solution u on an interval (T_-, T_+), two points $T_0, T_1 \in (T_-, T_+)$, and a sequence of initial data, indexed by l, converging to the initial data of u at T_0, then the corresponding solutions u_l have the property that they exist until T_1 for l large enough and the initial data of u_l at T_1 converge to the initial data of u at T_1. Let us be more precise.

Proposition 9.17. *Let $1 \le N, n \in \mathbb{Z}$, g be a C^∞ N, n-admissible metric and f be a C^∞ N, n-admissible non-linearity. Consider the initial value problem*

$$g^{\mu\nu} \partial_\mu \partial_\nu u = f, \tag{9.30}$$
$$u(T_0, \cdot) = U_0, \tag{9.31}$$
$$\partial_t u(T_0, \cdot) = U_1. \tag{9.32}$$

Assume $u \in C^\infty[(T_-, T_+) \times \mathbb{R}^n, \mathbb{R}^N]$ is the solution to this equation corresponding to initial data $U_0, U_1 \in C_0^\infty(\mathbb{R}^n, \mathbb{R}^N)$. Assume furthermore that there is a sequence of functions $U_{0,l}, U_{1,l} \in C_0^\infty(\mathbb{R}^n, \mathbb{R}^N)$ such that

$$\lim_{l \to \infty} [\|U_{0,l} - U_0\|_{H^{k+1}} + \|U_{1,l} - U_1\|_{H^k}] = 0,$$

for some $k > n/2 + 1$. Let $u_l \in C^\infty[(T_{l,-}, T_{l,+}) \times \mathbb{R}^n, \mathbb{R}^N]$ be the solution to (9.30)–(9.32) *with U_0, U_1 replaced by $U_{0,l}, U_{1,l}$. Let $T_1 \in (T_-, T_+)$. Then there is an l_0 such that for $l \ge l_0$, $T_1 \in (T_{l,-}, T_{l,+})$. Furthermore,*

$$\lim_{l \to \infty} [\|u_l(T_1, \cdot) - u(T_1, \cdot)\|_{H^{k+1}} + \|\partial_t u_l(T_1, \cdot) - \partial_t u(T_1, \cdot)\|_{H^k}] = 0.$$

Remark 9.18. When we say *the solution*, we take it for granted that the existence intervals (T_-, T_+) and $(T_{l,-}, T_{l,+})$ are maximal.

Proof. Without loss of generality, we can assume $T_1 > T_0$ and we can restrict our attention to a fixed compact subset of (T_-, T_+) containing T_1 in its interior. Let $g = g[u]$, $f = f[u]$, $g_l = g[u_l]$ and $f_l = f[u_l]$. Taking the difference of the equations (9.30) for u and u_l, we obtain

$$g_l^{\mu\nu} \partial_\mu \partial_\nu v_l = F_l,$$

where

$$v_l = u_l - u, \quad F_l = (g^{\mu\nu} - g_l^{\mu\nu})\partial_\mu\partial_\nu u + f_l - f.$$

Note that there is a $G \in C^\infty[(T_-, T_+) \times \mathbb{R}^{nN+2N+n}, \mathbb{R}^N]$ such that the first condition of a C^∞ N, n-admissible non-linearity is satisfied if one restricts one's attention to compact subintervals of (T_-, T_+), such that $G(t, x, 0) = 0$ for $(t, x) \in (T_-, T_+) \times \mathbb{R}^n$ and such that

$$F_l(t, x) = G\{t, x, v_l(t, x), \partial_0 v_l(t, x), \dots, \partial_n v_l(t, x)\}.$$

As a consequence, an analogue of Lemma 6.17 is applicable so that if I is a compact subinterval of (T_-, T_+), then

$$\|F_l(t, \cdot)\|_{H^k} \le \sigma_I(m[v_l])[\|v_l(t, \cdot)\|_{H^{k+1}} + \|\partial_t v_l(t, \cdot)\|_{H^k}]$$

for all $t \in I$, where σ is an object similar to an N, n-admissible majorizer, the only difference being that in addition to depending on g and f, σ also depends on u, and the intervals I for which $\sigma_I[g, f, u]$ is defined are the compact subintervals of (T_-, T_+) (we shall below use the notation σ_I for objects that may be different but have the same properties). Let us define

$$\mathcal{E}_{l,j} = \frac{1}{2}\sum_{|\alpha|\le j}\int_{\mathbb{R}^n}[-g_l^{00}|\partial^\alpha\partial_0 v_l|^2 + g_l^{ij}\partial^\alpha\partial_i v_l \cdot \partial^\alpha\partial_j v_l + |\partial^\alpha v_l|^2]\,dx.$$

Note that this energy is equivalent to the H^{j+1} norm of $v_l(t, \cdot)$ plus the H^j norm of $\partial_t v_l(t, \cdot)$ due to our assumptions concerning g. We shall also use the notation $\mathcal{E}_l = \mathcal{E}_{l,0}$. By an argument similar to the derivation of (9.16), we get

$$\begin{aligned}\partial_t\mathcal{E}_l = \int_{\mathbb{R}^n}\Big[&- g_l^{\mu\nu}\partial_\mu\partial_\nu v_l \cdot \partial_t v_l - \partial_i(g_l^{0i})|\partial_t v_l|^2 - \frac{1}{2}(\partial_t g_l^{00})|\partial_t v_l|^2 \\ &- (\partial_i g_l^{ij})\partial_j v_l \cdot \partial_t v_l + \frac{1}{2}(\partial_t g_l^{ij})\partial_j v_l \cdot \partial_i v_l + v_l \cdot \partial_t v_l\Big]\,dx \\ \le{}& C\|F_l(t, \cdot)\|_2\mathcal{E}_l^{1/2} + \sigma_I(m[v_l])\mathcal{E}_l \le \sigma_I(m[v_l])\mathcal{E}_l,\end{aligned}$$

for $t \in I$, where I is a compact subinterval of (T_-, T_+) and we have used the fact that all derivatives of u are bounded on compact subsets of (T_-, T_+). Note that $\partial^\alpha v_l$ satisfies the equation

$$g_l^{\mu\nu}\partial_\mu\partial_\nu\partial^\alpha v_l = [g_l^{\mu\nu}\partial_\mu\partial_\nu, \partial^\alpha]v_l + \partial^\alpha F_l.$$

In the above argument, we thus get the same estimate for $\partial^\alpha v_l$ if we exchange F_l with $\partial^\alpha F_l + [\partial^\alpha, g_l^{\mu\nu}\partial_\mu\partial_\nu]v_l$. Summing these estimates, we get

$$\partial_t\mathcal{E}_{l,j} \le \sigma_I(m[v_l])\mathcal{E}_{l,j} + C\sum_{|\alpha|\le j}\|[\partial^\alpha, g_l^{\mu\nu}\partial_\mu\partial_\nu]v_l\|_2\mathcal{E}_{l,j}^{1/2}.$$

Note that $[\partial^\alpha, g_l^{\mu\nu}\partial_\mu\partial_\nu]v_l$ is a sum of terms of the form

$$(\partial^{\alpha_1}\partial_i g_l^{\mu\nu})\partial^{\alpha_2}\partial_\mu\partial_\nu v_l = [\partial^{\alpha_1}\partial_i(g_l^{\mu\nu} - g^{\mu\nu})]\partial^{\alpha_2}\partial_\mu\partial_\nu v_l + (\partial^{\alpha_1}\partial_i g^{\mu\nu})\partial^{\alpha_2}\partial_\mu\partial_\nu v_l,$$

up to numerical factors, where $|\alpha_1| + |\alpha_2| = |\alpha| - 1$. Using (6.22), we obtain

$$\|(\partial^{\alpha_1}\partial_i g_l^{\mu\nu})\partial^{\alpha_2}\partial_\mu\partial_\nu v_l\|_2 \le \sigma_I(m[v_l])\mathcal{E}_{l,j}^{1/2},$$

since $g_l^{00} = g^{00} = -1$. Thus

$$\partial_t\mathcal{E}_{l,j} \le \sigma_I(m[v_l])\mathcal{E}_{l,j}. \tag{9.33}$$

Let

$$\begin{aligned}\mathcal{A} = \{t \in [T_0, T_1] \ :\ & \text{there exists } l_1 \text{ such that for all } s \in [T_0, t],\ l \ge l_1, \\ & \text{it holds } t < T_{l,+} \text{ and } M_k[u_l](s) \le M_k[u](s) + 1\}.\end{aligned}$$

Note that $\mathcal{A}$ is non-empty since $T_0 \in \mathcal{A}$. Assume $t \in \mathcal{A}$. Then there is an l_1 such that $[T_0, t] \subset (T_{l,-}, T_{l,+})$ and

$$M_k[u_l](s) \le M_k[u](s) + 1 \tag{9.34}$$

for $s \in [T_0, t]$ and $l \ge l_1$. Since the existence time in Proposition 9.12 only depends on the norm of the data, there is an $\varepsilon > 0$ such that we can extend u_l an interval of length ε beyond t for all $l \ge l_1$. Furthermore, we get a bound on $M_k[u_l](s)$ which is uniform in l for $l \ge l_1$ and $s \in [T_0, t + \varepsilon]$. Using the equation, we get a uniform bound on $m[v_l](s)$ for $s \in [T_0, t + \varepsilon]$ and $l \ge l_1$. Using (9.33), we conclude that for l large enough, (9.34) holds on $[T_0, t + \varepsilon]$. Thus $\mathcal{A}$ is open. Let $t_i \in \mathcal{A}$ be such that $t_i \to T$. We wish to prove that $T \in \mathcal{A}$. Since $\mathcal{A}$ is connected and $T_0 \in \mathcal{A}$, we can assume $T > t_i$. Let l_i be such that (9.34) holds on $[T_0, t_i]$ for $l \ge l_i$. Due to Proposition 9.12, there is an $\varepsilon > 0$ such that if we specify initial data $u_l(t_i, \cdot)$, $\partial_t u_l(t_i, \cdot)$ at t_i where $l \ge l_i$, then we get existence up to $t_i + \varepsilon$ and a bound on $M_k[u_l](s)$ for $s \in [T_0, t_i + \varepsilon]$ independent of $l \ge l_i$. Fix an i such that $T - t_i < \varepsilon/2$. Then, for $l \ge l_i$, $[T_0, T + \varepsilon/2] \subset (T_{l,-}, T_{l,+})$ and we have a uniform bound on $M_k[u_l](s)$ and on $m[u_l](s)$ for $l \ge l_i$ and $s \in [T_0, T + \varepsilon/2]$. Using (9.33) again, we see that for l large enough, (9.34) holds on $[T_0, T + \varepsilon/2]$. Thus $\mathcal{A}$ is an open, closed and non-empty subset of $[T_0, T_1]$. Thus $T_1 \in \mathcal{A}$. Using the equation, we conclude that there is an l_1 such that $m[v_l](s)$ is uniformly bounded for $s \in [T_0, T_1]$ and $l \ge l_1$. Inserting this information into (9.33), integrating and letting $l \to \infty$, we get the desired conclusion. □

Part II

Background in geometry, global hyperbolicity and uniqueness

10 Basic Lorentz geometry

10.1 Manifolds

The purpose of this chapter and the next is to lay the foundations for a geometric uniqueness statement. The coefficients of the highest order derivatives in (9.4) are determined by a Lorentz metric, and the uniqueness should be expressed in terms of the causal structure of that metric.

We assume that the reader is familiar with elementary differential geometry as well as Lorentz geometry. However, we wish to recall the basic definitions. In particular we wish to write down precise definitions of orientability and of the relation between the orientation of a manifold and its boundary. The reference we have in mind as far as integration and orientability is concerned is [25], the reference concerning Lorentz geometry is [65] and for basic concepts of topology, we shall refer the reader to [8]. Let us start by defining the concept of a differentiable manifold.

Definition 10.1. An n-dimensional *differentiable manifold* is a second countable Hausdorff space M^n together with a collection of maps called "charts" such that:

- a chart is a homeomorphism $\phi\colon U \to U'$, where U is open in M^n and U' is open in $\mathbb{R}^n$;
- each point $x \in M^n$ is in the domain of some chart;
- for charts $\phi\colon U \to U'$ and $\psi : V \to V'$, $\phi \circ \psi^{-1}\colon \psi(U \cap V) \to \phi(U \cap V)$ is C^∞;
- the collection of charts is maximal with the above properties.

Recall that a topological space X is *Hausdorff* if for every pair of $x, y \in X$ such that $x \neq y$, there are open subsets U, V of X such that $x \in U$, $y \in V$ and $U \cap V = \emptyset$. If X is a topological space and $\mathcal{B}$ is a collection of subsets of X, then $\mathcal{B}$ is called a *basis* for the topology of X if the open sets are precisely the unions of members of $\mathcal{B}$. A topological space is said to be *second countable* if it has a countable basis for the topology. As a consequence of its definition, a differentiable manifold is paracompact, cf. Theorem 12.12 p. 38 of [8]. However, we shall not use this fact. Let us recall the definition of paracompactness. If $\mathcal{U}$ and $\mathcal{V}$ are open coverings of a set, then $\mathcal{U}$ is said to be a *refinement* of $\mathcal{V}$ if each element of $\mathcal{U}$ is a subset of some element of $\mathcal{V}$. A collection $\mathcal{U}$ of subsets of a topological space X is said to be *locally finite* if each point $x \in X$ has a neighbourhood N which meets, non-trivially, only a finite number of the members of $\mathcal{U}$. Finally, a Hausdorff space X is said to be *paracompact* if every open covering has an open, locally finite refinement. The point of paracompactness is that if $\mathcal{U}$ is an open covering of a differentiable manifold, then there is a smooth partition of unity subordinate to $\mathcal{U}$. Again, let us explain the terminology.

Definition 10.2. Let $\mathcal{U}$ be an open covering of a differentiable manifold M. Then a *smooth partition of unity* subordinate to this covering is a collection of smooth maps $f_\beta : M \to [0,1]$ for $\beta \in \mathcal{B}$ such that

- There is a locally finite open refinement $\{V_\beta : \beta \in \mathcal{B}\}$ of $\mathcal{U}$ such that $\mathrm{supp}(f_\beta) \subset V_\beta$ for all $\beta \in \mathcal{B}$;
- $\sum_\beta f_\beta(x) = 1$ for all $x \in M$.

In this definition $\mathrm{supp}(f_\beta)$ is the closure of the set of points for which f_β is non-zero, and $f : M \to N$ between differentiable manifolds is said to be smooth if $\phi \circ f \circ \psi^{-1}$ is smooth for all combination of charts ϕ and ψ on N and M respectively.

One can prove the existence of partitions of unity using only the assumption of paracompactness, but in the situation we are interested in, there is an easier proof.

Lemma 10.3. *Let M be a smooth manifold and let $\mathcal{U}$ be an open covering. Then there is a locally finite open covering V_i, $i = 1, 2, \ldots$ where the closures of the V_i are compact and contained in elements of $\mathcal{U}$ and in the domains of coordinate charts. Furthermore, there are $f_i \in C_0^\infty(V_i)$ such that $\sum_i f_i = 1$.*

Proof. Since M is second countable, there is a countable basis for the topology, say $\mathcal{B}$. Due to the existence of charts, each point $p \in M$ has a compact neighbourhood. Consequently, the collection of elements in $\mathcal{B}$ with compact closure, say U_n, $n = 1, 2, \ldots$ is a covering of M. Note that here we implicitly assume that the closure of a set contained in a compact set is compact. This is based on two facts. Namely that a closed set contained in a compact set is compact and that a compact set is closed. It is important to keep in mind that the latter statement *uses the fact that M is Hausdorff*. Define $K_1 = \bar{U}_1$. Given that K_n has been defined, define

$$K'_{n+1} = \bigcup_{m=1}^{n+1} \bar{U}_m \cup K_n.$$

The U_i form an open covering of the compact set K'_{n+1}. Thus there is a finite subcovering. Define K_{n+1} to be the union of K'_{n+1} and the closure of the elements of the finite subcovering. Then K_{n+1} is compact and contains K_n in its interior. We have thus constructed a sequence of compact sets K_n such that $K_n \subset \mathrm{int}\, K_{n+1}$ and the union of the K_n is M.

For each $p \in K_2$, let V_p be an open set containing p such that its closure is compact and contained in a coordinate chart and $\bar{V}_p \subseteq \mathrm{int}\, K_3 \cap U_\beta$ for some $U_\beta \in \mathcal{U}$. Let W_p be an open set containing p such that the closure of W_p is contained in V_p. Let $p_{2,i}$, $i = 1, \ldots, k_2$ be points such that $W_{p_{2,i}}$ cover K_2 and let $f_{2,i} \in C_0^\infty(V_{p_{2,i}})$, $i = 1, \ldots, k_2$ be such that $0 \le f_{2,i} \le 1$ and $f_{2,i} = 1$ on $W_{p_{2,i}}$, cf. Proposition A.12. Given that we have a covering of K_n, let, for each $p \in K_{n+1} - \mathrm{int}\, K_n$, V_p be an open neighbourhood of p with compact closure contained in a coordinate neighbourhood such that

$$\bar{V}_p \subseteq (\mathrm{int}\, K_{n+2} - K_{n-1}) \cap U_\beta$$

for some $U_\beta \in \mathcal{U}$. Let W_p be a neighbourhood of p with compact closure contained in V_p. Let $p_{n+1,i}$, $i = 1, \dots, k_{n+1}$ be points such that $W_{p_{n+1,i}}$ cover $K_{n+1} - \operatorname{int} K_n$ and let $f_{n+1,i} \in C_0^\infty(V_{p_{n+1,i}})$, $i = 1, \dots, k_{n+1}$ be such that $0 \leq f_{n+1,i} \leq 1$ and $f_{n+1,i} = 1$ on $W_{p_{n+1,i}}$, cf. Proposition A.12.

Let us prove that $V_{p_{i,j}}$ for $i = 2, 3, \dots$ and $j = 1, \dots, k_i$ is a locally finite refinement of $\mathcal{U}$. By construction it is a refinement. Since $V_{p_{i_1,j_1}} \cap V_{p_{i_2,j_2}} \neq \emptyset$ implies that $|i_1 - i_2| \leq 2$, we conclude that the refinement is locally finite. Consequently, the sum

$$\phi = \sum_{i=2}^{\infty} \sum_{j=1}^{k_i} f_{i,j}$$

defines a smooth function, since each point of the manifold has a neighbourhood such that only a finite number of the terms are non-zero. Furthermore, $\phi \geq 1$, since each $p \in M$ belongs to some $W_{p_{i,j}}$. Letting $g_{i,j} = f_{i,j}/\phi$, we get the desired partition of unity. □

For future reference let us note that the above proof gives a construction of a nice open covering.

Corollary 10.4. *Let M be a differentiable manifold. Then there is a countable, locally finite open covering V_i, $i = 1, 2, \dots$ all the members of which are contained in coordinate charts. Furthermore, there are open sets U_i, $i = 1, 2, \dots$ such that $\bar{U}_i$ is compact and contained in V_i and the U_i also form an open covering.*

We assume the reader is familiar with the standard constructions of the tangent bundle, cotangent bundle, etc. We shall say that a differentiable manifold is *oriented* if there is an atlas obeying the additional condition that for any pair of charts $\phi\colon U \to U'$ and $\psi\colon V \to V'$, with $U \cap V \neq \emptyset$, the determinant of the derivative of $\phi \circ \psi^{-1}$ is positive. We shall call such an atlas an oriented atlas. Given an *ordered* collection of n linearly independent tangent vectors $v_1, \dots, v_n$ at a point p of M, we say that it is *positively oriented*, if the matrix A whose components A_{ij} are given by $v_i = \sum_j A_{ij} \partial_j|_p$, where $\phi = (x^1, \dots, x^n)$ is a chart belonging to the oriented atlas, has the property that $\det A > 0$.

In order to define the concept of a manifold with boundary, let us introduce

$$H^n = \{x \in \mathbb{R}^n : x^n \geq 0\}, \quad \partial H^n = \{x \in \mathbb{R}^n : x^n = 0\}.$$

The latter set is referred to as the *boundary* of H^n. If $f\colon H^n \to \mathbb{R}^n$, we say that it is smooth if it can be extended to be a smooth map from $\mathbb{R}^n$ to $\mathbb{R}^n$. We say that M is a *differentiable manifold with boundary* if M is a second countable Hausdorff space which is covered by charts mapping open subsets of M either to open subsets of $\mathbb{R}^n$ or to open subsets of H^n. Again, we require smoothness and maximality in the same way as we did in the definition of differentiable manifolds. Given that M is a differentiable manifold with boundary, we define the boundary of M, denoted ∂M, to be the set of points $p \in M$ such that there is a chart (U, ϕ) with $p \in U$, $\phi\colon U \to U'$ where U' is an open subset of H^n and $\phi(p) \in \partial H^n$. One can check that if (V, ψ) is a chart and $p \in V$, then $\psi : V \to V'$ with V' an open subset of H^n and $\psi(p) \in \partial H^n$. Note that ∂M is an $n-1$ dimensional differentiable manifold.

Given that M is an oriented manifold with boundary, let us define an orientation on ∂M. First of all, given a tangent vector at $p \in \partial M$ which is not tangent to ∂M, say v, then we can write $v = v^i \partial_i|_p$ with respect to coordinates $\phi = (x^1, \dots, x^n)$ where $\phi : U \to H^n$. If $v^n > 0$, we say that v is *inward pointing* and if $v^n < 0$ we say that v is *outward pointing*. One can check that this is not dependent on the choice of coordinates and that the following definition yields a well defined orientation.

Definition 10.5. Let M be an oriented differentiable manifold with boundary. Then the *induced* orientation on ∂M is defined by requiring that $(v_1, \dots, v_{n-1}) \in T_p(\partial M)$ be positively oriented if and only if $(u, v_1, \dots, v_{n-1})$ is a positively oriented basis for $T_p M$ whenever $u \in T_p M$ is outward pointing.

Let us define integration on a manifold. Our basic reference for integration, as well as orientability is [25]. Given a continuous n-form ω which has compact support within a coordinate chart (U, ϕ), we define

$$\int_M \omega = \int_{\phi(U)} h \circ \phi^{-1} \, d\mu,$$

where h is a function such that

$$\omega = h dx^1 \wedge \cdots \wedge dx^n$$

and μ is the Lebesgue measure on $\mathbb{R}^n$. That this definition is independent of coordinates follows from the standard properties of forms and the change of variables formula, cf. Theorem 7.26 of [79]. Note, however, that for this to hold it is essential that the manifold be oriented. The integral of a general continuous n-form with compact support is then obtained by using a partition of unity. Furthermore, one can prove Stokes' theorem.

Theorem 10.6. *Let M be an oriented n-manifold with boundary ∂M. Let ω be a smooth $(n-1)$-form on M having compact support. Then*

$$\int_M d\omega = \int_{\partial M} \omega.$$

We shall be interested in applying this result in the case of a Lorentz manifold, cf. Section 10.2. Let (M, g) be an oriented $n+1$ dimensional Lorentz manifold. One can then define the *volume form* ε by

$$\varepsilon = \sqrt{-\det g} \, dx^0 \wedge \cdots \wedge dx^n, \tag{10.1}$$

where $\det g$ is the determinant of the matrix $g(\partial_\mu, \partial_\nu)$. By the standard properties of forms, this definition is independent of the choice of coordinates, so that ε is well defined globally. Given a smooth k-form ω and a smooth vector field ξ, we define

$$i_\xi \omega(v_{1p}, \dots, v_{k-1p}) = \omega(\xi_p, v_{1p}, \dots, v_{k-1p})$$

for $v_{1p}, \dots, v_{k-1p} \in T_pM$. Thus $i_\xi\omega$ is a smooth $k-1$ form. Given a smooth vector field ξ, we define

$$\operatorname{div}\xi = \nabla_\mu \xi^\mu = \frac{\partial \xi^\mu}{\partial x^\mu} + \Gamma^\mu_{\mu\nu}\xi^\nu$$

in coordinates, cf. Section 10.2, where

$$\Gamma^\alpha_{\beta\gamma} = \frac{1}{2}g^{\alpha\delta}(\partial_\beta g_{\gamma\delta} + \partial_\gamma g_{\beta\delta} - \partial_\delta g_{\beta\gamma}) \tag{10.2}$$

are the Christoffel symbols and ∇ is the Levi-Civita connection associated with g. We wish to prove the following result.

Lemma 10.7. *Let (M, g) be an $(n+1)$-dimensional oriented Lorentz manifold with volume form ε given by* (10.1). *If ξ is a smooth vector field, then*

$$d(i_\xi\varepsilon) = (\operatorname{div}\xi)\varepsilon.$$

Proof. Concerning the wedge product, all we need to know in the following is that if $v_1, \dots, v_n$ are linearly independent vectors in an n-dimensional vector space and $v^1, \dots, v^n$ are the duals, then

$$v^{i_1} \wedge \dots \wedge v^{i_k}(v_{i_1}, \dots, v_{i_k}) = 1.$$

Let us compute, for $\mu = 0, \dots, n$,

$$\begin{aligned} i_\xi\varepsilon(\partial_0, \dots, \widehat{\partial_\mu}, \dots, \partial_n) &= \varepsilon(\xi, \partial_0, \dots, \widehat{\partial_\mu}, \dots, \partial_n) \\ &= \xi^\mu \varepsilon(\partial_\mu, \partial_0, \dots, \widehat{\partial_\mu}, \dots, \partial_n) \\ &= (-1)^\mu \xi^\mu \varepsilon(\partial_0, \dots, \partial_n) \\ &= (-1)^\mu \xi^\mu \sqrt{-\det g}, \end{aligned}$$

where we *do not* sum over μ and the hat signifies omission. By the properties of forms, we see that

$$i_\xi\varepsilon = \sum_{\mu=0}^{n} (-1)^\mu \xi^\mu \sqrt{-\det g}\, dx^0 \wedge \dots \wedge \widehat{dx^\mu} \wedge \dots \wedge dx^n.$$

Let us compute $d(i_\xi\varepsilon)$ at a point p, assuming that $(x^0, \dots, x^n)$ are geodesic normal coordinates at p, cf. pp. 71–73 of [65] (see also Subsection 10.2.5). Then $\partial_\alpha g_{\beta\gamma}(p) = 0$ for all α, β, γ, so that, at p,

$$d(i_\xi\varepsilon) = \sum_{\mu,\nu=1}^{n} (-1)^\mu \frac{\partial \xi^\mu}{\partial x^\nu} \sqrt{-\det g}\, dx^\nu \wedge dx^0 \wedge \dots \wedge \widehat{dx^\mu} \wedge \dots \wedge dx^n = \sum_{\mu=0}^{n} \frac{\partial \xi^\mu}{\partial x^\mu}\varepsilon.$$

Since the right-hand side equals $\operatorname{div}\xi\ \varepsilon$ at p, we get the desired result. □

The idea is of course to integrate this relation using Stokes' theorem, but before we do so, we need to sort out what $i_\xi \varepsilon$ is when restricted to the boundary. In order to be able to obtain a nice expression, we shall need to assume that the metric, when restricted to the boundary, is either Riemannian or Lorentzian. If the first case occurs, we shall say that the boundary is *spacelike* and if the second case occurs, we shall say that the boundary is *timelike*. As a consequence, the outward pointing normal to the boundary, N, is not null, i.e., $\langle N, N\rangle \neq 0$, where we use the notation $g(N,N) = \langle N, N\rangle$, so that we can assume it to be a unit vector (± 1). Let $e_1, \dots, e_n$ be an oriented orthonormal basis of $T_p(\partial M)$. Then $N, e_1, \dots, e_n$ is an orthonormal basis

$$i_\xi \varepsilon(e_1, \dots, e_n) = \frac{\langle \xi, N\rangle}{\langle N, N\rangle}\varepsilon(N, e_1, \dots, e_n) = \frac{\langle \xi, N\rangle}{\langle N, N\rangle},$$

where the last equality is due to our choice of orientation of the boundary. Thus

$$i_\xi \varepsilon_M = \frac{\langle \xi, N\rangle}{\langle N, N\rangle}\varepsilon_{\partial M}$$

on ∂M, where ε_M is the volume form of M and $\varepsilon_{\partial M}$ is the volume form of ∂M. To conclude, we have the following result.

Lemma 10.8. *Assume (M, g) is an $(n+1)$-dimensional oriented Lorentz manifold with boundary, assume that the boundary is spacelike or timelike and let ξ be a smooth vector field with compact support. Then, if ε_M and $\varepsilon_{\partial M}$ are the volume forms of M and ∂M respectively and N is the outward pointing unit normal to ∂M,*

$$\int_M \operatorname{div}\xi\, \varepsilon_M = \int_{\partial M} \frac{\langle \xi, N\rangle}{\langle N, N\rangle}\, \varepsilon_{\partial M}. \tag{10.3}$$

Remark 10.9. A similar result is of course true in the Riemannian setting.

10.2 Lorentz geometry

In this section we are concerned with the basic properties of Lorentz manifolds, i.e., manifolds with a Lorentz metric. We refer the reader to [65] for the proofs of the statements made.

10.2.1 Lorentz metrics. A *Lorentz metric* g is a symmetric non-degenerate covariant 2-tensor field on M such that at every point p of M, there is a basis for T_pM, say $e_0, \dots, e_n$ such that $g(e_\mu, e_\nu)$ are the components of the standard Minkowski metric $\eta = \operatorname{diag}(-1, 1, \dots, 1)$. We call a couple (M, g), where M is a manifold and g is a Lorentz metric on M a Lorentz manifold. Note that we can use the metric to define an isometry between vector fields and one-forms. In fact, if X is a vector field, we obtain a one-form according to the formula

$$X^\flat(Y) = \langle Y, X\rangle,$$

where we write $\langle Y, X\rangle$ instead of $g(X, Y)$. Conversely, given a one form η on M, there is a unique vector field $\eta^\sharp$ such that

$$\eta(Y) = \langle Y, \eta^\sharp\rangle, \tag{10.4}$$

cf. Proposition 10, p. 60 of [65]. Needless to say, this construction also works for individual vectors and covectors. Note that with respect to any coordinates x, $g_{\mu\nu} = g(\partial_{x^\mu}, \partial_{x^\nu})$ are the components of an invertible matrix. We shall use the notation $g^{\mu\nu}$ for the components of the inverse. These are the components of a contravariant 2-tensor field on M. This tensor field, say h, can be characterized more invariantly by the condition that

$$h(\eta, \omega) = g(\eta^\sharp, \omega^\sharp).$$

10.2.2 Covariant differentiation. Let (M, g) be a Lorentz manifold. Then there is a unique metric and torsion free connection, called the Levi-Civita connection, cf. Theorem 11, p. 61 of [65]. We shall denote it ∇ unless otherwise stated. Recall that the condition that ∇ be metric is the requirement that

$$X\langle V, W\rangle = \langle \nabla_X V, W\rangle + \langle V, \nabla_X W\rangle$$

hold for all vector fields X, V, W. Furthermore, ∇ is said to be torsion free if

$$\nabla_X Y - \nabla_Y X = [X, Y]$$

for all vector fields X, Y. It is of interest to extend ∇ to apply to tensor fields as well. We shall use the notation $\mathcal{T}_s^r(M)$ for smooth tensorfields on M, covariant of order s and contravariant of order r. One can extend ∇ by requiring it to be $\mathbb{R}$-linear, to satisfy

$$\nabla(A \otimes B) = \nabla A \otimes B + A \otimes \nabla B,$$

and to commute with contractions, i.e., by requiring ∇_V to be a tensor derivation for every vector field V, cf. Definition 11, p. 43 of [65]. That this is possible follows from Theorem 15, p. 45 of [65], see also Definition 16, p. 64. We shall most of the time use the notation of [65], but sometimes it will be convenient to use the notation of [87]. Therefore we wish to relate the different notations here, as well as fix our conventions concerning curvature. If $T \in \mathcal{T}_s^r(M)$, then

$$(\nabla_\alpha T^{\beta_1\ldots\beta_r}_{\alpha_1\ldots\alpha_s})X^\alpha Y_1^{\alpha_1}\ldots Y_s^{\alpha_s}\theta_{1\beta_1}\ldots\theta_{r\beta_r} = (\nabla_X T)(Y_1,\ldots,Y_s,\theta_1,\ldots,\theta_r)$$

defines what we mean by $\nabla_\alpha T^{\beta_1\ldots\beta_r}_{\alpha_1\ldots\alpha_s}$. Note that the right-hand side can be computed by using the fact that ∇ commutes with contractions in order to obtain

$$\begin{aligned}(\nabla_X T)(Y_1,\ldots,Y_s,\theta_1,\ldots,\theta_r) = {} & X[T(Y_1,\ldots,Y_s,\theta_1,\ldots,\theta_r)] \\ & - T(\nabla_X Y_1,\ldots,Y_s,\theta_1,\ldots,\theta_r) - \cdots \\ & \cdots - T(Y_1,\ldots,Y_s,\theta_1,\ldots,\nabla_X\theta_r).\end{aligned}$$

In particular

$$(\nabla_\alpha g_{\mu\nu})X^\alpha Y^\mu Z^\nu = X[g(Y,Z)] - g(\nabla_X Y, Z) - g(Y, \nabla_X Z) = 0,$$

due to the fact that the Levi-Civita connection is metric. Thus $\nabla_\alpha g_{\mu\nu} = 0$. Applying a vector field X to the identity (10.4) and using the metricity of ∇, one obtains

$$(\nabla_X \eta)^\sharp = \nabla_X \eta^\sharp.$$

Using this observation, one can argue similarly to the proof that $\nabla_\alpha g_{\mu\nu} = 0$ in order to obtain $\nabla_\alpha g^{\mu\nu} = 0$. In order to see what we obtain in terms of coordinates, let (U, x) be a coordinate system and define $\Gamma^\alpha_{\mu\nu}$ by the condition that

$$\nabla_{\partial_\mu}\partial_\nu = \Gamma^\alpha_{\mu\nu}\partial_\alpha \tag{10.5}$$

(if one wishes to be formal, one can here view (U, g) as a Lorentz manifold). One can then compute, using the Koszul formula for instance, cf. Theorem 11, p. 61 of [65], that

$$\Gamma^\alpha_{\mu\nu} = \frac{1}{2}g^{\alpha\beta}(\partial_\mu g_{\nu\beta} + \partial_\nu g_{\mu\beta} - \partial_\beta g_{\mu\nu}).$$

Furthermore,

$$\nabla_{\partial_\mu} dx^\nu(\partial_\alpha) = -dx^\nu(\nabla_{\partial_\mu}\partial_\alpha) = -\Gamma^\nu_{\mu\alpha},$$

i.e.,

$$\nabla_{\partial_\mu} dx^\nu = -\Gamma^\nu_{\mu\alpha} dx^\alpha.$$

With respect to coordinates, we can thus compute

$$\begin{aligned}\nabla_\alpha T^{\beta_1\dots\beta_r}_{\alpha_1\dots\alpha_s} &= (\nabla_{\partial_\alpha} T)(dx^{\beta_1},\dots,dx^{\beta_r},\partial_{\alpha_1},\dots,\partial_{\alpha_s})\\ &= \partial_\alpha T^{\beta_1\dots\beta_r}_{\alpha_1\dots\alpha_s} - \Gamma^\mu_{\alpha\alpha_1} T^{\beta_1\dots\beta_r}_{\mu\alpha_2\dots\alpha_s} - \dots - \Gamma^\mu_{\alpha\alpha_s} T^{\beta_1\dots\beta_r}_{\alpha_1\alpha_2\dots\alpha_{s-1}\mu}\\ &\quad + \Gamma^{\beta_1}_{\alpha\mu} T^{\mu\beta_2\dots\beta_r}_{\alpha_1\alpha_2\dots\alpha_s} + \dots + \Gamma^{\beta_r}_{\alpha\mu} T^{\beta_1\dots\beta_{r-1}\mu}_{\alpha_1\alpha_2\dots\alpha_s}.\end{aligned}$$

Let us define $S^{\beta_2\dots\beta_r}_{\alpha_2\dots\alpha_s} = T^{\alpha_1\beta_2\dots\beta_r}_{\alpha_1\dots\alpha_s}$. By the above formula, we see explicitly that

$$\nabla_\alpha S^{\beta_2\dots\beta_r}_{\alpha_2\dots\alpha_s} = \nabla_\alpha T^{\alpha_1\beta_2\dots\beta_r}_{\alpha_1\dots\alpha_s}.$$

This is simply the coordinate manifestation of the fact that ∇ commutes with contractions. Similarly, we see directly that the covariant derivative of the tensor product of two tensorfields is given by the expected formula with respect to coordinates. Note that one can use $g^{\mu\nu}$ and $g_{\mu\nu}$ to raise and lower indices of tensor fields. Since $\nabla_\alpha g_{\mu\nu} = 0$ and $\nabla_\alpha g^{\mu\nu} = 0$, these operations commute with covariant differentiation. In the following, we shall use the notation

$$\nabla_\alpha(T^{\alpha_1\beta_2\dots\beta_r}_{\alpha_1\dots\alpha_s}) = \nabla_\alpha S^{\beta_2\dots\beta_r}_{\alpha_2\dots\alpha_s},$$

and similarly in other cases, but since ∇ commutes with contractions it is clear that one need not distinguish between, for instance, $\nabla_\alpha(T^{\alpha_1\beta_2\dots\beta_r}_{\alpha_1\dots\alpha_s})$ and $\nabla_\alpha T^{\alpha_1\beta_2\dots\beta_r}_{\alpha_1\dots\alpha_s}$.

Let $T = \nabla X$ and let us compute

$$\begin{aligned}(\nabla_\alpha \nabla_\beta X^\gamma) Y^\alpha Z^\beta \theta_\gamma &= Y[T(Z,\theta)] - T(\nabla_Y Z, \theta) - T(Z, \nabla_Y \theta) \\ &= Y[\theta(\nabla_Z X)] - \theta(\nabla_{\nabla_Y Z} X) - \nabla_Y \theta(\nabla_Z X) \\ &= \theta(\nabla_Y \nabla_Z X - \nabla_{\nabla_Y Z} X).\end{aligned}$$

Since the Levi-Civita connection is torsion free, we conclude that

$$(\nabla_\alpha \nabla_\beta X^\gamma - \nabla_\beta \nabla_\alpha X^\gamma) Y^\alpha Z^\beta \theta_\gamma = \theta(\nabla_Y \nabla_Z X - \nabla_Z \nabla_Y X - \nabla_{[Y,Z]} X).$$

Let us introduce the Riemannian curvature tensor R by

$$R_{YZ} X = \nabla_Y \nabla_Z X - \nabla_Z \nabla_Y X - \nabla_{[Y,Z]} X. \tag{10.6}$$

Note that we have the opposite sign to that of [65] p. 74. That this object defines a tensor field follows from Lemma 35, p. 74 of [65]. In order to have the same conventions as in [87], cf. (3.2.11), p. 38, we wish to have

$$(\nabla_\alpha \nabla_\beta - \nabla_\beta \nabla_\alpha) X^\gamma = -R_{\alpha\beta\delta}{}^\gamma X^\delta.$$

If we let $\theta = W^\flat$ for some vector field W and use the convention $R_{\alpha\beta\delta}{}^\gamma g_{\gamma\mu} = R_{\alpha\beta\delta\mu}$, we get

$$R_{\alpha\beta\delta\mu} Y^\alpha Z^\beta X^\delta W^\mu = -(\nabla_\alpha \nabla_\beta X^\gamma - \nabla_\beta \nabla_\alpha X^\gamma) Y^\alpha Z^\beta \theta_\gamma = -\langle R_{YZ} X, W\rangle.$$

With respect to a coordinate system, we thus see that

$$R_{\alpha\beta\delta\mu} = -\langle R_{\partial_\alpha \partial_\beta} \partial_\delta, \partial_\mu \rangle. \tag{10.7}$$

This tensor has several symmetries, cf. Proposition 36, p. 75 of [65]. In particular,

$$R_{\alpha\beta\delta\mu} = -R_{\beta\alpha\delta\mu} = R_{\delta\mu\alpha\beta} = -R_{\alpha\beta\mu\delta} \tag{10.8}$$

and

$$R_{\alpha\beta\delta\mu} + R_{\beta\delta\alpha\mu} + R_{\delta\alpha\beta\mu} = 0.$$

The last equality is referred to as the first Bianchi identity and can also be written

$$R_{[\alpha\beta\delta]\mu} = 0,$$

where we use the notation

$$T^{\beta_1\dots\beta_k}_{[\alpha_1\dots\alpha_j]\alpha_{j+1}\dots\alpha_l} = \frac{1}{j!} \sum_{\sigma\in S_j} \operatorname{sign}\sigma\, T^{\beta_1\dots\beta_k}_{\alpha_{\sigma(1)}\dots\alpha_{\sigma(j)}\alpha_{j+1}\dots\alpha_l},$$

where S_j is the set of permutations on $\{1,\dots,j\}$, $\operatorname{sign}\sigma$ equals 1 if σ is even and -1 if σ is odd. This process is called antisymmetrizing over $\alpha_1,\dots,\alpha_l$. Similarly, we define

$$T^{\beta_1\dots\beta_k}_{(\alpha_1\dots\alpha_j)\alpha_{j+1}\dots\alpha_l} = \frac{1}{j!} \sum_{\sigma\in S_j} T^{\beta_1\dots\beta_k}_{\alpha_{\sigma(1)}\dots\alpha_{\sigma(j)}\alpha_{j+1}\dots\alpha_l}.$$

This process is called symmetrizing over $\alpha_1, \dots, \alpha_l$. These examples can be generalized in a natural way. We shall sometimes wish to exclude one of the indices in the antisymmetrization or symmetrization. The notation we use for exclusion is to write an absolute value sign around the index, or indices, we wish to exclude. For instance

$$A_{[\alpha|\beta|\gamma]} = \frac{1}{2}(A_{\alpha\beta\gamma} - A_{\gamma\beta\alpha}).$$

Note that if $f \in C^\infty(M)$, then

$$(\nabla_\alpha \nabla_\beta f) X^\alpha Y^\beta = \nabla_X(\nabla_Y f) - \nabla_{\nabla_X Y} f = XY(f) - \nabla_X Y(f).$$

Consequently,

$$(\nabla_\alpha \nabla_\beta f - \nabla_\beta \nabla_\alpha f) X^\alpha Y^\beta - [X,Y]f - (\nabla_X Y - \nabla_Y X) f = 0,$$

since the connection is torsion free. In other words, $\nabla_\alpha \nabla_\beta f = \nabla_\beta \nabla_\alpha f$. Since ∇ commutes with contraction, we have

$$\begin{aligned}\nabla_\alpha \nabla_\beta (\omega_\gamma X^\gamma) &= \nabla_\alpha [(\nabla_\beta \omega_\gamma) X^\gamma + \omega_\gamma \nabla_\beta X^\gamma] \\ &= (\nabla_\alpha \nabla_\beta \omega_\gamma) X^\gamma + (\nabla_\beta \omega_\gamma) \nabla_\alpha X^\gamma + (\nabla_\alpha \omega_\gamma) \nabla_\beta X^\gamma + \omega_\gamma \nabla_\alpha \nabla_\beta X^\gamma.\end{aligned}$$

Subtracting the same expression with α and β interchanged and using the fact that $\omega_\gamma X^\gamma$ is a function, we obtain

$$0 = -R_{\alpha\beta\delta}{}^{\gamma} X^\delta \omega_\gamma + X^\gamma (\nabla_\alpha \nabla_\beta - \nabla_\beta \nabla_\alpha) \omega_\gamma.$$

This can be rewritten

$$0 = [(\nabla_\alpha \nabla_\beta - \nabla_\beta \nabla_\alpha)\omega_\gamma - R_{\alpha\beta\gamma}{}^{\delta} \omega_\delta] X^\gamma.$$

In other words,

$$(\nabla_\alpha \nabla_\beta - \nabla_\beta \nabla_\alpha)\omega_\gamma = R_{\alpha\beta\gamma}{}^{\delta} \omega_\delta. \tag{10.9}$$

Similarly, one can compute

$$\begin{aligned}(\nabla_\alpha \nabla_\beta - \nabla_\beta \nabla_\alpha) T^{\beta_1 \dots \beta_k}_{\alpha_1 \dots \alpha_l} &= R_{\alpha\beta\alpha_1}{}^{\gamma} T^{\beta_1 \dots \beta_k}_{\gamma \dots \alpha_l} + \dots + R_{\alpha\beta\alpha_l}{}^{\gamma} T^{\beta_1 \dots \beta_k}_{\alpha_1 \dots \gamma} \\ &\quad - R_{\alpha\beta\gamma}{}^{\beta_1} T^{\gamma \dots \beta_k}_{\alpha_1 \dots \alpha_l} - \dots - R_{\alpha\beta\gamma}{}^{\beta_k} T^{\beta_1 \dots \gamma}_{\alpha_1 \dots \alpha_l}.\end{aligned} \tag{10.10}$$

We define the Ricci tensor to be the contraction

$$R_{\alpha\beta} = R_{\alpha\gamma\beta}{}^{\gamma}. \tag{10.11}$$

Note that due to the symmetries, we have

$$R_{\alpha\beta} = R^{\gamma}{}_{\alpha\gamma\beta}.$$

Using (10.10), we have

$$(\nabla_\alpha\nabla_\beta - \nabla_\beta\nabla_\alpha)\nabla_\gamma\omega_\mu = R_{\alpha\beta\gamma}{}^{\nu}\nabla_\nu\omega_\mu + R_{\alpha\beta\mu}{}^{\nu}\nabla_\gamma\omega_\nu.$$

On the other hand

$$\nabla_\alpha(\nabla_\beta\nabla_\gamma\omega_\mu - \nabla_\gamma\nabla_\beta\omega_\mu) = \omega_\nu\nabla_\alpha R_{\beta\gamma\mu}{}^{\nu} + R_{\beta\gamma\mu}{}^{\nu}\nabla_\alpha\omega_\nu.$$

If we antisymmetrize over α, β, γ in the last two equations, then the resulting expressions are such that the two left-hand sides equal. Consequently,

$$R_{[\alpha\beta\gamma]}{}^{\nu}\nabla_\nu\omega_\mu + R_{[\alpha\beta|\mu|}{}^{\nu}\nabla_{\gamma]}\omega_\nu = \omega_\nu\nabla_{[\alpha} R_{\beta\gamma]\mu}{}^{\nu} + R_{[\beta\gamma|\mu|}{}^{\nu}\nabla_{\alpha]}\omega_\nu$$

The first expression on the left-hand side vanishes due to the first Bianchi identity. The second term on the left-hand side equals the second term on the right-hand side. Since the equality holds for arbitrary ω, we get

$$\nabla_{[\alpha} R_{\beta\gamma]\mu}{}^{\nu} = 0.$$

This is referred to as the second Bianchi identity. Due to the symmetries of R, we can write this as

$$\nabla_\alpha R_{\beta\gamma\mu}{}^{\nu} + \nabla_\gamma R_{\alpha\beta\mu}{}^{\nu} + \nabla_\beta R_{\gamma\alpha\mu}{}^{\nu} = 0.$$

Putting $\nu = \gamma$, we get

$$\nabla_\alpha R_{\beta\mu} + \nabla_\gamma R_{\alpha\beta\mu}{}^{\gamma} - \nabla_\beta R_{\alpha\mu} = 0.$$

Contracting this equation with $g^{\alpha\mu}$, we obtain

$$2\nabla_\alpha R_\beta{}^{\alpha} - \nabla_\beta R = 0,$$

where we (by abuse of notation) use R to denote the *scalar curvature*, i.e., $R = R_{\mu\nu}g^{\mu\nu}$. This can also be written

$$\nabla^\alpha G_{\alpha\beta} = 0, \tag{10.12}$$

where

$$G_{\alpha\beta} = R_{\alpha\beta} - \frac{1}{2}g_{\alpha\beta}R$$

is the *Einstein tensor.*

10.2.3 Coordinate expressions for curvature. Let us derive some coordinate expressions for curvature that we shall use later. Due to (10.7), we have

$$R_{\alpha\beta\delta\mu} = -\langle R_{\partial_\alpha\partial_\beta}\partial_\delta, \partial_\mu\rangle.$$

Recalling that $\Gamma^\alpha_{\mu\nu}$ is defined by (10.5) and the formula (10.6), one can compute that

$$R_{\mu\beta\rho}{}^{\delta} = g^{\delta\alpha}R_{\mu\beta\rho\alpha} = \partial_\beta\Gamma^\delta_{\mu\rho} - \partial_\mu\Gamma^\delta_{\beta\rho} + \Gamma^\alpha_{\mu\rho}\Gamma^\delta_{\beta\alpha} - \Gamma^\alpha_{\beta\rho}\Gamma^\delta_{\mu\alpha}.$$

Putting $\beta = \delta$, we get the Ricci curvature

$$R_{\mu\rho} = \partial_\nu \Gamma^\nu_{\mu\rho} - \partial_\mu \Gamma^\nu_{\nu\rho} + \Gamma^\alpha_{\mu\rho}\Gamma^\nu_{\alpha\nu} - \Gamma^\alpha_{\nu\rho}\Gamma^\nu_{\alpha\mu}.$$

Let us first sort out the highest order terms. Let us introduce the notation

$$\Gamma_{\alpha\beta\gamma} = \frac{1}{2}[\partial_\alpha g_{\gamma\beta} + \partial_\gamma g_{\alpha\beta} - \partial_\beta g_{\alpha\gamma}],$$

so that

$$\Gamma^\mu_{\alpha\beta} = g^{\mu\gamma}\Gamma_{\alpha\gamma\beta}.$$

Note also that $\Gamma_{\alpha\gamma\beta} = \Gamma_{\beta\gamma\alpha}$. We have

$$\partial_\nu \Gamma^\nu_{\mu\rho} - (\partial_\nu g^{\nu\beta})\Gamma_{\mu\beta\rho} + g^{\nu\beta}\partial_\nu \Gamma_{\mu\beta\rho}$$

and

$$\partial_\mu \Gamma^\nu_{\nu\rho} = (\partial_\mu g^{\nu\beta})\Gamma_{\nu\beta\rho} + g^{\nu\beta}\partial_\mu \Gamma_{\nu\beta\rho}.$$

However,

$$\begin{aligned} g^{\nu\beta}[\partial_\nu \Gamma_{\mu\beta\rho} - \partial_\mu \Gamma_{\nu\beta\rho}] &= -\frac{1}{2}g^{\nu\beta}\partial_\nu\partial_\beta g_{\mu\rho} \\ &\quad + \frac{1}{2}g^{\nu\beta}[\partial_\nu\partial_\rho g_{\mu\beta} + \partial_\mu\partial_\beta g_{\nu\rho} - \partial_\mu\partial_\rho g_{\nu\beta}] \\ &= -\frac{1}{2}g^{\nu\beta}\partial_\nu\partial_\beta g_{\mu\rho} + \frac{1}{2}g^{\beta\nu}[\partial_\mu \Gamma_{\beta\rho\nu} + \partial_\rho \Gamma_{\beta\mu\nu}] \\ &= -\frac{1}{2}g^{\nu\beta}\partial_\nu\partial_\beta g_{\mu\rho} + \nabla_{(\mu}\Gamma_{\rho)} + \Gamma^\alpha_{\mu\rho}\Gamma_\alpha \\ &\quad - \frac{1}{2}(\partial_\mu g^{\beta\nu})\Gamma_{\beta\rho\nu} - \frac{1}{2}(\partial_\rho g^{\beta\nu})\Gamma_{\beta\mu\nu}. \end{aligned}$$

Here we have used the notation

$$\nabla_\mu \Gamma_\rho = \partial_\mu \Gamma_\rho - \Gamma^\alpha_{\mu\rho}\Gamma_\alpha$$

and

$$\Gamma_\alpha = g^{\mu\nu}\Gamma_{\mu\alpha\nu}.$$

Note that this is abuse of notation, since Γ_ρ are not the components of a covector. Adding up the above, we obtain

$$\begin{aligned} \partial_\nu \Gamma^\nu_{\mu\rho} - \partial_\mu \Gamma^\nu_{\nu\rho} &= -\frac{1}{2}g^{\nu\beta}\partial_\nu\partial_\beta g_{\mu\rho} + \nabla_{(\mu}\Gamma_{\rho)} + \Gamma^\alpha_{\mu\rho}\Gamma_\alpha \\ &\quad - \frac{1}{2}(\partial_\mu g^{\beta\nu})\Gamma_{\beta\rho\nu} - \frac{1}{2}(\partial_\rho g^{\beta\nu})\Gamma_{\beta\mu\nu} \\ &\quad + (\partial_\nu g^{\nu\beta})\Gamma_{\mu\beta\rho} - (\partial_\mu g^{\nu\beta})\Gamma_{\nu\beta\rho}. \end{aligned}$$

Inserting this expression in our expression for Ricci and using the equations

$$\partial_\gamma g_{\alpha\beta} = \Gamma_{\alpha\beta\gamma} + \Gamma_{\beta\alpha\gamma}, \quad \partial_\gamma g^{\alpha\beta} = -g^{\alpha\mu} g^{\beta\nu} \partial_\gamma g_{\mu\nu},$$

one sees that one can express all terms except $-\frac{1}{2} g^{\nu\beta} \partial_\nu \partial_\beta g_{\mu\rho}$ and $\nabla_{(\mu}\Gamma_{\rho)}$ as a sum of terms of the form

$$g^{\alpha\beta} g^{\gamma\delta} \Gamma_{***} \Gamma_{***},$$

where the stars are a suitable permutation of $\{\alpha, \beta, \gamma, \delta, \mu, \rho\}$. In the end, we get

$$R_{\mu\rho} = -\frac{1}{2} g^{\alpha\beta} \partial_\alpha \partial_\beta g_{\mu\rho} + \nabla_{(\mu}\Gamma_{\rho)} + g^{\alpha\beta} g^{\gamma\delta} [\Gamma_{\alpha\gamma\mu}\Gamma_{\beta\delta\rho} + \Gamma_{\alpha\gamma\mu}\Gamma_{\beta\rho\delta} + \Gamma_{\alpha\gamma\rho}\Gamma_{\beta\mu\delta}]. \tag{10.13}$$

Alternately, we can write

$$R_{\mu\rho} = -\frac{1}{2} g^{\alpha\beta} \partial_\alpha \partial_\beta g_{\mu\rho} + \nabla_{(\mu}\Gamma_{\rho)} + \Gamma^{\eta}_{\lambda\mu} g_{\eta\delta} g^{\lambda\gamma} \Gamma^{\delta}_{\rho\gamma} + 2\Gamma^{\lambda}_{\delta\eta} g^{\delta\gamma} g_{\lambda(\mu} \Gamma^{\eta}_{\rho)\gamma}.$$

The last expression should be compared with (2.29) of [41].

10.2.4 Basic causality concepts. We shall say that a vector $v \in T_pM$, $v \neq 0$, is *timelike*, *null* or *spacelike* if $g(v,v) < 0$, $g(v,v) = 0$ or $g(v,v) > 0$ respectively. The 0 vector is defined to be spacelike. If v is either timelike or null it is said to be *causal*. A *curve* in a manifold M is a smooth map $\alpha: I \to M$, where I is an interval. If $\langle \alpha', \alpha' \rangle < 0$ on all of I, we say that α is *timelike*. If $\alpha' \neq 0$ and $\langle \alpha', \alpha' \rangle = 0$ on all of I, we say that α is *null*. If $\alpha' \neq 0$ and $\langle \alpha', \alpha' \rangle \leq 0$ on all of I, we say that α is *causal*. At each $p \in M$ the set of timelike vectors forms two disjoint cones, referred to as time cones. Let τ be a function which assigns, to each point $p \in M$ a timecone τ_p in T_pM. We say that τ is smooth if for every $p \in M$ there is a smooth vector field V on some neighbourhood U of p such that $V_q \in \tau_q$ for every $q \in U$. Such a τ is called a *time-orientation* of M. If M admits a time-orientation, we say that M is *time orientable*. Note that these definitions make sense even when M is not second countable. However, if we assume second countability, we can patch together the local vector fields ensured by time-orientability into a global one, cf. Lemma 32, p. 145 of [65]. Thus, assuming that M is second countable, time orientability is equivalent to the existence of a smooth timelike vector field T (a vector field T is said to be timelike if $\langle T, T \rangle < 0$ at all points). From now on we shall assume all the Lorentz manifolds we consider to be time oriented. We shall furthermore assume them to be oriented and connected. The last two assumptions are there to ensure that the volume form is well defined, even though that is in many cases strictly speaking not necessary, and to ensure the existence of a complete Riemannian metric, see Lemma 11.1 below. If $v \in TM$ is a causal vector, i.e., if $v \neq 0$ and $\langle v, v \rangle \leq 0$, then $\langle v, T \rangle$ is either positive or negative. If it is negative, we say that v is *future pointing* and if it is positive we say that v is *past pointing*. A causal (timelike, null) curve α is said to be future pointing if α' is future pointing at all points. The concept past pointing causal (timelike, null) curve is defined analogously. We say that $p \ll q$ if there is a future pointing timelike curve in M from

p to q, that $p < q$ if there is a future pointing causal curve in M from p to q. Finally $p \leq q$ means $p = q$ or $p < q$. Given a subset A of M, define

$$I^+(A) = \{p \in M : q \ll p \text{ for some } q \in A\},$$
$$J^+(A) = \{p \in M : q \leq p \text{ for some } q \in A\}.$$

Analogously,

$$I^-(A) = \{p \in M : p \ll q \text{ for some } q \in A\},$$
$$J^-(A) = \{p \in M : p \leq q \text{ for some } q \in A\}.$$

The set $I^+(A)$ is called the *chronological future* of A, $J^+(A)$ is called the *causal future* of A and similarly $I^-(A)$ and $J^-(A)$ are referred to as the chronological and causal past of A respectively. The sets $I^+(A)$ and $I^-(A)$ are always open, cf. Lemma 3, p. 403 of [65]. Concerning $J^+(A)$ and $J^-(A)$ there are no general statements. It is however of interest to note that we have the following relations:

$$I^+(A) = I^+(I^+(A)) = I^+(J^+(A)) = J^+(I^+(A)) \subseteq J^+(J^+(A)) = J^+(A), \tag{10.14}$$

cf. Corollary 1, p. 402 of [65].

10.2.5 Geodesics. A *geodesic* is a curve $\gamma: I \to M$ such that $\gamma'' = 0$, cf. p. 65 of [65] for the notation. Alternately, γ is a curve which in local coordinates satisfies the equation

$$\frac{d^2\gamma^\mu}{dt^2} + \Gamma^\mu_{\alpha\beta}\frac{d\gamma^\alpha}{dt}\frac{d\gamma^\beta}{dt} = 0,$$

where $\gamma^\mu = x^\mu \circ \gamma$ for local coordinates $x: U \to \mathbb{R}^{n+1}$ and $\Gamma^\mu_{\alpha\beta}$ are the Christoffel symbols, cf. (10.2). Given $v \in T_pM$ there is consequently a unique geodesic γ satisfying $\gamma(0) = p$ and $\gamma'(0) = v$, cf. Proposition 24, p. 68 of [65]. We define the exponential map at a point $p \in M$, denoted $\exp_p$, to be the map taking $v \in T_pM$ to $\gamma_v(1)$, where γ_v is the geodesic satisfying $\gamma_v(0) = p$ and $\gamma_v'(0) = v$. Then $\exp_p$ is defined on an open subset of T_pM. Furthermore, there is an open subset U' of T_pM such that $\exp_p: U' \to M$ is a diffeomorphism onto its image, cf. Proposition 30, p. 71 of [65]. If U' is *starshaped*, i.e., if $v \in U'$ implies $tv \in U'$ for $t \in [0,1]$, and $\exp_p: U' \to U$ is a diffeomorphism, then U is said to be a *normal neighbourhood* of p. One can then construct a *normal coordinate system* on U, cf. p. 72 of [65], so that $g_{\mu\nu}(p)$ are the components of the Minkowski metric and $\partial_\alpha g_{\mu\nu}(p) = 0$ for all indices α, μ, ν. An open set in a Lorentz manifold is said to be *convex* if it is a normal neighbourhood of each of its points. Due to Proposition 7, p. 130 of [65], each point has a convex neighbourhood.

10.2.6 Global hyperbolicity. Lorentz manifolds in the generality we have discussed them up till now allow many pathologies one wishes to exclude. For instance, all compact Lorentz manifolds admit closed timelike curves, cf. Lemma 10, p. 407 of

[65]. From a physical point of view, this means that there are observers that can travel into their past. A Lorentz manifold which does not admit a closed timelike curve is said to satisfy the *chronology condition*. However, one often wants to demand more than that. The *strong causality condition* is said to hold at $p \in M$ provided that given any neighbourhood U of p there is a neighbourhood $V \subseteq U$ of p such that every causal curve segment with endpoints in V lies entirely in U. A Lorentz manifold is said to be *globally hyperbolic* if the strong causality condition holds at each of its points and if for each pair $p < q$, the set $J(p,q) = J^+(p) \cap J^-(q)$ is compact. The assumption that a Lorentz manifold M be globally hyperbolic has many consequences, e.g., if $p, q \in M$ and $p < q$, then there is a causal geodesic from p to q such that no causal curve from p to q can have greater length, cf. Proposition 19, p. 411 of [65]. Furthermore, the causality relation $\leq$ is closed on M, i.e., if $p_n \to p$, $q_n \to q$ and $p_n \leq q_n$, then $p \leq q$, cf. Lemma 22, p. 412 of [65].

10.2.7 Cauchy surfaces. A subset A of a Lorentz manifold (M, g) is said to be *achronal* if there is no pair of points $p, q \in A$ that can be connected by a timelike curve and it is said to be *acausal* if no pair of points in A can be connected by a causal curve. Given an achronal subset A of M, the *future Cauchy development* of A is the set $D^+(A)$ of all points p of M such that every past inextendible causal curve through p meets A. The past Cauchy development $D^-(A)$ is defined analogously and we write $D(A) = D^+(A) \cup D^-(A)$. A *Cauchy (hyper)surface* in M is a subset S that is met exactly once by every inextendible timelike curve in M. Recall that a piecewise smooth curve $\gamma\colon [0, a) \to M$ is *extendible* if it has a continuous extension $\tilde{\gamma}\colon [0, a] \to M$. Extendibility for curves of the form $\gamma\colon (a, 0] \to M$ and $\gamma\colon (a, b) \to M$ is defined similarly. A curve which is not extendible is called *inextendible*. Due to Lemma 29, p. 415 of [65], a Cauchy hypersurface S is a closed achronal topological hypersurface and is met by every inextendible causal curve. Furthermore, due to Corollary 39, p. 422 of [65], if M has a Cauchy surface, then it is globally hyperbolic. Assume M has a Cauchy surface S. Let $p \in M$ and let γ be an inextendible timelike geodesic through p (note that if a geodesic is continuously extendible, it is extendible as a geodesic, cf. Lemma 8, p. 130 of [65]). Then γ intersects S exactly once by definition. Thus p is in one of the sets S, $I^+(S)$ and $I^-(S)$. Note also that these sets are disjoint due to the definition of a Cauchy hypersurface. Furthermore, $J^{\pm}(S)$ and $I^{\mp}(S)$ are disjoint, cf. (10.14). As a consequence, $J^{\pm}(S) = M - I^{\mp}(S)$, so that $J^{\pm}(S)$ are closed sets. We have $D^+(S) \subseteq J^+(S)$ since $p \in I^-(S)$ clearly implies $p \notin D^+(S)$. By its definition, it is clear that $S \subseteq D^+(S)$. Let $p \in I^+(S)$. Let γ be a past inextendible causal curve through p. Then it has to intersect S by the properties of a Cauchy hypersurface, so that $p \in D^+(S)$. We conclude that $J^+(S) = S \cup I^+(S) \subseteq D^+(S)$. In other words $J^+(S) = D^+(S)$. Reversing time orientation, we get the same conclusion with $+$ replaced by $-$, so that $D(S) = M$.

10.2.8 Technical observations. When solving Einstein's equations, the following technical observations will be of use.

Lemma 10.10. *Let (M, g) be a Lorentz manifold and assume it admits a smooth spacelike Cauchy hypersurface S. If $U \subseteq M$ is open, $q \in J^+(S)$ and $J^-(q) \cap J^+(S) \subseteq U$, then if $q_i \in J^+(S)$ are such that $q_i \to q$, we have $J^-(q_i) \cap J^+(S) \subseteq U$ for i large enough. If $q_i \leq q$, $q \in I^+(S)$ and $q_i \to q$, then the closure of the union of the $J^-(q_i) \cap J^+(S)$ is $J^-(q) \cap J^+(S)$.*

Proof. In order to prove the first statement, let p be such that there is a future directed timelike curve from q to p. Then $J^-(p) \cap J^+(S)$ is compact and $J^-(p)$ contains q in its interior, cf. Lemma 3, p. 403 and Lemma 40, p. 423 of [65]. Since $q_i \to q$, $J^-(q_i) \subseteq J^-(p)$ for i large enough. Assuming the desired statement is not true, there is a subsequence q_{i_k} and points $r_{i_k} \in J^-(q_{i_k}) \cap J^+(S)$ such that $r_{i_k} \notin U$. Since the r_{i_k} are in the compact set $J^-(p) \cap J^+(S) - U$ for k large enough, we can assume that they converge to a point r. Then $r \in J^-(q) \cap J^+(S) - U$, a contradiction. To prove the second statement, let $p \in J^-(q)$. If $p \in I^+(S)$, let $p_k \to p$ be such that $p_k \in J^+(S)$ is in the timelike past of p. Then q is in the timelike future of p_k. Thus there is an i_k such that q_{i_k} is in the timelike future of p_k. Thus all the p_k are in the union of the $J^-(q_i)$, so that p is in the closure of the union of $J^-(q_i) \cap J^+(S)$. Assume $p \in S \cap J^-(q)$. Let $p_k \ll p$ be such that $p_k \to p$, let i_k be such that q_{i_k} is in the timelike future of p_k and let γ_k be a timelike curve from p_k to q_{i_k}. Denote the point of intersection between γ_k and S by p'_k. Since $p'_k \in J^-(q) \cap J^+(S)$, which is compact, we can choose a subsequence so that it converges to, say, r. Since $p_k \leq p'_k$ and p_k converges to p, we conclude that $p \leq r$. Since $p \in S$ and S is a spacelike Cauchy hypersurface, we have to have $p = r$. The conclusion follows. □

11 Characterizations of global hyperbolicity

The main purpose of the present chapter is to prove that a globally hyperbolic spacetime has a smooth spacelike Cauchy hypersurface and to prove the existence of so-called time functions, a concept we shall define below. The basic references for the material presented here are [4], [5], [6] and [43]. In particular, the proof of the existence of a time function is taken from the first three references.

11.1 Existence of a Cauchy hypersurface

In the end we wish to prove that a globally hyperbolic Lorentz manifold has a smooth spacelike Cauchy hypersurface S and that it is diffeomorphic to $\mathbb{R} \times S$. In the proof of the latter statement, the following technical observation will be useful.

Lemma 11.1. *Let M be a smooth connected manifold. Then there exists a complete Riemannian metric on M.*

Remark 11.2. An alternate proof is obtained by Whitney's embedding theorem, Theorem 10.8, p. 92 of [8], and the Hopf–Rinow theorem, Theorem 21, p. 138 of [65]. The reason is that due to the former result, M can be embedded into $\mathbb{R}^{2n+1}$ and the image under the embedding is a *closed* subset. Pulling back the standard Euclidean metric on $\mathbb{R}^{2n+1}$ by the embedding, we get a Riemannian metric g on M. Let d be the corresponding topological metric. The image of a closed and bounded subset with respect to d under the embedding is a closed and bounded subset of $\mathbb{R}^{2n+1}$, i.e. a compact set. Thus closed and bounded subsets with respect to d are compact. By the Hopf–Rinow theorem g is then complete.

Proof. By a standard partition of unity argument there is a Riemannian metric h on M, cf. Lemma 25, p. 140 of [65]. Let us construct a smooth proper function $\rho\colon M \to \mathbb{R}$, i.e., a smooth function with the property that the inverse image of every compact set is compact. Let ϕ_i, $i = 1, 2, \dots$ be the smooth partition of unity constructed in Lemma 10.3. Then all the functions ϕ_i have compact support. Define

$$\rho = \sum_{i=1}^{\infty} i\phi_i .$$

This is a smooth function and if $\rho(x) \in [-i, i]$, then $\phi_j(x) \neq 0$ for some $j = 1, \dots, i$, since $\rho(x) \geq i + 1$ otherwise. Consequently

$$\rho^{-1}([-i, i]) \subseteq \bigcup_{j=1}^{i} \operatorname{supp} \phi_j ,$$

which is a compact set. Since any compact subset of $\mathbb{R}$ is contained in $[-i, i]$ for some i, we conclude that ρ is proper (since the inverse image is closed by continuity and contained in a compact set). Let

$$g = h + d\rho \otimes d\rho .$$

The Riemannian metric g induces a topological metric d. Due to the Hopf–Rinow theorem, Theorem 21, p. 138 of [65], to prove that g is complete, all we need to prove is that closed and bounded subsets of M with respect to d are compact. Let $\gamma\colon [a,b] \to M$ be a smooth curve. Then the length of γ with respect to g is

$$L(\gamma) = \int_a^b \left[h(\dot\gamma,\dot\gamma) + \left(\frac{d}{dt}\rho\circ\gamma\right)^2\right]^{1/2} dt$$
$$\geq \int_a^b \left|\frac{d}{dt}\rho\circ\gamma\right| dt \geq |\rho\circ\gamma(b) - \rho\circ\gamma(a)|.$$

In particular,

$$|\rho(p) - \rho(q)| \leq d(p,q).$$

Thus if C is a closed and bounded set with respect to d, $\rho(C)$ is contained in a compact set. Since ρ is proper, C is compact. Thus g is complete. □

Proposition 11.3. *Let (M,g) be a connected, time oriented Lorentz manifold and assume that there is a smooth spacelike Cauchy hypersurface S. Then M is diffeomorphic to $\mathbb{R}\times S$. Furthermore, if S' is another smooth spacelike Cauchy hypersurface, then S' and S are diffeomorphic. Finally, if S is a Cauchy hypersurface, then S is a topological manifold and M is homeomorphic to $\mathbb{R}\times S$.*

Remark 11.4. The statement in the case that S is not smooth requires the use of invariance of domain, but we shall not use that result in what follows.

Proof. Let us start with the smooth setting. Since M is time oriented, there is a smooth timelike vector field T_1. Let h be a complete Riemannian metric on M and $T = T_1/\|T\|_h$, where $\|T_1\|_h$ is the norm of the vector field T_1 with respect to the Riemannian metric h. Then T is a complete vector field, since if (t_-, t_+) is the maximal existence interval of an integral curve through $p \in M$ and, say, $t_+ < \infty$, then the integral curve restricted to $[0,t_+)$ is contained in a compact set, since the length of the curve restricted to $[0,t_+)$ (with respect to h) is bounded by t_+. This contradicts the maximality of the existence interval since it allows one to extend the solution beyond the maximal existence interval, cf. for instance [25] Proposition 4.33, p. 55. Similarly, we must have $t_- = -\infty$. Thus the flow Φ is defined on all of $\mathbb{R}\times M$ and is a smooth function, cf. Theorem 4.26 of [25], p. 54. Define a smooth map

$$f\colon \mathbb{R}\times S \to M$$

by $f(t,x) = \Phi(t,x)$. Then f is smooth and injective. The reason for the latter property is that $f(t_1,x_1) = f(t_2,x_2)$ implies $x_1 = \Phi(t_2 - t_1, x_2)$. Due to the achronality of S, we get the conclusion that $t_2 = t_1$ and thus that $x_2 = x_1$. Let $(t_0,x_0) \in \mathbb{R}\times S$. We wish to prove that f_* is surjective at this point. Let $g(p) = \Phi(-t_0, p)$. Then g is a diffeomorphism of M and f_* is a surjective map if and only if $(g\circ f)_*$ is a surjective map. The image of $(g\circ f)_*$ at (t_0,x_0) contains the tangent space of S at x_0 and T_{x_0}. We conclude that f_* is surjective at all points (t_0,x_0) whence f is a local diffeomorphism. Since it is injective, we conclude that it is a diffeomorphism onto its image. To prove surjectivity, let $p \in M$ and let γ be the maximal integral curve of T through p. Since T is a complete vector field, $\gamma\colon \mathbb{R} \to M$. If γ were inextendible, it would have to intersect S, and consequently p would be in the image of f. In order to obtain a contradiction, let us therefore assume γ is extendible. This means that there is a $q \in M$ such that $\gamma(t) \to q$ as $t \to \infty$ (or $t \to -\infty$). Let $\xi\colon \mathbb{R} \to M$ be the integral curve of T through q. Then

$$\xi(t) = \Phi(t,q) = \lim_{s\to\infty} \Phi[t,\gamma(s)] = \lim_{s\to\infty} \gamma(t+s) = q.$$

In other words, $T_q = \xi'(0) = 0$, a contradiction. Let $\pi\colon \mathbb{R} \times S \to S$ be defined by $\pi(t,s) = s$. If S' is another smooth spacelike Cauchy hypersurface, we get a map $r\colon S' \to S$ defined by

$$r(x) = \pi \circ f^{-1}(x).$$

Then r is smooth. Note that r maps a point $x \in S'$ to the point of intersection between the integral curve of T through x and S. By symmetry, the inverse of r is smooth, so that S and S' are diffeomorphic.

Let us now assume that S is a Cauchy hypersurface, not necessarily spacelike or smooth. Then S is a topological manifold, cf. Lemma 29, p. 415 of [65], and we get a continuous map $f\colon \mathbb{R} \times S \to M$ defined by $f(t,x) = \Phi(t,x)$, with Φ as above. For the same reason as above, f is injective and due to the fact that all integral curves of T have to intersect S, it is surjective. By invariance of domain, see e.g. Corollary 19.9, p. 235 of [8], we conclude that f is open. Thus the inverse is continuous, so that f is a homeomorphism. □

Our next goal is to prove that globally hyperbolic Lorentz manifolds admit Cauchy hypersurfaces. The argument is due to Robert Geroch, cf. [43]. We shall need the following lemma.

Lemma 11.5. *Let (M,g) be a globally hyperbolic Lorentz manifold and let K_1, $K_2 \subseteq M$ be two compact subsets. Then $J^+(K_1) \cap J^-(K_2)$ is compact.*

Proof. Note that a subset K of M is compact if and only if it is sequentially compact, i.e., if and only if every sequence $x_n \in K$ has a convergent subsequence. The reason is that M admits a Riemannian metric, cf. Lemma 11.1, so that the topology of M is metrizable. A subset K of M is then a metric space in its own right and due to Theorem 9.4, p. 25 of [8], for metric spaces, compactness is equivalent to sequential compactness. If the intersection is empty, there is nothing to prove. Assume $q_n \in J^+(K_1) \cap J^+(K_2)$ is a sequence of points. We wish to prove that there is a convergent subsequence. By assumption there are $p_{n,i} \in K_i$ such that $p_{n,1} \leq q_n \leq p_{n,2}$. Since the sets K_i are compact there are subsequences of the sequences $p_{n,i}$ that converge to, say, $p_i \in K_i$. Thus we can assume, without loss of generality, that $p_{n,i} \to p_i$. Let $r_i \in M$, $i = 1,2$ be such that $r_1 \ll p_1$ and $p_2 \ll r_2$. Then $r_1 \ll q_n \ll r_2$ for n large enough due to the fact that $I^+(r_1)$ and $I^-(r_2)$ are open and (10.14). Consequently, $q_n \in J^+(r_1) \cap J^-(r_2)$ for n large enough. Since this set is compact, due to global hyperbolicity, there is a subsequence, which we shall also denote q_n, such that $q_n \to q$. Since the relation $\leq$ is closed on globally hyperbolic Lorentz manifolds, we conclude that $p_1 \leq q \leq p_2$. In other words, $q \in J^+(K_1) \cap J^+(K_2)$. The lemma follows. □

Theorem 11.6. *Let (M, g) be an oriented, time oriented Lorentz manifold which is globally hyperbolic. Then there is a continuous, onto function $\tau : M \to \mathbb{R}$ which is strictly increasing along any causal curve and if $\gamma : (t_-, t_+) \to M$ is an inextendible causal curve, then $\tau[\gamma(t)] \to \pm\infty$ as $t \to t_\pm \mp$. Furthermore, similar statements hold for past inextendible and future inextendible causal curves. In particular, for every $a \in \mathbb{R}$, $\tau^{-1}(a)$ is an acausal Cauchy hypersurface.*

Proof. Let us start by constructing a measure on M. Let h be a Riemannian metric on M. Let $\phi_i, i = 1, 2, \ldots$ be the smooth partition of unity constructed in Lemma 10.3. Given i, let (U, x) be coordinates with $\operatorname{supp} \phi_i \subset U$ and let m_i be defined by

$$m_i = \int \phi_i \, \mu_h,$$

where

$$\mu_h = \sqrt{\det h} dx^1 \wedge \cdots \wedge dx^n$$

and $\det h$ is the determinant of the matrix with components $h(\partial_i, \partial_j)$. We can then define a smooth n-form ω on M by

$$\omega = \sum_{i=1}^{\infty} \frac{1}{m_i 2^i} \phi_i \mu_h.$$

We can use this form to define a linear functional Λ on the continuous complex-valued functions with compact support $C_c(M)$, by

$$\Lambda(f) = \int_M f \, \omega.$$

Due to the Riesz representation theorem, cf. Theorem 2.14, p. 40 of [79], we conclude the existence of a positive measure μ and a σ-algebra of subsets of M, say $\mathcal{A}$, such that $\mathcal{A}$ contains all the Borel sets, μ is complete and

$$\int f \, d\mu = \int_M f \, \omega$$

for all $f \in C_c(M)$. Let

$$g_i = \sum_{j=1}^{i} \phi_j.$$

Then $g_j \to 1$ everywhere and it is an increasing sequence of smooth functions. Due to Lebesgue's monotone convergence theorem, p. 21 of [79], we obtain

$$\int_M d\mu = \lim_{j \to \infty} \int_M g_j \, d\mu = 1.$$

Note also that if U is an open, non-empty set, then $\mu(U) > 0$. The reason is that if K is a compact set with non-empty interior contained in U, there is a $\phi \in C_0^\infty(U)$ which equals 1 on K such that $0 \leq \phi \leq 1$. Then

$$0 < \int_M \phi\,\omega = \int_M \phi\, d\mu \leq \int_U d\mu = \mu(U).$$

Let $p \in M$. We wish to prove that if $C_p^+ = J^+(p) - I^+(p)$ then

$$\mu(C_p^+) = 0. \tag{11.1}$$

Assume $q \in C_p^+$. Then there is a causal curve from p to q. Due to the global hyperbolicity of M, there is then a causal geodesic connecting p and q such that there is no causal curve connecting p and q of greater length. If this geodesic were not a null geodesic, $q \in I^+(p)$, a contradiction. Thus there is a $v \in T_pM$ which is null and such that $\exp_p(v) = q$. Let N be the future directed null vectors in T_pM such that $\exp_p$ is defined on them. If M is an n dimensional manifold, this is an $n-1$ dimensional submanifold of T_pM. Furthermore, $\exp_p$ defines a smooth map from N to M and C_p, except for p, is in the image of this map. One way to prove that (11.1) holds is to note that all the image points of $\exp_p$ restricted to N are critical values, so that Sard's theorem implies (11.1). However, in the present setting, less sophisticated techniques suffice. Let us proceed in as follows. Let $v \in N$. Then there is a neighbourhood W of v and coordinates (y, W) such that $w \in W \cap N$ if and only if $w \in W$ and $y^n(w) = 0$. Let V be a neighbourhood of v whose closure is compact and contained in W and is such that $\bar{V} \cap N$ is compact. Assume furthermore that $\exp_p(\bar{V})$ is contained in a coordinate chart (U, x) of M. Note that $K_2 = \exp_p(\bar{V} \cap N)$ coincides with the image of the hyperplane $y^n = 0$ intersected with $y(\bar{V})$, let us call this set K_1, under the map $\exp_p \circ y^{-1}$. For any $\varepsilon > 0$, we can cover K_1 with a finite number of balls so that the sum of the Euclidean volume of these balls is less than ε and the radius of the balls is less than ε. For ε small enough, the image of the closure of all the balls under $\exp_p \circ y^{-1}$ is contained in U. Due to the form of the measure μ, the measure of $\exp_p(\bar{V} \cap N)$ with respect to μ is bounded by a real constant times the ordinary Lebesgue measure of $x \circ \exp_p(\bar{V} \cap N)$. Since the balls are contained in a fixed compact set K_3 for ε small enough, and the derivative of $x \circ \exp_p \circ y^{-1}$ is bounded on K_3, there is a real constant α_0 such that the image of a ball of radius r_0 is contained in a ball of radius $\alpha_0 r_0$. The Lebesgue measure of the union of the image of the balls is therefore bounded by $\alpha_0^n \varepsilon$. Since $\varepsilon > 0$ was arbitrary and since $x \circ \exp_p(\bar{V} \cap N)$ is contained in the union of the image of the balls, we conclude that the measure of $\exp_p(\bar{V} \cap N)$ is zero. Since we can find a countable number of such open sets V covering N, we conclude that (11.1) holds. Similarly,

$$\mu[J^-(p) - I^-(p)] = 0.$$

Let us define two functions

$$f_-(p) = \mu[J^-(p)], \quad f_+(p) = \mu[J^+(p)].$$

Let us prove that these functions are continuous. Let us start by proving that if $p_j \to p$ and $p_j \ll p$, then $f_-(p_j) \to f_-(p)$. Since $J^-(p_j) \subseteq J^-(p)$, we know that $f_-(p_j) \le f_-(p)$. However, by the above observation

$$f_-(p) = \int_M \chi_{I^-(p)}\, d\mu$$

and similarly for p_j. Furthermore

$$\lim_{j\to\infty} \chi_{I^-(p_j)} = \chi_{I^-(p)}$$

everywhere, because if $q \in I^-(p)$ then $q \in I^-(p_j)$ for j large enough. By Lebesgue's dominated convergence theorem, p. 26 of [79], we conclude that $f_-(p_j) \to f_-(p)$. Assume now that $p_j \to p$ and $p \ll p_j$. Say that $q \notin J^-(p)$. Then, for j large enough, $q \notin J^-(p_j)$. In order to prove this statement, assume the opposite. Then there is an infinite subsequence of the p_j, say p_{j_k}, such that $q \le p_{j_k}$. Since the relation $\le$ is closed on globally hyperbolic manifolds, we conclude that $q \le p$, a contradiction. As a conclusion, we have

$$\lim_{j\to\infty} \chi_{J^-(p_j)} = \chi_{J^-(p)}$$

everywhere. Consequently, we can apply Lebesgue's dominated convergence theorem again in order to conclude that $f_-(p_j) \to f_-(p)$. Finally, let $p_j \to p$ be any sequence of points converging to p. Let $\varepsilon > 0$. Then there are points q_1, q_2 such that $q_1 \ll p \ll q_2$ and

$$f_-(p) \le f_-(q_2) \le f_-(p) + \varepsilon, \quad f_-(p) - \varepsilon \le f_-(q_1) \le f_-(p).$$

For j large enough, $q_1 \ll p_j \ll q_2$ so that

$$f_-(p) - \varepsilon \le f_-(q_1) \le f_-(p_j) \le f_-(q_2) \le f_-(p) + \varepsilon.$$

The continuity of f_- follows. The fact that f_+ is continuous follows by time reversal. Note that $f_+(p), f_-(p) > 0$ for all $p \in M$, since $J^+(p)$ and $J^-(p)$ contain open sets for every $p \in M$.

Let us prove that f_- is strictly increasing along causal curves. Assume $p < q$. Then there is a neighbourhood U of p that does not contain q. By the strong causality condition, there is an open neighbourhood V of p, contained in U, such that every causal curve with endpoints in V is contained in U. Assume $q \in J^-(p)$. This means that there is a causal curve from q to p. But by assumption, there is a causal curve from p to q. Thus we have constructed a causal curve with endpoints in V which leaves U, in contradiction to strong causality. If $p < q$, we conclude that $q \notin J^-(p)$. Since the relation $\le$ is closed on globally hyperbolic spacetimes, $J^-(p)$ is closed. Thus there is an open neighbourhood U of q which does not intersect $J^-(p)$. Since $J^-(q) \cap U$ contains an open neighbourhood and $J^-(p) \subseteq J^-(q)$, we conclude that $f_-(q) > f_-(p)$. Similarly, if $p < q$, then $f_+(q) < f_+(p)$.

The next statement we wish to prove is that if $\gamma\colon [0,a) \to M$ is a future inextendible causal curve, then $f_+ \circ \gamma(t)$ converges to zero as $t \to a-$. Let $K \subseteq M$ be a compact set. We wish to prove that for t large enough, $K_t = J^+[\gamma(t)] \cap J^-(K)$ is empty. Note that K_t is compact for every t due to Lemma 11.5. Let $C = K_0 \cup \{\gamma(0)\}$. Then C is a compact set and γ starts in C. Due to Lemma 13, p. 408 of [65], we conclude that there is a t_0 such that for $t > t_0$, $\gamma(t) \notin C$. Since $\gamma(t) \in J^+[\gamma(0)]$ for all $t \geq 0$, we conclude that $\gamma(t) \notin J^-(K)$ for $t > t_0$. Thus K_t is empty for $t > t_0$. Given i, let

$$C_i = \bigcup_{j=1}^{i} \operatorname{supp} \phi_j.$$

Note that $\mu(M - C_i) \leq 2^{-i}$. Since for t large enough $J^+[\gamma(t)] \cap C_i$ is empty, we conclude that for t large enough, $f_+[\gamma(t)] \leq 2^{-i}$. Thus $f_+[\gamma(t)]$ converges to zero as $t \to a-$. Similarly, if $\gamma\colon (b,0] \to M$ is a past inextendible causal curve, then $f_-[\gamma(t)] \to 0$ as $t \to b+$.

Let us define the function $\tau\colon M \to \mathbb{R}$ by

$$\tau(p) = \ln \frac{f_-(p)}{f_+(p)}.$$

This is a continuous function with the property that it is strictly increasing along causal curves. Furthermore, if $\gamma\colon (a,b) \to M$ is an inextendible causal curve, then $\tau \circ \gamma(t)$ converges to ∞ as $t \to a-$ and converges to $-\infty$ as $t \to b+$. □

11.2 Basic constructions

The purpose of the present section is to construct the basic functions that will later be used to construct smooth time functions.

Lemma 11.7. *Let $U \subseteq \mathbb{R}^n$ be an open set and $V \subseteq U$ be an open subset. Assume $f \in C^\infty(U)$ is such that $f(x) > 0$ for $x \in V$ and $f(x) = 0$ for $x \in \partial V \cap U$. Let*

$$g(x) = \begin{cases} \exp[-1/f(x)], & x \in V, \\ 0, & x \notin V. \end{cases}$$

Then $g \in C^\infty(U)$.

Proof. If $x \in V$ or if $x \in U - \bar{V}$, then g is smooth in a neighbourhood of x. If $x \in \partial V \cap U$ and $x_j \to x$, then $g(x_j) \to 0$, so that g is continuous. Any derivative of g for $x \in V$ can be written as a sum of terms which consist of a smooth factor times a factor of the form

$$r = \frac{1}{f^k} \exp\left[-\frac{1}{f}\right].$$

Since $e^{-1/t} \leq k!t^k$ for any $t > 0$ and non-negative integer k, we conclude that this object converges to zero as x tends to the boundary of V. We can thus extend r to be zero in the complement of V. What we need to prove is that this sort of object is differentiable at the boundary and that the derivative is zero. For $x \in \partial V$, $v \in \mathbb{R}^n$ and $h \in \mathbb{R}$, $h \neq 0$, we need to prove that

$$\frac{r(x+hv)-r(x)}{h}$$

converges to zero as $h \to 0$. Since $r(x) = 0$, this object is zero unless $x + hv \in V$, so let us assume that that is the case. We get

$$\left|\frac{r(x+hv)}{h}\right| \leq (k+2)!\frac{f^2(x+hv)}{|h|}.$$

Since $f(x) = 0$ and f is smooth, it is clear that this object converges to zero as $h \to 0$. Thus r is differentiable at every boundary point. Since it is differentiable away from the boundary as well, we conclude that it is differentiable. The above arguments prove that g is continuous and that if $g \in C^k(U)$ with all derivatives of order less than or equal to k equal zero in the complement of V, then the same is true with k replaced by $k+1$. The lemma follows. □

We are interested in the following set up which is described in detail on pp. 127-128 of [65]. Let (M, g) be a Lorentz manifold and let U be a normal neighbourhood of a point $p \in M$. Let U' be the corresponding neighbourhood of the origin in T_pM under $\exp_p$. Let $\tilde{q}: TM \to \mathbb{R}$ be defined by $\tilde{q}(v) = g(v, v)$ and $q: U \to \mathbb{R}$ be defined by $q = \tilde{q} \circ \exp_p^{-1}$. Let V be the image of the future directed timelike vectors in U' under $\exp_p$. Then V is an open subset of U and $q(p) < 0$ for $p \in V$. Thus we can define

$$f(x) = \begin{cases} \exp[1/q(x)], & x \in V, \\ 0, & x \notin V. \end{cases} \tag{11.2}$$

This is a smooth function on U by Lemma 11.7. Before we compute the gradient of this function, let us introduce some more terminology. Let $\widetilde{P}$ be the vector field on T_pM defined by the condition that $\widetilde{P}_v$ is the vector v based at v. Let P be this vector field transferred to U by means of $\exp_p$. Then $\operatorname{grad} q = 2P$, cf. Corollary 3, p. 128 of [65]. Furthermore

$$\operatorname{grad} f = -\frac{1}{q^2} f \operatorname{grad} q$$

whenever $f \neq 0$, so that $\operatorname{grad} f$ is past directed timelike on V and zero everywhere else.

Lemma 11.8. *Let (M, g) be a globally hyperbolic Lorentz manifold and let S be an acausal Cauchy hypersurface. Fix a $p \in S$ and a convex neighbourhood U of p. Then there is a smooth function $h_p: M \to [0, \infty)$ such that*

- $h_p(p) = 1$;

- $\operatorname{supp} h_p$ *is compact and contained in* U;
- *if* $r \in J^-(S)$ *and* $h_p(r) \neq 0$, *then* $\operatorname{grad} h_p(r)$ *is past pointing timelike.*

Remark 11.9. The argument does not hold if S is merely a Cauchy hypersurface, it needs to be acausal. The reason is that the proof requires the existence of an $r \ll p$ such that $J^+(r) \cap J^-(S)$ is contained in U. Taking S to be a Cauchy hypersurface in Minkowski space which has a null part and fixing a convex neighbourhood at a point of the null part which does not include the entire null part, one sees that there is no such r.

Proof. Let $\gamma\colon (-\varepsilon, \varepsilon) \to M$ be a future directed timelike curve with $\gamma(0) = p$. We wish to prove that for $t < 0$ close enough to 0, $K_t = J^+[\gamma(t)] \cap J^-(S)$ is contained in U. The sets K_t are compact, cf. Lemma 40, p. 423 of [65] with plus signs replaced by minus signs and vice versa. Furthermore, $s < t$ implies $K_t \subset K_s$. Assume there is no $t < 0$ such that $K_t \subset U$. Then there is an increasing sequence $s_j \to 0-$, $j = 1, 2, \dots$ and $p_j \in K_{s_j} \cap U^c$. Since $p_j \in K_{s_1}$ for all j, there is a subsequence, which we shall also denote p_j, converging to $r \in U^c \cap K_{s_1}$. Since $\gamma(s_j) \leq p_j$ and the relation $\leq$ is closed, $p \leq r$. Since $p \neq r$, we have $p < r$. But $r \in J^-(S)$ and $p \in S$, so that we obtain a contradiction to the acausality of S. We conclude that for $t < 0$ small enough, $K_t \subset U$. Fix such a t and let $o = \gamma(t)$. Since U is a convex neighbourhood of o, we can construct the function f given by (11.2) with the base point p replaced by o. Then $f \in C^\infty(U)$ has a past directed timelike gradient wherever it is non-zero. Let K be a compact subset contained in U containing K_t in its interior. There is a $\phi \in C_0^\infty(U)$ such that $\phi = 1$ on K and $\phi(r) \in [0, 1]$ for all $r \in U$, cf. Proposition A.12. Consider the function $H = \phi f$. It is a smooth function with compact support in U and can thus be considered to be a smooth function on M. If $r \in J^-(S)$ and $H(r) \neq 0$, then $H = f$ in a neighbourhood of r, so that $\operatorname{grad} H$ is past pointing timelike. The only thing missing is the normalization. Note that $H(p) > 0$ since $o \ll p$ and $\phi(p) = 1$. Then $h_p = H/H(p)$ has the desired properties. □

Proposition 11.10. *Let* (M, g) *be an oriented, time oriented, connected and globally hyperbolic Lorentz manifold and let* S *be an acausal Cauchy hypersurface. Given an open neighbourhood* W *of* S, *there is a smooth function* $h\colon M \to [0, \infty)$ *such that*

- $\operatorname{supp} h \subseteq W$;
- $h(p) > 1/2$ *for* $p \in S$;
- *the gradient of* h *is timelike and past pointing on* $h^{-1}[(0, \infty)] \cap J^-(S)$.

Proof. Fix a complete Riemannian metric on M, which exists due to Lemma 11.1, and let d be the associated topological distance. Then the closed ball of radius $\rho < \infty$ around a point p, $\bar{B}_\rho(p)$, is compact due to the Hopf–Rinow theorem, cf. Theorem 21, p. 138 of [65]. We shall denote the open ball by $B_\rho(p)$, with the convention that $B_0(p) = \emptyset$. Fix a point $p \in M$ and define, for $l = 1, 2, \dots$, the compact sets

$$K_l = \bar{B}_l(p) - B_{l-1}(p), \quad R_l = K_l \cap S.$$

For each $r \in S$ fix a convex set U_r which contains r, has diameter < 1 with respect to d and is contained in W. By Lemma 11.8, there is a smooth function h_r with the properties stated in that lemma. Let $V_r = h_r^{-1}[(1/2, \infty)]$. Given l, there is a finite number of $r_{l,1}, \dots, r_{l,k_l} \in R_l$ such that the corresponding $V_{l,i} = V_{r_{l,i}}$ form an open cover of R_l. Let us use the notation $U_{r_{l,i}} = U_{l,i}$. Note that if $|l - m| \geq 3$, then $U_{l,i}$ and $U_{m,j}$ have empty intersection since both have diameter < 1 and $d(r_{l,i}, r_{m,j}) \geq 2$. Thus the sum

$$h = \sum_{l=1}^{\infty} \sum_{i=1}^{k_l} h_{r_{l,i}}$$

defines a smooth function since each point has a neighbourhood in which all but a finite number of the terms vanish. If $x \in S$, then $h(x) > 1/2$, since $x \in V_{l,i}$ for some l, i. Furthermore, the support of h is the union of the supports of $h_{r_{l,i}}$ since the covering $U_{l,i}$ is locally finite. Thus $\operatorname{supp} h \subseteq W$. If $x \in h^{-1}[(0, \infty)] \cap J^-(S)$, then $\operatorname{grad} h_{r_{l,i}}(x)$ is timelike and past pointing for every $h_{r_{l,i}}$ such that $h_{r_{l,i}}(x) > 0$ and it is zero for every $h_{r_{l,i}}$ such that $h_{r_{l,i}}(x) = 0$ (the reason for this is that x is a minimum for $h_{r_{l,i}}$). Since $h_{r_{l,i}}(x) > 0$ for some l, i, we conclude that $\operatorname{grad} h$ is past pointing and timelike at x. The proposition follows. □

11.3 Smooth time functions

A *time function* on a Lorentz manifold (M, g) is a function that is strictly increasing along any causal curve. A *temporal* function is a smooth function with the property that its gradient is everywhere past directed and timelike. The purpose of the present section is to prove that given a globally hyperbolic Lorentz manifold there is a temporal function defined on it, all of whose level sets are Cauchy hypersurfaces. The argument is taken from [5]. Below, we shall assume that all Lorentz manifolds are oriented, connected, time oriented and globally hyperbolic. Due to Theorem 11.6, there is thus a function τ with the properties stated in that theorem. We shall use the notation $S_t = \tau^{-1}(t)$.

Definition 11.11. Fix $t_- < t_a < t < t_b < t_+$ and define $S_\pm = S_{t_\pm}$ and $S = S_t$. We shall say that $\sigma : M \to \mathbb{R}$ is a *temporal step function around* t, compatible with the outer extremes t_-, t_+ and the inner extremes t_a, t_b if it satisfies

- $\operatorname{grad} \sigma$ is timelike and past pointing in $V = \{p \in M : \operatorname{grad} \sigma(p) \neq 0\}$;
- $\sigma(p) \in [-1, 1]$ for all $p \in M$;
- $\sigma(p) = -1$ for $p \in J^-(S_-)$ and $\sigma(p) = 1$ for $p \in J^+(S_+)$;
- $S_{t'} \subset V$ for all $t' \in (t_a, t_b)$.

Lemma 11.12. *Fix $t_-, t, t_+ \in \mathbb{R}$ such that $t_- < t < t_+$. Then there exists an open set U such that*

$$J^-(S_t) \subset U \subset I^-(S_{t_+}),$$

and a function $h^+ : M \to \mathbb{R}$ with $h^+ \geq 0$, and support contained in $I^+(S_{t_-})$ such that

- *if* $p \in U$ *and* $h^+(p) > 0$, *then* $\operatorname{grad} h^+(p)$ *is timelike and past pointing*;
- $h^+(p) > 1/2$ *for* $p \in J^+(S_t) \cap U$.

Proof. Take h^+ to be the function h constructed in Proposition 11.10 with $S = S_t$ and $W = I^-(S_{t_+}) \cap I^+(S_{t_-})$. For $x \in S_t$, $h(x) > 1/2$ and $\operatorname{grad} h(x)$ is past pointing timelike. Let V_x be an open neighbourhood of x in which the same conditions hold and define U to be the union of $I^-(S_t)$ and the sets V_x for $x \in S_t$. □

Lemma 11.13. *Fix* $t, t_+ \in \mathbb{R}$ *such that* $t < t_+$ *and let* $U \subset I^-(S_{t_+})$ *be an open neighbourhood of* $J^-(S_t)$. *Then there exists a smooth function* $h^-\colon M \to \mathbb{R}$ *with* $-1 \leq h^-(p) \leq 0$ *for all* $p \in M$, *such that*

- $\operatorname{supp} h^- \subset U$;
- *if* $\operatorname{grad} h^-(p) \neq 0$ *at* $p \in U$, *then* $\operatorname{grad} h^-(p)$ *is timelike and past pointing;*
- $h^-(p) = -1$ *for* $p \in J^-(S_t)$.

Proof. Reverse time orientation and construct h as in Proposition 11.10 with $S = S_t$ and $W = U$. We then get a smooth function $h\colon M \to [0, \infty)$ such that $\operatorname{supp} h \subset U$, if $h(p) > 0$ for $p \in J^+(S_t)$, then $\operatorname{grad} h(p)$ is future pointing timelike and finally, $h(p) > 1/2$ for $p \in S_t$. Let $h_1 = -h$ and let $\phi\colon \mathbb{R} \to \mathbb{R}$ be a smooth function such that $\phi(t) = -1$ for $t \leq -1/2$, $\phi'(t) > 0$ for $t \in (-1/2, 0)$, $\phi(t) = 0$ for $t \geq 0$ and $\phi(t) \in [-1, 0]$ for all t. In order to prove that there is such a function ϕ, let f be the function defined in Lemma A.9 and $\psi(t) = f(t + 1/2) f(-t)$. Then ψ is a smooth function, $\psi \geq 0$, $\psi(t) > 0$ for $t \in (-1/2, 0)$ and $\psi(t) = 0$ for $t \notin (-1/2, 0)$. Let

$$\phi(t) = \int_{-1/2}^{t} \psi(s)\,ds \left[\int_{-1/2}^{0} \psi(s)\,ds\right]^{-1} - 1.$$

Define $h^- = \phi \circ h_1$ on $J^+(S_t)$ and $h^-(p) = -1$ for $p \in J^-(S_t)$. Then h^- has the desired properties. □

Proposition 11.14. *Fix* $t_-, t, t_+ \in \mathbb{R}$ *such that* $t_- < t < t_+$. *Then there is a smooth function* $\sigma\colon M \to \mathbb{R}$ *satisfying the first three conditions of Definition* 11.11 *and the condition that* S_t *is contained in* $\{p \in M : \operatorname{grad} \sigma(p) \neq 0\}$.

Proof. According to Lemma 11.12 there is an open set U and a smooth function h^+ with the properties stated in that lemma. Given this U, construct h^- as in Lemma 11.13. Then $h^+ - h^- > 1/2$ on U, so that we can define

$$\sigma = 2\frac{h^+}{h^+ - h^-} - 1$$

on U and σ to be 1 on $M - \operatorname{supp} h^-$. Compute

$$\operatorname{grad} \sigma = 2\frac{h^+ \operatorname{grad} h^- - h^- \operatorname{grad} h^+}{(h^+ - h^-)^2},$$

which, at a given point, is past pointing timelike or zero. □

Corollary 11.15. *Fix $t \in \mathbb{R}$ and let $t_\pm = t \pm 1$. Assume t_a, t_b are such that $t_- < t_a < t < t_b < t_+$ and that K is a compact subset contained in $\tau^{-1}([t_a, t_b])$. Then there is a smooth function $\sigma : M \to \mathbb{R}$ satisfying the first three conditions of Definition* 11.11 *and, furthermore, K is contained in $\{p \in M : \operatorname{grad} \sigma(p) \neq 0\}$.*

Proof. For each $s \in [t_a, t_b]$, let σ_s be the function constructed in Proposition 11.14 with t_-, t, t_+ replaced by $t-1$, s, $t+1$ and let V_s be the set on which $\operatorname{grad} \sigma_s$ is non-zero. Then the sets V_s constitute an open covering of K. Let $s_1, \dots, s_k$ be such that V_{s_i} is an open covering of K and define

$$\sigma = \frac{1}{k} \sum_{i=1}^{k} \sigma_{s_i}.$$

Then σ has the desired properties. □

Before proceeding to add an infinite number of timelike vectors, let us note the following.

Lemma 11.16. *Let $\{v_i\}$ be a sequence of timelike vectors and assume that they are all future or all past pointing. If the sum $v = \sum_{i=1}^{\infty} v_i$ is well defined, then v is timelike.*

Proof. Since $\sum_{i=2}^{\infty} v_i$ is causal and future or past pointing according to the nature of the v_i, and since the sum of a future pointing causal vector and a future pointing timelike vector is a future pointing timelike vector, the statement follows. □

Theorem 11.17. *Fix $t \in \mathbb{R}$ and let $t_\pm = t \pm 1$. Given t_a, $t_b \in \mathbb{R}$ such that $t_- < t_a < t < t_b < t_+$, there is a temporal step function around t, compatible with the outer extremes t_-, t_+ and the inner extremes t_a, t_b.*

Proof. Choose a sequence G_j, $j = 1, \dots,$ of open sets such that $\bar{G}_j$ is compact, $\bar{G}_j \subset G_{j+1}$ and M is contained in the union of the G_j. Let

$$K_j = \bar{G}_j \cap J^+(S_{t_a}) \cap J^-(S_{t_b}).$$

On the compact sets K_j, we get a smooth function σ_j with the properties stated in Corollary 11.15. The idea is to take the sum of all these functions. The problem is then to get convergence. Let U_i and V_i be the open coverings whose existence is guaranteed by Corollary 10.4, and let (V_i, x_i) be coordinates. Let $A_j > 1$ be constants such that for each $1 \leq i \leq j$ and $0 \leq m \leq j$,

$$\left| \frac{\partial^m \sigma_j}{\partial x_i^{l_1} \dots \partial x_i^{l_m}} \right| < A_j$$

on $\bar{U}_i$ for $l_1, \dots, l_m \in \{1, \dots, n\}$. Define

$$\sigma = \sum_{j=1}^{\infty} \frac{1}{2^j A_j} \sigma_j. \tag{11.3}$$

Since the series converges absolutely, it defines a continuous function. To prove differentiability of this object, let $p \in M$. Then $p \in U_i$ for some i. To prove that σ is C^l, let $j > i, l$. Then $\sigma_j/(2^j A_j)$ and all its derivatives of order $\leq l$ with respect to the coordinates x_i are bounded by 2^{-j}. Thus the series (11.3) and all its partial derivatives up to order l converges uniformly. Thus σ is smooth. What remains to be done is to fix the normalization of σ. We have $\sigma(p) = \sigma_- < 0$ for $p \in J^-(S_-)$ and $\sigma(p) = \sigma_+ > 0$ for $p \in J^+(S_+)$. Let $\psi : \mathbb{R} \to \mathbb{R}$ be such that $\psi(t) = -1$ for $t \leq \sigma_-$, $\psi(t) = 1$ for $t \geq \sigma_+$, $\psi'(t) > 0$ for $t \in (\sigma_-, \sigma_+)$ and $\psi(t) \in [-1, 1]$ for all $t \in \mathbb{R}$. Then $\psi \circ \sigma$ is the desired function. □

Theorem 11.18. *Let (M, g) be an oriented, connected, time oriented and globally hyperbolic Lorentz manifold. Then there is a smooth function $\mathcal{T} : M \to \mathbb{R}$ such that* grad $\mathcal{T}$ *is timelike and past pointing on all of M and for each inextendible causal curve $\gamma\colon (t_-, t_+) \to M$, $\mathcal{T}[\gamma(t)] \to \pm\infty$ as $t \to t_\pm \mp$. In particular, each hypersurface $\mathcal{T}^{-1}(t)$ is a smooth spacelike Cauchy hypersurface.*

Proof. Let $t_k = k/2$, $t_{k,\pm} = t_k \pm 1$, $t_{k,a} = t_k - 1/2$ and $t_{k,b} = t_k + 1/2$. Let σ_k be the function constructed in Theorem 11.17 with t replaced by t_k, $t_\pm$ replaced by $t_{k,\pm}$ and t_a, t_b replaced by $t_{k,a}, t_{k,b}$ respectively. Note that each $p \in M$ is such that $\tau(p) \in (t_{k,a}, t_{k,b})$ for some k. Define

$$\mathcal{T} = \sigma_0 + \sum_{k=1}^{\infty} (\sigma_{-k} + \sigma_k). \tag{11.4}$$

Note that if $k \geq 3$, then $(\sigma_k + \sigma_{-k})(p) = 0$ if $-k/2 + 1 \leq \tau(p) \leq k/2 - 1$. Consequently, for any $p \in M$, there is a neighbourhood of p such that only a finite number of the terms in the sum (11.4) are non-zero in that neighbourhood. Thus (11.4) defines a smooth function. Since σ_k has a past directed timelike gradient at p if $\tau(p) \in [t_{k,a}, t_{k,b}]$, we conclude that the gradient of $\mathcal{T}$ is everywhere timelike and past directed. Assume now that $\gamma\colon (t_-, t_+ \to M$ is a future directed inextendible causal curve. Note first that $\mathcal{T} \circ \gamma$ is strictly monotonically increasing. Since γ has to intersect every Cauchy hypersurface S_t, we conclude that given $m \geq 1$, there is an $s_m \in (t_-, t_+)$ such that $\tau[\gamma(s_m)] = m$. Let $l = 2(m + 1)$. By the above observation, $(\sigma_{-k} + \sigma_k)[\gamma(s_m)] = 0$ for $k \geq l$. Thus

$$\mathcal{T}[\gamma(s_m)] = \Big[\sigma_0 + \sum_{k=1}^{l} (\sigma_k + \sigma_{-k})\Big][\gamma(s_m)].$$

Since $m \geq 1$, $\sigma_0[\gamma(s_m)] = 1$. Furthermore,

$$(\sigma_k + \sigma_{-k})[\gamma(s_m)] \geq 0$$

for all $k \geq 1$, since $\sigma_{-k}[\gamma(s_m)] = 1$ and $\sigma_k[\gamma(s_m)] \geq -1$ for all $k \geq 1$. Finally, for $1 \leq k \leq 2(m-1)$, we have

$$(\sigma_k + \sigma_{-k})[\gamma(s_m)] = 2.$$

Adding up the above observations, we conclude that $\mathcal{T}[\gamma(s_m)] \geq 4(m-1)+1$ and consequently, $\mathcal{T}[\gamma(s)] \geq 4(m-1)+1$ for $s \in [s_m, t_+)$. Thus $\mathcal{T}[\gamma(s)] \to \infty$ as $t \to t_+-$. The argument to prove that $\mathcal{T}[\gamma(s)] \to -\infty$ as $t \to t_-+$ is similar. □

Corollary 11.19. *Let (M, g) be an oriented, connected, time oriented and globally hyperbolic Lorentz manifold. Then M has a smooth spacelike Cauchy surface S and M is diffeomorphic to $\mathbb{R} \times S$.*

Proof. Theorem 11.18 yields a smooth spacelike Cauchy hypersurface and Proposition 11.3 the desired diffeomorphism. □

11.4 Smooth temporal functions adapted to Cauchy hypersurfaces

The purpose of the present section is to prove that given a spacelike Cauchy hypersurface S, there is a smooth temporal function $\mathcal{T}$ with the properties listed in Theorem 11.18 such that $\mathcal{T}^{-1}(0) = S$. The arguments are taken from [6]. We begin with a lemma.

Lemma 11.20. *Let (M, g) be a globally hyperbolic time oriented Lorentz manifold and let S be a Cauchy hypersurface. Then $I^-(S)$ and $I^+(S)$, regarded as spacetimes, are globally hyperbolic.*

Proof. If we regard $I^+(S)$ as a spacetime, we conclude that the strong causality condition holds on it, and that

$$J^+[p, I^+(S)] \cap J^-[q, I^+(S)] = J^+(p) \cap J^-(q)$$

for $p, q \in I^+(S)$. Since the latter set is compact and contained in $I^+(S)$, the statement follows. □

Proposition 11.21. *Let (M, g) be a connected, oriented, time oriented globally hyperbolic Lorentz manifold and let S be an acausal Cauchy hypersurface. Then there is a continuous onto function $\hat{\tau}: M \to \mathbb{R}$ such that*

- $S = \hat{\tau}^{-1}(0)$;
- *$\hat{\tau}$ is smooth and has past directed timelike gradient on $M - S$;*
- *each $\hat{\tau}^{-1}(t)$, $t \in \mathbb{R} - \{0\}$ is a smooth spacelike Cauchy hypersurface;*
- *$\hat{\tau}$ is a timefunction.*

Remark 11.22. Recall that a timefunction is a function which is strictly increasing along all future directed causal curves.

Proof. Since $I^\pm(S)$ are time oriented globally hyperbolic Lorentz manifolds, there are temporal functions $\mathcal{T}_\pm$ on them with properties as stated in Theorem 11.18 (note that $I^\pm(S)$ are connected, cf. Proposition 11.3). Define

$$\hat{\tau}(p) = \begin{cases} \exp[\mathcal{T}_+(p)], & p \in I^+(S), \\ 0, & p \in S, \\ -\exp[-\mathcal{T}_-(p)], & p \in I^-(S). \end{cases}$$

Let us first prove that for each $t \in \mathbb{R} - \{0\}$, $\hat{\tau}^{-1}(t)$ is a smooth spacelike Cauchy hypersurface. Let γ be an inextendible timelike curve in M. Then γ has to intersect S and consequently $I^{\pm}(S)$. Thus γ restricted to $\gamma^{-1}[I^{\pm}(S)]$ is an inextendible timelike curve in $I^{\pm}(S)$. Thus if S_1 is a Cauchy hypersurface in one of $I^{\pm}(S)$, γ has to intersect S_1. Assume γ intersects S_1 twice. Assuming for example that S_1 is a Cauchy hypersurface in $I^+(S)$, one then gets a future directed timelike curve in $I^+(S)$ intersecting S_1 twice, contradicting the fact that S_1 was a Cauchy hypersurface of $I^+(S)$. Since $\hat{\tau}^{-1}(t)$ is a smooth spacelike Cauchy hypersurface in $I^{\pm}(S)$ for $t \neq 0$, depending on the sign of t, we conclude that $\hat{\tau}^{-1}(t)$ is a smooth spacelike Cauchy hypersurface in M for $t \neq 0$.

To prove continuity, let $p_k \to p$ with $p_k, p \in M$. If $p \notin S$, it is clear that $\hat{\tau}(p_k) \to \hat{\tau}(p)$, so let us assume $p \in S$. Let $l \geq 1$ be an integer, $S_{l,+} = \mathcal{T}_+^{-1}(-l)$ and $S_{l,-} = \mathcal{T}_-^{-1}(l)$. Then $U = I^-(S_{l,+}) \cap I^+(S_{l,-})$ is an open set containing S. For k large enough, we thus have $p_k \in U$ so that $\hat{\tau}(p_k) \in [-e^{-l}, e^{-l}]$. Consequently $\hat{\tau}(p_k) \to 0$, so that $\hat{\tau}$ is continuous.

The fact that $\hat{\tau}$ is onto follows from the fact that $\mathcal{T}_{\pm}$ are onto. That $S = \hat{\tau}^{-1}(0)$ is clear from the definition. What remains to be proved is that it is a timefunction. Let γ be a causal curve. On $\gamma^{-1}(M - S)$, $\hat{\tau} \circ \gamma$ is strictly increasing due to the fact that $\mathcal{T}_{\pm}$ have past directed timelike gradients. Due to the acausality of S, $\gamma^{-1}(S)$ is a point. The statement follows. □

Theorem 11.23. *Let (M, g) be an oriented, time oriented, connected and globally hyperbolic Lorentz manifold and let S be an acausal Cauchy hypersurface. Then there is a smooth onto function $\tilde{\tau}: M \to \mathbb{R}$ such that*

- $S = \tilde{\tau}^{-1}(0)$;
- *$\tilde{\tau}$ has past directed timelike gradient on $M - S$;*
- *each $\tilde{\tau}^{-1}(t)$, $t \in \mathbb{R} - \{0\}$ is a smooth spacelike Cauchy hypersurface;*
- *$\tilde{\tau}$ is a timefunction.*

Remark 11.24. The only improvement in comparison with Proposition 11.21 is the smoothness of $\tilde{\tau}$.

Proof. For each $k \geq 1$, let $\phi_k^{\pm} \in C^{\infty}(\mathbb{R})$ be such that

$$\begin{aligned} &\phi_k^+(t) = 0 \ \text{ for } t \leq 1/k, && \phi_k^+(t) = t \ \text{ for } t \geq 2, && \frac{d\phi_k^+}{dt}(t) > 0 \ \text{ for } t > 1/k \\ &\phi_k^-(t) = 0 \ \text{ for } t \geq -1/k, && \phi_k^-(t) = t \ \text{ for } t \leq -2, && \frac{d\phi_k^-}{dt}(t) > 0 \ \text{ for } t < -1/k. \end{aligned}$$

Choose constants $C_k \geq 1$ such that

$$\left|\frac{d^m\phi_k^+}{dt^m}\right| + \left|\frac{d^m\phi_k^-}{dt^m}\right| \leq C_k$$

for all $m = 1, \ldots, k$. Define $\tau_k^{\pm} = \phi_k^{\pm} \circ \hat{\tau}$, where $\hat{\tau}$ is the function whose existence is guaranteed by Proposition 11.21. Then grad $\tau_k^{\pm}$ is timelike and past directed whenever it is non-zero. Define

$$\tilde{\tau} = \sum_{k=1}^{\infty} \frac{1}{2^k C_k} (\tau_k^+ + \tau_k^-).$$

Then $\tilde{\tau}$ is smooth, $S = \tilde{\tau}^{-1}(0)$ and the gradient of $\tilde{\tau}$ is past directed and timelike on $M - S$, since the sum of timelike vectors is timelike by Lemma 11.16. Since there is a constant $c_0 > 0$ such that $\tilde{\tau}(p) = c_0 \hat{\tau}(p)$ for all p such that $|\hat{\tau}(p)| \geq 2$, $\tilde{\tau}^{-1}(t)$ are smooth spacelike Cauchy hypersurfaces for $|t| \geq 2c_0$. Let γ be an inextendible causal curve, then $\tilde{\tau} \circ \gamma$ is strictly increasing since $\gamma^{-1}(S)$ is a point. Since γ has to intersect $\tilde{\tau}^{-1}(t)$ for every $|t| \geq 2c_0$, $\tilde{\tau} \circ \gamma$ is onto, so that $\tilde{\tau}$ is onto. Thus γ intersects each set $\tilde{\tau}^{-1}(t)$ exactly once. Since $\tilde{\tau}^{-1}(t)$ is a smooth spacelike hypersurface for $t \neq 0$, all the statements follow. □

Lemma 11.25. *Let (M, g) be a connected, oriented, time oriented globally hyperbolic Lorentz manifold, let S be a smooth spacelike Cauchy hypersurface and let W be an open neighbourhood of S. Then there are smooth functions $h_{\pm}$ on M such that*

- $\pm h_{\pm} \geq 0$ *and* $\operatorname{supp} h_{\pm} \subset W \cup J^{\pm}(S)$;
- $h_{\pm} = \pm 1$ *on* $J^{\pm}(S)$;
- *if* grad $h_{\pm}(p) \neq 0$, *then it is past directed timelike.*

Proof. Construct h_- exactly as in Lemma 11.13 with U replaced by $W \cup J^-(S)$. Define $-h_+$ to be the same function for the Lorentz manifold with opposite time orientation. □

Proposition 11.26. *Let (M, g) be a connected, oriented, time oriented globally hyperbolic Lorentz manifold and let S be a smooth spacelike Cauchy hypersurface. Let $\tilde{\tau}$ be a function with the properties stated in Theorem 11.23 and define $S_i = \tilde{\tau}^{-1}(i)$, $i = -1, 0, 1$. Then $S_0 = S$. There is a smooth function $\tau_0 \colon M \to \mathbb{R}$ such that*

- grad τ_0 *is timelike and past directed in* $V = \{p \in M : \text{grad}\, \tau_0 \neq 0\}$;
- $\tau_0(p) \in [-1, 1]$ *for all* $p \in M$;
- $\tau_0(p) = \pm 1$ *for* $p \in J^{\pm}(S_{\pm 1})$, *so that* $V \subseteq \tilde{\tau}^{-1}[(-1, 1)]$;
- $S = \tau_0^{-1}(0) \subset V$.

Proof. In order to construct the function τ_0, we shall use the concept of a normal neighbourhood of a submanifold introduced on pp. 197–200 of [65]. The set of vectors perpendicular to S can be viewed as a manifold, let us call it NS, and let Z be the subset of zero vectors in NS. Then NS is a manifold and Z is a submanifold diffeomorphic to S. We can define $\exp^{\perp}$ on an open subset of NS by $\exp^{\perp}(v) = \gamma_v(1)$, where γ_v is the unique geodesic going through p with initial velocity v, assuming that $v \in T_pM$.

Then $\exp^{\perp}$ is a smooth function defined on an open subset of NS. A neighbourhood W of S is said to be *normal* if it is the diffeomorphic image under $\exp^{\perp}$ of an open neighbourhood of Z in NS. Due to Proposition 26, p. 200, of [65], S has such a normal neighbourhood. Let us call it W. Let T be the future oriented timelike normal to S. Then for every $r \in W$, there is a unique $p \in S$ and $t \in \mathbb{R}$ such that $\exp^{\perp}(tT_p) = r$. Let $h_S(r) = t + 1$. Then h_S is a smooth function on W. Let us prove that the gradient is timelike, at least in a neighbourhood of S. Let $p \in S$ and let U be an open neighbourhood of p in S such that there are coordinates $\phi\colon U \to \mathbb{R}^n$, and let us assume that $\phi(p) = 0$. Define a map ψ which takes $(t, x^1, \dots, x^n)$ to $\exp^{\perp}(tT_r)$, where $r = \phi^{-1}(x^1, \dots, x^n)$, wherever it is defined. Then ψ is a smooth map from an open subset of $\mathbb{R}^{n+1}$ into M. Note that the image of $T_0\mathbb{R}^{n+1}$ under ψ_* is T_pM. Thus we can assume ψ to be a diffeomorphism onto its image by reducing the domain. Letting x^μ, $\mu = 0, \dots, n$ be the coordinates defined by ψ^{-1}, we see that $\partial_0|_r = T_r$ for $r \in S$ in the image of ψ and that $\partial_i|_r$ is tangent to S for $r \in S$. In particular, $g_{\mu\nu} = \langle \partial_\mu, \partial_\nu \rangle$ has the property that $g_{00} < 0$, $g_{0i} = 0$ for $i = 1, \dots, n$ and g_{ij}, $i, j = 1, \dots, n$ are the components of a positive definite matrix. Consequently

$$\langle \operatorname{grad} h_S, \operatorname{grad} h_S \rangle = \langle \operatorname{grad} t, \operatorname{grad} t \rangle = \langle \operatorname{grad} x^0, \operatorname{grad} x^0 \rangle = g^{00} < 0,$$

so that the gradient of h_S is timelike on an open neighbourhood of p. That it is past oriented follows from the construction. For every $p \in S$, there is thus an open neighbourhood of p such that h_S has a past directed timelike gradient on that open neighbourhood. Thus, by restricting W, we can assume h_S to be smooth and have past directed timelike gradient on W.

Note that $h_S(p) = 1$ for $p \in S$. Without loss of generality, we can assume $W \subseteq \tilde{\tau}^{-1}[(-1, 1)]$. Furthermore, we can assume $h_S > 0$ on W. Let $h_\pm$ be the functions constructed in Lemma 11.25. Define

$$h^+ = h_S h_+.$$

Then h^+ is smooth on $W \cup J^-(S)$, by defining it to be zero on $J^-(S) - W$. Furthermore, its gradient is past directed timelike when it does not vanish. Define

$$\tau_0 = 2\frac{h^+}{h^+ - h_-} - 1$$

and extend it to be 1 on $J^+(S) - W$. Then the gradient of τ_0 is past directed timelike when it is non-zero and the desired properties hold. □

Theorem 11.27. *Let (M, g) be an oriented, time oriented, connected and globally hyperbolic Lorentz manifold and let S be a smooth spacelike Cauchy hypersurface. Then there is a smooth function $\mathcal{T}$ on M which has past directed timelike gradient everywhere and satisfies the property that $\mathcal{T}^{-1}(t)$ is a Cauchy hypersurface for every $t \in \mathbb{R}$. Furthermore $\mathcal{T}^{-1}(0) = S$ and for every inextendible causal curve $\gamma\colon (t_-, t_+) \to M$, $\mathcal{T}[\gamma(t)] \to \pm\infty$ as $t \to t_\pm \mp$.*

Proof. Let $\mathcal{T} = \tilde{\tau} + \tau_0$, where τ_0 is the function constructed in Proposition 11.26 and $\tilde{\tau}$ was the function constructed in Theorem 11.23. Then $\mathcal{T}$ has past directed timelike gradient everywhere and $\mathcal{T}^{-1}(0) = S$. Since $\tilde{\tau}$ tends to $\pm\infty$ along inextendible causal curves and τ_0 is bounded, the desired behaviour of $\mathcal{T}$ along inextendible causal curves follows. This in its turn implies that $\mathcal{T}^{-1}(t)$ is a Cauchy hypersurface for every $t \in \mathbb{R}$. □

11.5 Auxiliary observations

The purpose of this section is to make some observations concerning globally hyperbolic Lorentz manifolds that will be of use when proving uniqueness of constant mean curvature hypersurfaces, cf. Chapter 18. The following result is essentially a direct consequence of Theorem 11.27, cf. [5].

Lemma 11.28. *Let (M, g) be an oriented, time oriented, connected and globally hyperbolic Lorentz manifold and let S be a smooth spacelike Cauchy hypersurface. Then there is a diffeomorphism $f : \mathbb{R} \times S \to M$ such that $f(\{t\} \times S)$ is a Cauchy hypersurface for every $t \in \mathbb{R}$ and $f(0, p) = p$. Furthermore, there is a timelike vector field T on M such that $(\pi_2 \circ f^{-1})^{-1}(p)$ is the image of the maximal integral curve of T through p, where π_2 is projection to the second factor.*

Proof. Due to Theorem 11.27, there is a smooth surjective function $\mathcal{T} : M \to \mathbb{R}$ with the following properties

- it has a past directed timelike gradient,
- $\mathcal{T}^{-1}(0) = S$,
- for every inextendible causal curve $\gamma : (t_-, t_+) \to M$, $\mathcal{T}[\gamma(t)] \to \pm\infty$ as $t \to t_\pm \mp$,
- $\mathcal{T}^{-1}(t)$ is a Cauchy hypersurface for all $t \in \mathbb{R}$.

Let $T = -\operatorname{grad} \mathcal{T}$. Then T is a smooth future directed timelike vector field. Let $\Phi : \mathcal{D} \to M$ be the associated flow. The domain $\mathcal{D}$ is an open subset of $\mathbb{R} \times M$ such that if $I_p \times \{p\} = \mathbb{R} \times \{p\} \cap \mathcal{D}$, then I_p is the maximal existence interval of the integral curve of T through p, cf. the bottom of p. 29 of [65]. Furthermore, Φ is a smooth map. Let us prove that if $\gamma : I \to M$ is an integral curve of T, where $I = (t_-, t_+)$ is the maximal existence interval, then γ is inextendible. Assume γ is extendible to the future. If $t_+ < \infty$, we can use Lemma 56, p. 30 of [65] to conclude that I is not maximal, a contradiction. If $t_+ = \infty$, we can use the argument presented in the proof of Proposition 11.3 to conclude that T has to be zero somewhere, a contradiction. The argument in the other time direction is similar. Thus γ is inextendible. Let $\mathcal{D}_S = \mathbb{R} \times S \cap \mathcal{D}$ and let $\phi : \mathcal{D}_S \to M$ be the restriction of Φ to $\mathcal{D}_S$. Note that $\mathcal{D}_S$ is an open subset of $\mathbb{R} \times S$. Let us prove that ϕ is bijective. Let $p \in M$ and let γ be the integral curve of T through p. Then γ is an inextendible timelike curve and thus has to intersect S. Consequently, p is in the image of ϕ. Assume $\phi(t_1, p_1) = \phi(t_2, p_2)$. If t_1 and t_2 had different signs, there would be a timelike curve intersecting S twice, contradicting global hyperbolicity. Assume $0 \leq t_1 \leq t_2$. Then $p_1 = \phi(t_2 - t_1, p_2)$. If $t_2 > t_1$, we get a timelike curve from S to itself, a contradiction. Thus $t_1 = t_2$ so that $p_1 = p_2$. The argument for $t_1 \leq t_2 \leq 0$ is similar. Let $(t, p) \in \mathcal{D}_S$ and assume $t \geq 0$. We wish to prove that

$$\phi_*|_{T_{(t,p)}\mathbb{R}\times S} : T_{(t,p)}\mathbb{R} \times S \to T_{\phi(t,p)}M \tag{11.5}$$

is surjective. If $t = 0$, this is obvious, so let us assume $t > 0$. By the properties of the flow, there is an open set $U \subset M$ containing p and an $\varepsilon > 0$ such that Φ is defined on $(-\varepsilon, t + \varepsilon) \times U$. Furthermore, $\Phi(t, \cdot)$ is a diffeomorphism from $\Phi[(-\varepsilon, \varepsilon) \times U]$ to $\Phi[(t - \varepsilon, t + \varepsilon) \times U]$. Note that for $(s, q) \in (t - \varepsilon, t + \varepsilon) \times U \cap S$, we have

$$\phi(s, q) = \Phi[t, \phi(s - t, q)]$$

Thus $\Phi[-t, \phi(s, q)] = \phi(s - t, q)$. Viewing this as a map taking $(s, q) \in \mathcal{D}_S$ to a point on the manifold, we can consider the push forward of this map at (t, q). Since the push forward of ϕ at $(0, q)$ is surjective, we conclude that (11.5) has to be surjective. To conclude, ϕ is bijective and a local diffeomorphism. Thus ϕ is a diffeomorphism.

Consider $\psi : \mathcal{D}_S \to \mathbb{R} \times S$ defined by

$$\psi(t, p) = (\mathcal{T} \circ \Phi(t, p), p).$$

Due to the properties of $\mathcal{T}$, one can check that ψ is injective, surjective and that ψ_* is always surjective. Thus ψ is a diffeomorphism. We define $f := \phi \circ \psi^{-1}$. Then, by definition, $\mathcal{T} \circ f(t, p) = t$, and f has the desired properties. □

Proposition 11.29. *Let (M, g) be a connected, oriented, time oriented and globally hyperbolic Lorentz manifold. Assume (M, g) has a smooth, compact, spacelike and achronal hypersurface Σ. Then Σ is a Cauchy hypersurface.*

Proof. Since (M, g) is globally hyperbolic, we know that there is a smooth spacelike Cauchy hypersurface S in M. Furthermore, we know that there is a diffeomorphism $f : \mathbb{R} \times S \to M$ with the properties stated in Lemma 11.28. Let $\rho : M \to S$ be defined by $\rho = \pi_2 \circ f^{-1}$, where π_2 is the projection to the second factor. Note that ρ is a smooth map which takes open sets to open sets. Let ϱ denote the restriction of ρ to Σ. Since Σ is compact, we conclude that ϱ maps closed sets to closed sets. That ϱ is injective follows from the fact that if it were not, there would be a timelike curve intersecting Σ twice, cf. Lemma 11.28, in contradiction with the assumptions. Note that the image of ϱ is closed and non-empty. We wish to prove that it is open. By an argument similar to one presented in the proof of Lemma 11.28, there is, given a $p \in \Sigma$, an open set $U \subseteq \Sigma$ containing p and an $\varepsilon > 0$ such that the image of $(-\varepsilon, \varepsilon) \times U$ under Φ is an open subset of M, where Φ is the flow of T, cf. Lemma 11.28. Then

$$\varrho(U) = \rho\{\Phi[(-\varepsilon, \varepsilon) \times U]\},$$

but the latter set is open by the above observations. Consequently, ϱ takes open sets to open sets. The image of Σ under ϱ is thus non-empty, open and closed. Since M is connected and diffeomorphic to $\mathbb{R} \times S$, S is connected. Thus $\varrho\colon \Sigma \to S$ is an injective and surjective map which takes closed sets to closed sets. Consequently, the inverse is defined and continuous. Thus ϱ is a homeomorphism. Note in particular that S has to be compact. Let us define $h\colon S \to \mathbb{R}$ by $h(p) = \pi_1 \circ f^{-1} \circ \varrho^{-1}$, where π_1 denotes projection onto the first factor. Since ϱ is the restriction of ρ to a subset, we have $\rho \circ \varrho^{-1}(p) = p$ for all $p \in S$. Thus, keeping the definition of ρ in mind, we have, for $p \in S$,

$$\begin{aligned}[h(p), p] &= [\pi_1 \circ f^{-1} \circ \varrho^{-1}(p), \rho \circ \varrho^{-1}(p)] \\ &= [\pi_1 \circ f^{-1} \circ \varrho^{-1}(p), \pi_2 \circ f^{-1} \circ \varrho^{-1}(p)] = f^{-1} \circ \varrho^{-1}(p).\end{aligned}$$

Let

$$\mathcal{A} = \{(t, p) \in \mathbb{R} \times S : h(p) = t\}.$$

Then the above computation demonstrates that $f(\mathcal{A}) = \Sigma$.

Since h is a continuous function on a compact set, it attains a maximum and a minimum. Let t_1 and t_2 be such that $t_1 < h(p) < t_2$ for all $p \in S$. Let γ be an inextendible timelike curve in M. Let s_i, $i = 1, 2$ be such that $\gamma(s_i) \in f(\{t_i\} \times S)$. Consider

$$r(s) = h \circ \pi_2 \circ f^{-1}[\gamma(s)] - \pi_1 \circ f^{-1}[\gamma(s)].$$

We have $r(s_1) > 0$ and $r(s_2) < 0$. Since $r(s)$ is continuous and γ is defined on an interval, there has to be an s_0 between s_1 and s_2 such that $r(s_0) = 0$. Then $f^{-1}[\gamma(s_0)] \in \mathcal{A}$ so that $\gamma(s_0) \in \Sigma$. In other words γ intersects Σ. Since we already know that it intersects Σ at most once, we conclude that Σ is a Cauchy hypersurface. □

It is of some interest to note that the assumptions of Proposition 11.29 can be weakened. In fact, it is not necessary to assume achronality of Σ; connectedness is enough. However, to prove this fact, one needs to appeal to intersection theory, something we wish to avoid here. Nevertheless, let us state the result, which we shall not use in what follows.

Proposition 11.30. *Let (M, g) be a connected, oriented, time oriented and globally hyperbolic Lorentz manifold. Assume (M, g) has a smooth, compact, connected and spacelike hypersurface Σ. Then Σ is a Cauchy hypersurface.*

Remark 11.31. That it is necessary to demand compactness follows by considering a hyperboloid in Minkowski space. That it is necessary to demand connectedness is also quite clear.

Proof. Proposition 7, p. 465 of [4] implies that if there is a Cauchy hypersurface S in M which does not intersect Σ, then Σ is achronal. To prove that there is such a hypersurface, let f be as in Lemma 11.28 and consider $\pi_1 \circ f^{-1}(\Sigma)$. This is a compact subset of the real numbers. Consequently, there is a t_0 such that $t_0 > \pi_1 \circ f^{-1}(p)$ for all $p \in \Sigma$. Thus $f(\{t_0\} \times S)$ is a Cauchy hypersurface in M which does not intersect Σ. Proposition 11.29 yields the conclusion of the proposition. □

12 Uniqueness of solutions to linear wave equations

In this chapter, we wish to prove a geometric uniqueness statement for linear wave equations on a globally hyperbolic Lorentz manifold. In particular, consider the equation

$$\Box_g u + Xu + \kappa u = f,$$

where $\Box_g u = \nabla_\alpha \nabla^\alpha u$, ∇ is the Levi Civita connection associated with g, X is a smooth vector field and κ, f are smooth functions on M. Say that we have two solutions u_1, u_2 to this equation and say that u_1 and u_2 and their normal derivatives coincide on a subset Ω of a spacelike Cauchy hypersurface. The purpose of the present chapter is to prove that under such circumstances, $u_1 = u_2$ on $D(\Omega)$.

In Subsection 2.2.2 we gave an outline of the proof of uniqueness in the case of Minkowski space. The purpose of the first section of the present chapter is to provide the technical tools necessary for establishing that one has roughly speaking the same picture in suitable convex neighbourhoods in the general case. In particular, we need to know that the intersection of the past light cone of a suitable point and the Cauchy hypersurface is a manifold in the context of interest. This is in part achieved by Lemma 12.1. Consider the region above the $t = 0$ hypersurface and below the hyperbola in Figure 2.1. We shall wish to apply (10.3) to this region. Unfortunately, this formula does not apply, since the region is not a manifold with boundary. To overcome this problem, one can subtract the intersection of the hyperbola and the $t = 0$ hypersurface from the region of interest and apply the result to the thus obtained region. However, then we cannot assume the vector field of interest to have compact support. Thus we need to multiply it with a suitable cut off function. This is achieved by a construction carried out in a tubular neighbourhood of the intersection mentioned above, and the purpose of Lemma 12.3 is to construct the needed tubular neighbourhood.

The purpose of the remaining lemmas of Section 12.1 is to ensure that some elementary aspects of the intuitively obvious setup described in the Minkowski setting hold in the general case as well, given one is prepared to restrict one's attention to a small enough convex neighbourhood.

12.1 Preliminary technical observations

Let M be an n-dimensional differentiable manifold. We refer to a subset N of M as a k-dimensional submanifold, $k < n$, if for every $p \in N$, there is a chart (U, ϕ) with $p \in U$ and $\phi = (x^1, \dots, x^n)$ such that $q \in U \cap N$ if and only if $x^{k+1}(q) = \dots = x^n(q) = 0$. Due to the definition, we see that N can be given the structure of a differentiable manifold. There are other, equivalent, definitions of the concept of a submanifold, cf. pp. 15–18 of [65]. We shall need the following observation.

Lemma 12.1. *Let N_1 and N_2 be two n dimensional spacelike submanifolds of an $(n+1)$-dimensional Lorentz manifold (M, g). Assume that for each $p \in N_1 \cap N_2$,* *any two normals to N_1 and N_2 at p are linearly independent. Then $N_1 \cap N_2$ is an $n-1$-dimensional submanifold of M.*

Remark 12.2. The result is of course a trivial special case of the fact that the intersection of transversal submanifolds is a submanifold, cf. Theorem 7.7, p. 84 of [8].

Proof. Let $p \in N_1 \cap N_2$. Since N_i is a submanifold, there are charts (U_i, ϕ_i) such that $\phi_i = (x_i^0, \dots, x_i^n)$ has the property that $U_i \cap N_i$ equals the set of $p \in U_i$ such that $x_i^n(p) = 0$. Let $U = U_1 \cap U_2$ and $f_i = x_i^n$. Then $N_1 \cap N_2 \cap U$ corresponds to the set of $p \in U$ such that $f_i(p) = 0$, $i = 1, 2$. Note that a normal of N_i is given by grad f_i in U. If f is a smooth function, grad f is defined to be the vector field such that for any vector X

$$\langle \text{grad } f, X \rangle = X(f).$$

Choose coordinates $(\tilde{U}, y)$ such that $g_{\mu\nu}|_p = \eta_{\mu\nu}$, where $\eta_{\mu\nu}$ are the components of the Minkowski metric and

$$\text{grad } f_1|_p = \alpha \frac{\partial}{\partial y^0}\bigg|_p$$

$$\text{grad } f_1|_p = \beta \frac{\partial}{\partial y^0}\bigg|_p + \gamma \frac{\partial}{\partial y^1}\bigg|_p,$$

where α, β and γ are all non-zero (we have here used the fact that the normals are both timelike). Recall that

$$\frac{\partial f_i}{\partial y^j} = D_j(f_i \circ y^{-1}) \circ y.$$

For the sake of simplicity, let us assume $y(p) = 0$. Consider the smooth function $f : y(\tilde{U}) \to \mathbb{R}^2$ defined by

$$f = (f_1 \circ y^{-1}, f_2 \circ y^{-1}).$$

Differentiating this function with respect to the first two variables leads to a 2×2 matrix which is invertible. By the implicit function theorem there exists an open set $V \subseteq \mathbb{R}^{n+1}$ containing 0 and an open set $W \subseteq \mathbb{R}^{n-1}$ containing zero such that for every $w \in W$ there is a unique $z \in \mathbb{R}^2$ such that $(z, w) \in V$ and $f(z, w) = 0$. If this z is defined to be $g(w)$, then g is a smooth function on W, $g(0) = 0$ and $f[g(w), w] = 0$. Reduce the domain of definition of the coordinates y to be $U' = y^{-1}(\mathbb{R}^2 \times W \cap V)$. On U' we define the coordinates $v = (v^0, \dots, v^n)$ by

$$(v^0, \dots, v^n) = [y^0 - g_0(y^2, \dots, y^n), y^1 - g_1(y^2, \dots, y^n), y^2, \dots, y^n].$$

By considering $v \circ y^{-1}$, it is clear that v is a smooth map with a smooth inverse. Thus (U', v) is a coordinate chart. If $q \in U'$, $v^0(q) = 0$ and $v^1(q) = 0$, then $q \in N_1 \cap N_2$ by the above construction. If $q \in U' \cap N_1 \cap N_2$, we can conclude that $y(q) \in \mathbb{R}^2 \times W \cap V$ so that, by the above construction, $v^0(q) = 0$ and $v^1(q) = 0$. With respect to the chart (U', v), $N_1 \cap N_2$ is thus given by the condition $v^0 = v^1 = 0$. We conclude that $N_1 \cap N_2$ is an $n-1$-dimensional submanifold. □

We shall also need the following.

Lemma 12.3. *Assume M is a compact $n-2$-dimensional spacelike submanifold of an open subset $U \subseteq \mathbb{R}^n$ on which there is a Lorentz metric g. Assume that there are smooth maps $v, w\colon M \to \mathbb{R}^n$ such that for every $p \in M$, $v(p) = v_p$ and $w(p) = w_p$ are orthonormal non-null vectors normal to T_pM, where we identify T_pU with $\mathbb{R}^n$ using a fixed coordinate system on U. Define*

$$f\colon M \times \mathbb{R}^2 \to \mathbb{R}^n \quad by \quad f(p,t,s) = p + tv_p + sw_p.$$

Then f is smooth and there is an $\varepsilon > 0$ such that f, restricted to $M \times B_\varepsilon(0)$, is a diffeomorphism onto a neighbourhood of M.

Remark 12.4. Here $B_\varepsilon(0) = \{(t,s) \in \mathbb{R}^2 : t^2 + s^2 < \varepsilon^2\}$. The result is essentially a special case of the existence of tubular neighbourhoods, cf. Theorem 11.4, p. 93 of [8].

Proof. Let $p \in M$. Then there is a chart (V, x), where V is an open subset of U, such that $q \in M$ if and only if $x^{n-1}(q) = x^n(q) = 0$. Then $(V \cap M, y)$, where $y = (x^1, \dots, x^{n-2})$, is a chart. In these local coordinates, f can be considered to be a function $\hat{f}$ from $y(V \cap M) \times \mathbb{R}^2$ into $\mathbb{R}^n$ defined by

$$\begin{aligned}\hat{f}(z^1, \dots, z^{n-2}, t, s) = y^{-1}(z^1, \dots, z^{n-2}) &+ tv \circ y^{-1}(z^1, \dots, z^{n-2}) \\ &+ sw \circ y^{-1}(z^1, \dots, z^{n-2}).\end{aligned}$$

Since this function is smooth, we conclude that f is smooth. Computing the derivative with respect to $z^1, \dots, z^{n-2}$ at p we get the tangent space T_pM. Computing the derivative with respect to t and s at p, we get v_p and w_p respectively. By the inverse function theorem, there is a neighbourhood W of p in M and an $\varepsilon > 0$ such that f, restricted to $W \times B_\varepsilon(0)$, has a smooth inverse. By the compactness of M, there is a finite number of open neighbourhoods $W_1, \dots, W_m \subseteq M$ of points $p_1, \dots, p_m \in M$ and there are $\varepsilon_1, \dots, \varepsilon_m > 0$ such that f, restricted to $W_i \times B_{\varepsilon_i}(0)$, is a diffeomorphism onto an open subset of p_i with respect to the topology on U, and M is contained in the union of the W_i. Let $\varepsilon_a = \min\{\varepsilon_1, \dots, \varepsilon_m\}$. Then f, restricted to $W_i \times B_{\varepsilon_a}(0)$ is a diffeomorphism onto an open subset of p_i. Let us prove that for ε small enough, f, restricted to $M \times B_\varepsilon(0)$, is injective. Assume this is not the case. Then there are sequences $(p_l, t_l, s_l) \neq (q_l, \tau_l, \sigma_l)$ such that (t_l, s_l), $(\tau_l, \sigma_l) \to 0$ and $f(p_l, t_l, s_l) = f(q_l, \tau_l, \sigma_l)$. Due to the compactness of M, there is a subsequence l_k such that $p_{l_k} \to p$ for some $p \in M$. Since (t_l, s_l), $(\tau_l, \sigma_l) \to 0$, we conclude that $q_{l_k} \to p$. Consequently, there will sooner or later be a j such that both sequences are in $W_j \times B_{\varepsilon_a}(0)$. But then they have to equal by the above. We have a contradiction. Thus, for ε small enough, f is injective when restricted to $M \times B_\varepsilon(0)$. By the above observations we conclude that f, restricted to $M \times B_\varepsilon(0)$, is a diffeomorphism onto an open neighbourhood of M. □

Lemma 12.5. *Let (M, g) be an $(n+1)$-dimensional Lorentz manifold and let Σ be a smooth spacelike n-dimensional submanifold. If $p \in \Sigma$ there is a chart (U, x) with*

$p \in U$ and $x = (x^0, \dots, x^n)$ such that $q \in U \cap \Sigma$ if and only if $q \in U$ and $x^0(q) = 0$. Furthermore $\partial_{x^0}|_q$ is the future directed unit normal to Σ for $q \in \Sigma \cap U$. If we fix $\varepsilon > 0$ and let $g_{\mu\nu} = g(\partial_{x^\mu}, \partial_{x^\nu})$, then we can assume U to be such that $|g_{0i}| \leq \varepsilon$, $i = 1, \dots, n$ on U. If we let $a = g_{00}(p)$ and $b > 0$ be such that $g_{ij}(p)$, considered as a positive definite matrix, is bounded from below by b, we can assume that $g_{00} < a/2$ and that g_{ij}, considered as a positive definite matrix, is bounded from below by $b/2$.

Remark 12.6. One can of course construct the desired coordinates by using the concept of a normal neighbourhood of a submanifold, cf. the proof of Proposition 11.26, but that would be using more complicated methods than necessary.

Proof. Let (V, x) be a chart with $p \in V$ such that $q \in V \cap \Sigma$ if and only if $q \in V$ and $x^0(q) = 0$. Let $y^i = x^i|_{V \cap \Sigma}$, $i = 1, \dots, n$ and $y = (y^1, \dots, y^n)$. Define $V' = x^{-1}[x(V) \cap \mathbb{R} \times y(V \cap \Sigma)]$ and $\hat{x}\colon V' \to y(V \cap \Sigma)$ by $\hat{x} = (x^1, \dots, x^n)$ (the fact that the image of this map is in the right set follows from the definition of V'). Let $g_{\mu\nu} = g(\partial_{x^\mu}, \partial_{x^\nu})$. Then g_{ij}, $i, j = 1, \dots, n$ are the components of a positive definite matrix when restricted to $V \cap \Sigma$. Let ρ^{ij} be the components of the inverse of this matrix, which are smooth functions on $V \cap \Sigma$ and define $r^{ij} = \rho^{ij} \circ y^{-1} \circ \hat{x}$. Then r^{ij} are smooth functions on V' and on $V \cap \Sigma$, $r^{ij} g_{jl} = \delta^i_l$. Define $\alpha^i = r^{ij} g_{0j}$. Then α^i are smooth functions on V'. Define

$$z = (z^0, \dots, z^n) = (x^0, x^1 + \alpha^1 x^0, \dots, x^n + \alpha^n x^0).$$

Then z defines coordinates in a neighbourhood $W' \subseteq V'$ of p, since $\partial z^\alpha / \partial x^\beta$ are the components of an invertible matrix at a point where $x^0 = 0$. Furthermore $q \in W' \cap \Sigma$ if and only if $q \in W'$ and $z^0(q) = 0$. Let us compute, at a point $q \in W' \cap \Sigma$,

$$\left.\frac{\partial}{\partial z^i}\right|_q = \left.\frac{\partial}{\partial x^i}\right|_q, \quad i = 1, \dots, n, \qquad \left.\frac{\partial}{\partial z^0}\right|_q = \left.\frac{\partial}{\partial x^0}\right|_q - \alpha^i \left.\frac{\partial}{\partial x^i}\right|_q.$$

Thus

$$g(\partial_{z^0}|_q, \partial_{z^i}|_q) = g_{0i}(q) - (r^{lj} g_{0l} g_{ij})(q) = 0.$$

Define α_0 on $W' \cap \Sigma$ by

$$\alpha_0(q) = \pm[-\langle \partial_{z^0}|_q, \partial_{z^0}|_q \rangle]^{1/2}.$$

We choose the plus sign if ∂/∂_{z^0} is future oriented and the minus sign if it is past oriented. Note that α_0 is a smooth function on $\Sigma \cap W'$. Let $\hat{z} = (z^1|_{W' \cap \Sigma}, \dots, z^n|_{W' \cap \Sigma})$. Then $\hat{z}$ are coordinates on $\Sigma \cap W'$. Let $W = z^{-1}[z(W') \cap \mathbb{R} \times \hat{z}(W' \cap \Sigma)]$ and let $\bar{z}\colon W \to \hat{z}(W' \cap \Sigma)$ be defined by $\bar{z} = (z^1, \dots, z^n)$. Then $\alpha_1 = \alpha_0 \circ \hat{z}^{-1} \circ \bar{z}$ is a smooth function on W which coincides with α_0 on $\Sigma \cap W$. Let us define

$$w = (w^0, \dots, w^n) = [\alpha_1 z^0, z^1, \dots, z^n].$$

Then w are coordinates on W since $\alpha_1(q)$ can be computed if one knows $z^i(q)$, $i = 1, \dots, n$. Furthermore, $q \in W \cap \Sigma$ if and only if $q \in W$ and $w^0(q) = 0$. At a point $q \in \Sigma \cap W$, we get

$$\left.\frac{\partial}{\partial w^i}\right|_q = \left.\frac{\partial}{\partial z^i}\right|_q, \quad i = 1, \dots, n, \qquad \left.\frac{\partial}{\partial w^0}\right|_q = \alpha_0^{-1}(q) \left.\frac{\partial}{\partial z^0}\right|_q.$$

As a consequence $\partial/\partial_{w^0}|_q$ is the future directed unit normal to Σ for $q \in \Sigma$. All the statements of the lemma follow by restricting W suitably. □

Lemma 12.7. *Let (M, g) be an $(n+1)$-dimensional Lorentz manifold and let us assume that there is a smooth spacelike Cauchy hypersurface S. Then for every $p \in S$ there is a neighbourhood $U \ni p$ such that for every $q \in U$ which lies to the future of S, there are geodesic normal coordinates (V, ϕ) centered at q such that $J^-(q) \cap J^+(S)$ is compact and contained in V.*

Proof. Let (U, y) be coordinates around p as constructed in Lemma 12.5. We can assume, without loss of generality, that $\partial/\partial y^0$ is future pointing. Given a causal curve α in U, we can parameterize it by its y^0-component. The condition that the curve be causal and the bounds on the components of $g_{\mu\nu}$ given in the above lemma lead to the conclusion that the distance the curve can travel in the y^i directions for $i = 1, \dots, n$ is bounded by a constant times the distance it travels in the y^0 direction. By restricting the set U more and more, we see that the distance the causal curve can travel in the y^0 direction before intersecting S tends to zero. Therefore, the distance it can travel in the y^i directions, $i = 1, \dots, n$ tends to zero. Once the y^0 component of the curve is negative, we have reached the complement of $J^+(S)$ since S is a Cauchy surface (and we can never come back). Letting W be a convex neighbourhood of p, we see that for U small enough a neighbourhood contained in W, the desired properties hold. Note that the compactness follows from Lemma 40, p. 423 of [65]. □

12.2 Uniqueness of solutions to tensor wave equations

Let A be a tensorfield which is contravariant of order r and covariant of order s on a Lorentz manifold (M, g). Then we denote by $\Box_g A$ the tensor whose components are

$$(\Box_g A)^{\alpha_1 \dots \alpha_r}_{\beta_1 \dots \beta_s} = \nabla^\alpha \nabla_\alpha A^{\alpha_1 \dots \alpha_r}_{\beta_1 \dots \beta_s},$$

where ∇ is the Levi Civita connection associated with g. Alternately, one can define it to be a suitable contraction of $\nabla\nabla A$.

Lemma 12.8. *Let (M, g) be an $(n+1)$-dimensional Lorentz manifold and let us assume that there is a smooth spacelike Cauchy hypersurface S. Let p be a point to the future of S and assume that there are geodesic normal coordinates (V, ϕ) centered at p such that $J^-(p) \cap J^+(S)$ is compact and contained in V. Assume $u\colon V \to \mathbb{R}^l$ solves the equation*

$$\Box_g u + Xu + \kappa u = 0,$$

where X is an $l \times l$ matrix of smooth vector fields on V and κ is a smooth $l \times l$ matrix-valued function on V. Assume furthermore that u and grad u *vanish on $S \cap J^-(p)$. Then u and* grad u *vanish in $J^-(p) \cap J^+(S)$.*

Remark 12.9. The equation need only be satisfied in $J^-(p) \cap J^+(S)$. By time reversal, the analogous statement for p to the past of S holds.

Proof. The setup we are about to describe is taken from p. 127-128 of [65]. The exponential map exp gives a diffeomorphism of a neighbourhood $\widetilde{U}$ of the origin in T_pM to a neighbourhood U of p. On T_pM we have the function $\tilde{q}(v) = g(v, v)$ and we define $q = \tilde{q} \circ \exp_p^{-1}$ on U. Note that $\tilde{q}^{-1}(c)$ are hyperboloids for $c < 0$ and the family of hyperboloids ($c < 0$) foliate the interior of the light cones in T_pM. For $c < 0$, let us denote the component of $\tilde{q}^{-1}(c)$ corresponding to past directed timelike vectors by $\widetilde{Q}_c$. The image of the hyperboloids under exp are $q^{-1}(c)$. We shall use the notation Q_c for the component to the past of p and Q_0 for the image of the past directed null vectors under $\exp_p$. Let us denote the position vector field in T_pM by $\widetilde{P}$, i.e., this is the vector field which with $v \in T_pM$ associates the vector v (based at v). We denote the vector field transferred under the exponential map by P. As a consequence of the Gauss lemma, grad $q = 2P$, cf. Corollary 3, p. 128 of [65].

Let $D = J^-(p) \cap J^+(S)$ and $D_c = J^-(Q_c) \cap J^+(S)$. This should be compared with Figure 2.1 in which D is the triangle and its interior and D_c is the region below the hyperboloid and above the $t = 0$ hypersurface. Let us make some observations concerning these objects. First of all $Q_c \subset I^-(p)$ for $c < 0$ so that $J^-(Q_c) \subset J^-(p)$ for $c < 0$. Thus $D_c \subset D$ for $c < 0$ so that $D_c \subset V$. Let $q \in D_c$. Then there is a timelike curve from p to q in V. Due to Proposition 34, p. 147 of [65], the radial geodesic is the unique longest timelike curve from p to q in V. Since there is an $r \in Q_c$ such that $q \leq r \ll p$, we conclude that $q \in Q_{c_1}$ for some $c_1 \leq c$. Thus

$$D_c = \bigcup_{\gamma \leq c} Q_\gamma \cap J^+(S). \tag{12.1}$$

As a consequence, if $q \in Q_c \cap J^+(S)$, then $q \in \partial D_c$, the boundary of D_c. If $q \in J^-(Q_c) \cap S$, then considering a timelike curve through q and using the fact that S is a Cauchy hypersurface leads to the conclusion that $q \in \partial D_c$. Note that $I^-(Q_c) \cap I^+(S)$ is an open set contained in D_c. Since $Q_\gamma \subseteq I^-(Q_c)$ for $\gamma < c$ and since (12.1) holds, we see that

$$\begin{aligned} D_c &= Q_c \cap J^+(S) \cup I^-(Q_c) \cap J^+(S) \\ &= Q_c \cap J^+(S) \cup I^-(Q_c) \cap S \cup I^-(Q_c) \cap I^+(S). \end{aligned}$$

As a consequence, the interior of D_c is $I^-(Q_c) \cap I^+(S)$ and the boundary is $Q_c \cap J^+(S) \cup J^-(Q_c) \cap S$. Let us prove that if c_l is a sequence of negative numbers converging to zero, then

$$\operatorname{int} D \subseteq \bigcup_l D_{c_l} \subseteq D. \tag{12.2}$$

One of the inclusions is trivial. Let $q \in \operatorname{int} D$. Since int D is open, we conclude that there is a $t > 0$ such that $\gamma(t) \in D$. Thus $q \in I^-(p)$ so that $q \in Q_c$ for some $c < 0$ by arguments similar to ones given above. The conclusion follows.

Let ρ be any Riemannian metric on V and let d be the associated topological metric. Let $\varepsilon > 0$ and

$$R_\varepsilon = \{r \in S \cap D : d(r, Q_0) < \varepsilon\}, \quad d(r, Q_0) = \inf_{s \in Q_0} d(r, s).$$

Then R_ε is an open neighbourhood of $Q_0 \cap S$ in $S \cap D$. Let $L_\varepsilon = S \cap J^-(p) - R_\varepsilon$. Since $D = J^+(S) \cap J^-(p)$ is compact and S is closed, we conclude that $D \cap S = S \cap J^-(p)$ is compact. Since L_ε is a closed subset of $S \cap D$, we conclude that it is compact. Furthermore, for every $r \in L_\varepsilon$, $\exp_p^{-1} r$ is timelike. Consequently, $\exp_p^{-1} L_\varepsilon$ is a compact subset of the interior of the past lightcone in T_pM. Consequently, for $c < 0$ close enough to 0, Q_c does not intersect L_ε. The intersection of Q_c and S thus has to be in R_ε for $c < 0$ close enough to 0. Let T be a smooth unit normal to S. Then T has to be timelike. Let us prove that for $\varepsilon > 0$ small enough, P and T are linearly independent in $\bar{R}_\varepsilon$, the closure of R_ε. Since P and T are non-zero vector fields on $\bar{R}_\varepsilon$ and $\bar{R}_\varepsilon$ is compact, $\rho(T, T)$ and $\rho(P, P)$ are uniformly bounded away from 0 and ∞ on $\bar{R}_\varepsilon$. On the other hand $g(P, P) < 0$ tends to zero as ε tends to zero, but $g(T, T) < 0$ is uniformly bounded away from zero on $\bar{R}_\varepsilon$. Assume T and P to be linearly dependent at some point $r \in \bar{R}_\varepsilon$. Then there is an α_r such that $T_r = \alpha_r P_r$ so that

$$g(T_r, T_r) = \alpha_r^2 g(P_r, P_r), \quad \rho(T_r, T_r) = \alpha_r^2 \rho(P_r, P_r).$$

Due to the first equality, α_r has to tend to infinity as ε tends to zero. This contradicts the second inequality and our uniform bounds. Consequently, for $c < 0$ close enough to zero, every point in $Q_c \cap S$ is such that the normal to Q_c and S at that point are linearly independent. Since Q_c and S are smooth spacelike n-dimensional submanifolds, we conclude that the intersection is a smooth $n-1$ dimensional submanifold. To prove the compactness of $S \cap Q_c$, let $r_i \in S \cap Q_c$. Then $r_i \in D$, so that there is a subsequence converging to some point r. Since S is closed, $r \in S$ and since Q_c is the level set of a function, $r \in Q_c$.

Note that $D_c - S \cap Q_c$ can be considered to be a Lorentz manifold with boundary, the boundary being $Q_c \cap I^+(S) \cup S \cap I^-(Q_c)$. Let u be the solution assumed to exist in the statement of the lemma. Define

$$Q_{\alpha\beta} = \nabla_\alpha u \cdot \nabla_\beta u - \frac{1}{2} g_{\alpha\beta}(g^{\mu\nu} \nabla_\mu u \cdot \nabla_\nu u), \quad f = -\frac{1}{2} q, \quad N = -P.$$

Note that $\operatorname{grad} f = N$. Furthermore, we define

$$\xi_1^\alpha = g^{\alpha\gamma} Q_{\gamma\beta} N^\beta, \quad \eta = -e^{kf} |u|^2 N, \quad \xi = e^{kf} \xi_1,$$

where k is a constant to be determined. Note that in D_c, N is a future directed timelike vector field and that on $Q_c \cap I^+(S)$ it is the outward pointing normal relative to D_c. Compute

$$\nabla^\alpha Q_{\alpha\beta} = \Box_g u \cdot \nabla_\beta u.$$

Thus

$$\operatorname{div} \xi_1 = \Box_g u \cdot N(u) + Q_{\alpha\beta} \nabla^\alpha N^\beta.$$

Let us introduce the quantity

$$\mathcal{E} = \frac{1}{2}|u|^2 + \sum_{\alpha=0}^{n} |\partial_\alpha u|^2,$$

where the ∂_α are associated with the normal coordinates in V. Then

$$|\operatorname{div} \xi_1| \leq C\mathcal{E}$$

on D_c due to the equation. Compute

$$\begin{aligned}\operatorname{div} \xi &= e^{kf} \operatorname{div} \xi_1 + \xi_1(e^{kf}) \\ &= e^{kf} (\operatorname{div} \xi_1 + k\langle \operatorname{grad} f, \xi_1\rangle) \\ &= e^{kf} [\operatorname{div} \xi_1 + kQ(N, N)].\end{aligned}$$

We also have

$$\operatorname{div} \eta = -2e^{kf} u \cdot N(u) - e^{kf} |u|^2 \operatorname{div} N - ke^{kf} |u|^2 \langle N, N\rangle.$$

Due to the fact that N is timelike in all of D_c, we conclude that there is a constant $c_0 > 0$ and a constant C such that on D_c,

$$\operatorname{div} \eta \geq e^{kf} (kc_0|u|^2 - C\mathcal{E}).$$

For the same reason there is a constant $c_1 > 0$ such that on D_c,

$$c_0|u|^2 + Q(N, N) \geq c_1\mathcal{E}.$$

We conclude that

$$\operatorname{div} \eta + \operatorname{div} \xi \geq e^{kf} (kc_1 - C)\mathcal{E}.$$

For k large enough, it is clear that this object is positive and dominates $e^{kf}\mathcal{E}$. Let us compute

$$\frac{\langle N, \xi\rangle}{\langle N, N\rangle} = \frac{e^{kf} Q(N, N)}{\langle N, N\rangle}, \qquad \frac{\langle N, \eta\rangle}{\langle N, N\rangle} = -e^{kf} |u|^2.$$

Note that both these quantities are non-positive. If it were possible to apply (10.3), we would be done. In fact, applying (10.3) to $\xi + \eta$ with k large enough, the left-hand side is non-negative and the right-hand side is non-positive. Consequently, both have to equal zero, so that $u = 0$ in D_c. By (12.2), we conclude that $u = 0$ and $\operatorname{grad} u = 0$ in the interior of D, so that they equal zero in D by the smoothness of u.

The problem is of course that we are not allowed to apply (10.3). Let us apply Lemma 12.3. In order to do so, we need to construct two smooth vector fields normal to $S \cap Q_c$. Let T be a unit normal vector field to S. Then we can construct an orthonormal vector field by normalizing

$$P - \frac{\langle P, T\rangle}{\langle T, T\rangle} T.$$

These two vector fields can then be used as the vector fields assumed to exist in the statement of Lemma 12.3. As a conclusion, we get a smooth map

$$h\colon Q_c \cap S \times B_\varepsilon(0) \to V,$$

which is a diffeomorphism onto its image and contains an open neighbourhood of $Q_c \cap S$, assuming that $\varepsilon > 0$ is small enough. Let $\chi \in C_0^\infty(\mathbb{R}^2)$ be such that $\chi(x) = 1$ for $|x| \leq 1/2$ and $\chi(x) = 0$ for $|x| \geq 3/4$, and let $\chi_\delta(x) = \chi(x/\delta)$. For $\delta \leq \varepsilon$ we can consider the function

$$\psi_\delta : Q_c \cap S \times B_\varepsilon(0) \to \mathbb{R}, \quad \psi_\delta(p, x) = \chi_\delta(x),$$

to be a smooth function on V. Due to the existence of the diffeomorphism, the volume of the support of ψ_δ can be estimated by $C\delta^2$, where C is a constant independent of δ. For the same reason, we can estimate $|\partial_\alpha \psi_\delta| \leq C\delta^{-1}$, where ∂_α are the derivatives with respect to the normal coordinates. Let X be an arbitrary smooth vector field on V. Note that $D_c' = D_c - S \cap Q_c$ is a smooth manifold with boundary and that $(1-\psi_\delta)X$ can be considered as a smooth vector field with compact support on this manifold. Thus (10.3) is applicable to $(1 - \psi_\delta)X$. We have

$$\int_{D_c'} \operatorname{div}[(1-\psi_\delta)X]\,\mu_g = -\int_{D_c'} X^\alpha \partial_\alpha \psi_\delta\,\mu_g + \int_{D_c'} \operatorname{div} X\,\mu_g - \int_{D_c'} \psi_\delta \operatorname{div} X\,\mu_g,$$

where $\mu_g = \varepsilon_{D_c'}$. In the first term of the right-hand side, $|\partial_\alpha \psi_\delta| \leq C\delta^{-1}$, $|X^\alpha| \leq C$ and the volume of the set on which the integrand is non-zero is bounded by $C\delta^2$. Consequently, the first term converges to zero as $\delta \to 0+$. The third term also converges to zero. By Lebesgue's dominated convergence theorem, the boundary integral converges to what it should. Consequently, (10.3) holds on D_c, and the desired conclusion follows. □

Let us introduce the terminology that $A \in \mathcal{T}_s^r(M)$ if and only if A is a smooth tensorfield on M, contravariant of order r and covariant of order s.

Lemma 12.10. *Let (M, g) be an $(n+1)$-dimensional Lorentz manifold and let us assume that there is a smooth spacelike Cauchy hypersurface S. Let p be a point to the future of S and assume that there are geodesic normal coordinates (V, ϕ) centered at p such that $J^-(p) \cap J^+(S)$ is compact and contained in V. Assume $A \in \mathcal{T}_s^r(M)$, $B \in \mathcal{T}_{r+s}^{r+s+1}(M)$ and $C \in \mathcal{T}_{r+s}^{r+s}(M)$ satisfy the equation*

$$(\Box_g A)^{\alpha_1\ldots\alpha_r}_{\beta_1\ldots\beta_s} + B^{\alpha_1\ldots\alpha_r\gamma_1\ldots\gamma_{s+1}}_{\beta_1\ldots\beta_s\delta_1\ldots\delta_r} \nabla_{\gamma_1} A^{\delta_1\ldots\delta_r}_{\gamma_2\ldots\gamma_{s+1}} + C^{\alpha_1\ldots\alpha_r\gamma_1\ldots\gamma_s}_{\beta_1\ldots\beta_s\delta_1\ldots\delta_r} A^{\delta_1\ldots\delta_r}_{\gamma_1\ldots\gamma_s} = 0.$$

Assume furthermore that A and ∇A vanish on $S \cap J^-(p)$. Then A and ∇A vanish in $J^-(p) \cap J^+(S)$.

Remark 12.11. When we say that A vanishes at a point p, we mean that for any one-forms at p, say $\theta_1, \ldots, \theta_r$ and vectors at p, say $v_1, \ldots, v_s$, we have

$$A(\theta_1, \ldots, \theta_r, v_1, \ldots, v_s) = 0.$$

When we say that A vanishes on $S \cap J^-(p)$, we mean that A vanishes at q for every $q \in S \cap J^-(p)$. The conventions for ∇A are similar. It should reasonably be possible to prove similar uniqueness statements for equations where the unknown takes values in a vector bundle. However, we shall limit our attention to the above case here.

Proof. Writing out the equation on V in local coordinates, we see that it can be viewed as an equation of the same sort as the one dealt with in Lemma 12.8. □

Corollary 12.12. *Let (M, g) be a connected, oriented, time oriented, globally hyperbolic $(n+1)$-dimensional Lorentz manifold and let S be a smooth spacelike Cauchy hypersurface. Let $\Omega \subseteq S$. Assume $A \in \mathcal{T}_s^r(M)$, $B \in \mathcal{T}_{r+s}^{r+s+1}(M)$ and $C \in \mathcal{T}_{r+s}^{r+s}(M)$ satisfy the equation*

$$(\Box_g A)^{\alpha_1\ldots\alpha_r}_{\beta_1\ldots\beta_s} + B^{\alpha_1\ldots\alpha_r\gamma_1\ldots\gamma_{s+1}}_{\beta_1\ldots\beta_s\delta_1\ldots\delta_r}\nabla_{\gamma_1}A^{\delta_1\ldots\delta_r}_{\gamma_2\ldots\gamma_{s+1}} + C^{\alpha_1\ldots\alpha_r\gamma_1\ldots\gamma_s}_{\beta_1\ldots\beta_s\delta_1\ldots\delta_r}A^{\delta_1\ldots\delta_r}_{\gamma_1\ldots\gamma_s} = 0.$$

Assume furthermore that A and ∇A vanish on Ω. Then A and ∇A vanish in $D^+(\Omega)$.

Remark 12.13. The equation need only be satisfied in $D^+(\Omega)$. There is a similar statement concerning $D^-(\Omega)$.

Proof. Let t be a temporal function such that all its level sets are spacelike Cauchy hypersurfaces and $t^{-1}(0) = S$, cf. Theorem 11.27. Let us first prove that

$$D^+(\Omega) - \Omega \subseteq \overline{\text{int}[D^+(\Omega)]}. \tag{12.3}$$

Note that $O = I^-[D^+(\Omega)] \cap I^+(\Omega)$ is an open set which is contained in $D^+(\Omega)$. The reason for the latter statement is that if $p \in O$ and γ is a past inextendible causal curve through p, then there is a past inextendible causal curve from some point $q \in D^+(\Omega)$ that coincides with γ on some subinterval. Since $p \in I^+(\Omega)$, we can use the definition of $D^+(\Omega)$ and the fact that S is a Cauchy surface to conclude that γ has to intersect Ω. Thus $p \in D^+(\Omega)$. We conclude that $O \subseteq \text{int}[D^+(\Omega)]$. On the other hand, if $p \in D^+(\Omega) - \Omega$, we have $p \in I^+(\Omega)$ since an inextendible timelike geodesic through p has to intersect Ω, and the intersection point is certainly not p. Furthermore, there is a sequence $p_i \in I^-[D^+(\Omega)]$ such that p_i converge to p. Since $I^+(\Omega)$ is open, the points p_i have to belong to $I^+(\Omega)$ for i large enough. For i large enough, we conclude that $p_i \in O$. Thus

$$D^+(\Omega) - \Omega \subseteq \bar{O} \subseteq \overline{\text{int}[D^+(\Omega)]}.$$

We wish to prove that A and ∇A vanish on $D^+(\Omega)$. Due to the assumptions, (12.3) and the smoothness of A we conclude that it is enough to prove that A and ∇A vanish in the interior of $D^+(\Omega)$.

Let p be in the interior of $D^+(\Omega)$. Due to Lemma 40, p. 423 of [65], $K = J^-(p) \cap D^+(\Omega)$ is compact. Given an interval I and a real number t_0, let us introduce the notation

$$R_I = t^{-1}(I) \cap K, \quad S_{t_0} = t^{-1}(t_0).$$

Note that if I is a closed interval, R_I is compact. We shall use the notation R_{t_0} if the interval I coincides with the point t_0. Let t_0 be a real number, $\varepsilon > 0$, $I_\varepsilon = [t_0 - \varepsilon, t_0 + \varepsilon]$, assume that R_{t_0} is non-empty and that there is an open set U containing it. Then, we claim, there is an $\varepsilon > 0$ such that $R_{I_\varepsilon} \subseteq U$. Assume not. Then there are $r_i \in R_{I_{\varepsilon_i}}$

with $\varepsilon_i \to 0$ such that $r_i \notin U$. As a consequence $r_i \in R_{I_\varepsilon} \cap U^c$ for some fixed $\varepsilon > 0$. Since this set is compact, we conclude that there is a subsequence r_{i_k} converging to some $r \in R_{I_\varepsilon} \cap U^c$. By construction, we have to have $t(r) = t_0$ and $r \in K$ so that $r \in U$. Since $r \in U^c$, we have a contradiction.

Define T by $t(p) = T > 0$. Let $0 \le t_0 < T$. For every $q \in R_{t_0}$ there is an open neighbourhood U_q with the properties stated in Lemma 12.7. In this lemma, we take the Cauchy hypersurface to be S_{t_0}. Since R_{t_0} is compact, there is a finite subcovering, $U_{q_1}, \dots, U_{q_l}$. Let us denote the union by U. By the above observations there is an $\varepsilon > 0$ such that if $I_\varepsilon = [t_0, t_0 + \varepsilon]$, then $R_{I_\varepsilon} \subseteq U$. Assume that A and ∇A vanish on R_{t_0}. Let $q \in R_s$ for some $s \in [t_0, t_0 + \varepsilon]$. Then there are geodesic normal coordinates at q such that the conditions of Lemma 12.10 hold. We conclude that A and ∇A equal zero in R_s. Consider the set of $s \in [0, T)$ such that A and ∇A equal zero in R_τ for $\tau \in [0, s]$. By the assumptions this set is non-empty, by continuity it is closed and by the above argument it is open. We conclude that A and ∇A equal zero at p. The conclusion follows. □

Corollary 12.14. *Let (M, g) be a connected, oriented, time oriented, globally hyperbolic $(n+1)$-dimensional Lorentz manifold and let S be a smooth spacelike Cauchy hypersurface. Let $\Omega \subseteq S$ and let U be an open set containing $\overline{D^+(\Omega)}$. Assume $u\colon U \to \mathbb{R}^l$ solves the equation*

$$\Box_g u + Xu + \kappa u = 0,$$

where X is an $l \times l$ matrix of smooth vector fields on U and κ is a smooth $l \times l$ matrix-valued function on U. Assume furthermore that u and grad u *vanish on Ω. Then u and* grad u *vanish on $D^+(\Omega)$.*

Remark 12.15. The equation need only be satisfied in $D^+(\Omega)$. There is a similar statement concerning $D^-(\Omega)$.

Proof. The statement does, formally speaking, not follow from Corollary 12.12, but the proof is identical. □

12.3 Existence

So far we have only discussed the question of uniqueness. Let us address the question of existence. Before doing so, we need to carry out some preliminary constructions.

Lemma 12.16. *Let (M, g) be a connected, oriented, time oriented, globally hyperbolic $(n+1)$-dimensional Lorentz manifold and let S be a smooth spacelike Cauchy hypersurface. Let t be a temporal function constructed in Theorem* 11.27 *so that $t^{-1}(0) = S$ and let us use the notation $S_{t_0} = t^{-1}(t_0)$ for any real number t_0. If $p \in S_{t_0}$ there is an $\varepsilon > 0$ and open neighbourhoods U, W of p such that*

- *the closure of W is compact and contained in U,*

- *if $q \in W$ and $\tau \in [t_0-\varepsilon, t_0+\varepsilon]$, then $J^+(S_\tau)\cap J^-(q)$ is compact and contained in U;*

- *there are coordinates (U,ϕ) with $\phi = (x^0,\dots,x^n)$, where $x^0 = t$, such that if $g_{\mu\nu} = \langle \partial_\mu, \partial_\nu\rangle$ for $\mu,\nu = 0,\dots,n$, then there are positive constants a, b such that $g_{00} < -a$ and the matrix $g_\flat$ with components g_{ij} for $i,j = 1,\dots,n$ is positive definite with $g_\flat \geq b$, both inequalities holding on U;*

- *for any compact $K \subseteq U$, there is a smooth Lorentz matrix-valued function h on $\mathbb{R}^{n+1}$ such that $h_{\mu\nu} = g_{\mu\nu}\circ\phi^{-1}$ on $\phi(K)$ and such that there are positive constants a_1, b_1, c_1 with $h_{00} \leq -a_1$, $h_\flat \geq b_1$ and $|h_{\mu\nu}| \leq c_1$ on all of $\mathbb{R}^{n+1}$.*

Proof. Since $\operatorname{grad} t$ is timelike, we can construct coordinates (U,ϕ) at p with $\phi = (x^0,\dots,x^n)$ where $x^0 = t$, cf. Lemma 35, p. 20 of [65]. By modifying the coordinates in the same way as was done in the beginning of the proof of Lemma 12.5, we can assume that $g_{00} < 0$ uniformly in U, that g_{0i} is arbitrarily small and that g_{ij} are the components of a positive definite matrix. Since $q \in S_\tau \cap U$ if and only if $q \in U$ and $x^0(q) = \tau$, we obtain the first three properties stated in the lemma by an argument similar to the one given in the proof of Lemma 12.7, assuming that ε and W are small enough. In order to construct the function $h_{\mu\nu}$, let $\chi \in C_0^\infty[\phi(U)]$ be such that $\chi(r) = 1$ for all $r \in \phi(K)$. Define

$$h_{\mu\nu} = \chi g_{\mu\nu}\circ\phi^{-1} + (1-\chi)g_{\mu\nu}(p).$$

It is clear that $h_{00} < 0$ and that $h_\flat$ is positive definite. Thus $h_{\mu\nu}$ is a smooth Lorentz matrix-valued function due to Lemma 8.3. Furthermore, it is clear that the desired properties hold. □

Theorem 12.17. *Let (M,g) be a connected, oriented, time oriented, globally hyperbolic $(n+1)$-dimensional Lorentz manifold and let S be a smooth spacelike Cauchy hypersurface. Let t be the temporal function constructed in Theorem 11.27 so that $t^{-1}(0) = S$ and let us use the notation $S_{t_0} = t^{-1}(t_0)$ for any real number t_0. Assume $B \in \mathcal{T}^{r+s+1}_{r+s}(M)$, $C \in \mathcal{T}^{r+s}_{r+s}(M)$ and $E \in \mathcal{T}^r_s(M)$. Given smooth tensorfields $A_0, A_1 \in \mathcal{T}^r_s(M)$, there is a smooth tensorfield $A \in \mathcal{T}^r_s(M)$ solving the initial value problem*

$$\begin{aligned}(\Box_g A)^{\alpha_1\dots\alpha_r}_{\beta_1\dots\beta_s} + B^{\alpha_1\dots\alpha_r\gamma_1\dots\gamma_{s+1}}_{\beta_1\dots\beta_s\delta_1\dots\delta_r}\nabla_{\gamma_1}A^{\delta_1\dots\delta_r}_{\gamma_2\dots\gamma_{s+1}}&\\ + C^{\alpha_1\dots\alpha_r\gamma_1\dots\gamma_s}_{\beta_1\dots\beta_s\delta_1\dots\delta_r}A^{\delta_1\dots\delta_r}_{\gamma_1\dots\gamma_s} &= E^{\alpha_1\dots\alpha_r}_{\beta_1\dots\beta_s},\\ A(p) &= A_0(p),\\ \nabla_N A(p) &= A_1(p)\end{aligned} \tag{12.4}$$

for all $p \in S$, where N is the future directed unit normal to the hypersurfaces S_τ, i.e., $N = -\operatorname{grad} t/|\operatorname{grad} t|$.

Remark 12.18. It would perhaps have seemed more natural to require $A_0(p)$ and $A_1(p)$ to be tensors of the correct type for every $p \in S$ and furthermore A_0 and A_1, evaluated on the correct number of smooth vector fields and one-forms (on M restricted to S), to be smooth maps from S to the real numbers. However, the two definitions are equivalent.

Proof. Let us construct the solution to the future of S. Let us start by assuming that the initial data have support in some compact set $K_1 \subseteq S$ and that the support of E is contained in some compact set $K_2 \subseteq M$. Let $t_1 > 0$ and define R_{t_1} to be the set of q such that $0 \leq t(q) \leq t_1$. Note that this set is closed. Furthermore $K_3 = K_2 \cap R_{t_1}$ is compact. The union of $I^+(p)$ for $p \in I^-(S)$ is an open covering of $K_1 \cup K_3$. Since this set is compact, there is a finite number of points $p_1, \dots, p_l \in I^-(S)$ such that $K_1 \cup K_3 \subseteq \bigcup_{i=1}^{l} I^+(p_l)$. Note that the set $F = \bigcup_{i=1}^{l} J^+(p_i) \cap J^-(S_{t_1})$ is compact, due to Lemma 40, p. 423 of [65], and that if there is a solution in R_{t_1}, it has to equal zero in $R_{t_1} - F$ by uniqueness. We are left with the problem of defining A in the compact set $R_{t_1} \cap F$. Let us use the notation $F_\tau = F \cap S_\tau$. It is of interest to note that if U is an open set containing F_τ, then there is an $\varepsilon > 0$ such that $F_s \subseteq U$ for all $s \in [\tau - \varepsilon, \tau + \varepsilon]$. The argument is similar to an argument given in the proof of Corollary 12.12.

Let $0 \leq \tau < t_1$ and assume we either have a solution on R_τ or on R_s for every $0 \leq s < \tau$. We wish to construct a solution beyond τ. For every $p \in F_\tau$ there are neighbourhoods U_p, W_p and an ε_p with the properties stated in Lemma 12.16. By compactness, there is a finite number of points $p_1, \dots, p_l$ such that $W_{p_1}, \dots, W_{p_l}$ cover F_τ. Let $0 < \varepsilon \leq \min\{\varepsilon_{p_1}, \dots, \varepsilon_{p_l}\}$ be such that

$$F_s \subset \bigcup_{i=1}^{l} W_{p_i}$$

for all $s \in [\tau - \varepsilon, \tau + \varepsilon]$ and let s_1 be a point in the interval $[\tau - \varepsilon, \tau]$ such that there is a solution up to and including s_1. Let $p \in F_s$ for any $s \in [s_1, \tau + \varepsilon]$. Then $K_p = J^-(p) \cap J^+(S_{s_1})$ is compact and contained in one of the charts, say (U_{p_i}, ϕ). Let $\chi \in C_0^\infty(U_{p_i})$ be such that $\chi(q) = 1$ for all $q \in K_p$. Define $B' = \chi B$, $C' = \chi C$, $E' = \chi E$ and as initial data for A at S_{s_1}, we use $A_0' = (\chi A)|_{S_{s_1}}$ and $A_1' = (\chi \nabla_N A)|_{S_{s_1}}$. Note furthermore that we can extend the Lorentz metric $g \circ \phi^{-1}$ to a smooth Lorentz matrix-valued function $h_{\mu\nu}$ on $\mathbb{R}^{n+1}$ coinciding with $g_{\mu\nu} \circ \phi^{-1}$ on $\phi(K_p)$. If we replace all the objects occurring in (12.4) with their primed counterparts and replace the metric by $h_{\mu\nu}$, then we can consider this equation as an equation on $\mathbb{R}^{n+1}$. If we write it out in components, we see that it is of the same type as the one dealt with in Theorem 8.6. We conclude that there is a smooth global solution. Transferring this solution back to U_{p_i}, we see that we have a smooth solution to the original problem on K_p. For any point $q \in I^-(p) \cap J^+(S_{s_1}) = V_p$, we define A to be this solution. If $r \in V_p \cap V_q$, then we have two candidate definitions for A at r. Since $J^-(r) \cap J^+(S_{s_1})$ is contained in $V_p \cap V_q$, Corollary 12.12 ensures that the two possible definitions coincide. If we take the union of the V_p for $p \in F_s$, $s \in [s_1, \tau + \varepsilon]$, let us call it O_1, we get a set which contains F_s in its interior for any $s \in (s_1, \tau + \varepsilon)$. Let us define O_2 to be the set of q such that $s_1 \leq t(q) < \tau + \varepsilon$ for which $q \notin F$. Note that if $q \in O_2$ and $t(q) > s_1$ then there is an open neighbourhood of q contained in O_2 and similarly for O_1. In O_1 we have already defined A and ∇A and in O_2 we would like to define these objects to be zero. We need to check that these two definitions are consistent. If $q \in O_1 \cap O_2$ and $t(q) > s_1$, then A and $\nabla_N A$ vanish at $J^-(q) \cap S_{s_1}$ by the above observation and E vanishes in $J^-(q) \cap J^+(S_{s_1})$. Furthermore, there is an r such that $q \in V_r$ with $V_r \subseteq O_1$. By uniqueness, the solution defined on O_1 has to vanish at q.

To conclude, given a solution on R_τ or on R_s for any $s < \tau$, we get a solution on $R_{\tau+\varepsilon}$ for some $\varepsilon > 0$. Let $\mathcal{A}$ be the set of $s \in [0, \infty)$ such that there is a solution on R_s. By the above argument, this set is open, closed and non-empty. Consequently, we get a solution for all future times. By time reversal, we get a solution on all of M.

What remains is to remove the condition that E and the initial data have compact support. Let p be any point to the future of S. Then $K_p = J^-(p) \cap J^+(S)$ is a compact set. Let $\chi \in C_0^\infty(M)$ be such that $\chi(q) = 1$ for all $q \in K_p$. Define $E' = \chi E$, $A_0' = \chi A_0$ and $A_1' = \chi A_1$. Solve the equation (12.4) with E, A_0 and A_1 replaced by E', A_0' and A_1' respectively. This leads to a smooth tensorfield A'. Define A to equal A' in $I^-(p) \cap J^+(S) = V_p$. If $r \in V_p \cap V_q$, then for the same reasons as before, the two candidate solutions have to coincide in r. The theorem follows. □

Since the statement below is not, strictly speaking, a consequence of the above theorem, we write it down. The proof is however identical.

Theorem 12.19. *Let (M, g) be a connected, oriented, time oriented, globally hyperbolic $(n+1)$-dimensional Lorentz manifold and let S be a smooth spacelike Cauchy hypersurface. Let t be the temporal function constructed in Theorem 11.27 so that $t^{-1}(0) = S$ and let us use the notation $S_{t_0} = t^{-1}(t_0)$ for any real number t_0. Assume X is an $k \times k$ matrix of smooth vector fields on M, that κ is a smooth $k \times k$ matrix-valued function on M, that f is a smooth $\mathbb{R}^k$-valued function on M and that u_0, u_1 are smooth $\mathbb{R}^k$-valued functions on S. Then there is a unique smooth $u\colon M \to \mathbb{R}^k$ solving the initial value problem*

$$\begin{aligned} \Box_g u + Xu + \kappa u &= f, \\ u(p) &= u_0(p), \\ (Nu)(p) &= u_1(p) \end{aligned}$$

for all $p \in S$, where N is the future directed unit normal to the hypersurfaces S_τ, i.e., $N = -\operatorname{grad} t / |\operatorname{grad} t|$.

Part III

General relativity

13 The constraint equations

13.1 Introduction, equations

We are interested in Einstein's equations of general relativity, given by

$$G = T, \tag{13.1}$$

where G is the *Einstein tensor* of an $(n+1)$-dimensional Lorentz manifold (M, g) and T is the *stress energy tensor* of the matter. The Einstein tensor is defined by

$$G = \mathrm{Ric} - \frac{1}{2} S g,$$

where Ric is the Ricci tensor and S is the scalar curvature of (M, g). There are many possibilities for T, but we shall here only consider the case

$$T = d\phi \otimes d\phi - \left[\frac{1}{2} \langle \mathrm{grad}\, \phi, \mathrm{grad}\, \phi \rangle + V(\phi) \right] g, \tag{13.2}$$

where $\langle \cdot\, , \cdot \rangle := g$. We shall refer to this matter model as a *non-linear scalar field*, where the *potential* V is a given smooth function from $\mathbb{R}$ to itself and ϕ is a smooth function on M, referred to as the *scalar field.* Note that (13.1), in this case, is equivalent to

$$\mathrm{Ric} = d\phi \otimes d\phi + \frac{2}{n-1} V(\phi)\, g.$$

This equation should of course be coupled to a matter equation for ϕ, which is given by

$$\Box_g \phi - V'(\phi) = 0,$$

where $\Box_g \phi = \nabla^\alpha \nabla_\alpha \phi$. Note that this equation implies that the stress energy tensor is divergence free. Consequently, it ensures the compatibility of the choice of matter model with the Einstein equations. The system of equations of interest is thus

$$\mathrm{Ric} - d\phi \otimes d\phi - \frac{2}{n-1} V(\phi)\, g = 0, \tag{13.3}$$

$$\Box_g \phi - V'(\phi) = 0. \tag{13.4}$$

We shall refer to the system consisting of (13.3) and (13.4) as the *Einstein non-linear scalar field system.*

We would like to formulate an initial value problem for these equations. The question is then what the initial data should be. Let us discuss this issue informally in order to develop some intuition. Let us assume we have a solution (M, g, ϕ). We wish to view this solution as the development of initial data on some hypersurface Σ. If there are causal curves in Σ, the physical interpretation indicates that what happens at one point of Σ might affect what happens at another point of Σ; consider the ordinary wave equation in Minkowski space. To avoid this situation, we restrict our attention to spacelike hypersurfaces. As was mentioned in the outline, (13.3) and (13.4) constitute a system of wave equations when viewed in appropriate coordinates. Naively, one then expects it to be necessary to specify the metric, the scalar field and their first normal derivatives as initial data. However, the equations are diffeomorphism invariant: if we have a diffeomorphism $\psi : M \to M$, then $(M, \psi^* g, \phi \circ \psi)$ is also a solution; in fact, it should be considered as the same solution. Consequently, it seems natural to insist that the initial data be geometrically defined. The simplest piece of geometric information induced on Σ is of course the induced metric. Since we assume Σ to be spacelike, the induced metric is a Riemannian metric. Note, however, that if we view (13.3) and (13.4) as a system of wave equations (after an appropriate choice of coordinates), it would seem natural to specify all the metric components on the initial hypersurface, not just the components tangent to the initial hypersurface, which is the information the induced metric amounts to. In other words, the induced metric would naively seem to be less information than required. Another geometric quantity induced on Σ is the second fundamental form. Written down in local coordinates, this yields information concerning the normal derivative of some components of the metric, though, as with the case of the metric itself, not all the components one would naively wish to have. From a naive point of view, it would thus seem that, at the very minimum, the initial data concerning the metric would have to include the induced metric and the second fundamental form. The situation concerning the scalar field is more straightforward; since the equation for the scalar field is a wave equation, the natural quantities to specify are the induced function and normal derivative of the function on Σ. We refer the reader interested in a further discussion of these issues to [87] and [41].

To conclude, a suggestion would be to demand that the initial data consist of a smooth n-manifold Σ with a Riemannian metric g_0, a symmetric covariant 2-tensor k_0 and two functions ϕ_0 and ϕ_1, all assumed to be smooth. The problem would be to construct an $(n+1)$-manifold M with a Lorentz metric g, a smooth function ϕ, satisfying (13.3)–(13.4), and an embedding $i : \Sigma \to M$ such that if k is the second fundamental form of $i(\Sigma)$ in M and N is the future directed unit normal to $i(\Sigma)$ in M, then $i^*g = g_0$, $i^*k = k_0$, $i^*\phi = \phi_0$ and $i^*(N\phi) = \phi_1$.

Note that the preceding discussion was purely informal and does not justify calling the above mentioned objects initial data for the Einstein non-linear scalar field system. The only proper justification for calling the above objects initial data is that one can prove that given these objects, there is a corresponding development; in fact, a unique maximal globally hyperbolic development (a concept we shall define in Chapter 16).

The initial value problem as stated above cannot have a solution in all generality. The reason is that the Riemannian metric, the second fundamental form and the functions ϕ and $N\phi$ induced on $i(\Sigma)$ by g and ϕ are related by certain equations, assuming that (13.1) is satisfied. If the initial data do not satisfy the corresponding equations, it is clear that there will not be a development. In this chapter, we shall derive the corresponding *constraint equations*.

13.2 The constraint equations

We shall use the Gauß and Codazzi equations, as formulated in Theorem 5, p. 100, and Proposition 33, p. 115, of [65] respectively, in order to derive the constraint equations. Consequently, we shall use the notation of O'Neill in the current section. However, in the end we wish to formulate the equations as they are usually formulated in general relativity, and this requires a translation. To begin with, let us define the second fundamental form.

Definition 13.1. Let $(\bar{M}, \bar{g})$ be a time oriented Lorentz manifold, let M be a spacelike hypersurface in $(\bar{M}, \bar{g})$, let $i : M \to \bar{M}$ be the embedding and let N be a future directed unit timelike vector field such that for every $p \in M$, $\bar{g}(N_p, i_* v) = 0$ for every $v \in T_p M$. Then the second fundamental form of M is the covariant 2-tensor field k on M defined by

$$k(v, w) = \bar{g}(\bar{D}_{i_* v} N, i_* w),$$

for $v, w \in T_p M$, where $\bar{D}$ is the Levi-Civita connection of $(\bar{M}, \bar{g})$.

Remark 13.2. The second fundamental does not depend on the choice of N. In what follows, we shall not make a distinction between v and $i_* v$ for $v \in T_p M$.

Proposition 13.3. *Let $(\bar{M}, \bar{g})$ be a time oriented Lorentz manifold, let M be a spacelike hypersurface with induced metric g, let N be as in Definition* 13.1 *and let k be the second fundamental form of M. Then*

$$\bar{G}(N_p, N_p) = \frac{1}{2}[S - k_{ij}k^{ij} + (\mathrm{tr}_g\, k)^2](p), \tag{13.5}$$

$$\bar{G}(N_p, v) = [D^j k_{ji} - D_i(\mathrm{tr}_g\, k)]v^i, \tag{13.6}$$

where $\bar{G}$ is the Einstein tensor of $(\bar{M}, \bar{g})$, $p \in M$, $v \in T_p M$, D is the Levi-Civita connection and S the scalar curvature of (M, g).

Proof. As the proof will be based on results from [65], let us use the corresponding notation. First of all, we let $\langle \cdot , \cdot \rangle = \bar{g}$. A smooth map $X : M \to T\bar{M}$ such that $\pi \circ X(p) = p$, where $\pi : T\bar{M} \to \bar{M}$ is the standard projection, will be referred to as an $\bar{M}$ *vector field on* M. The restriction of a vector field on $\bar{M}$ to M is clearly an $\bar{M}$ vector field on M, and it is possible to divide an $\bar{M}$ vector field on M, say X, into its normal and tangential components, which we shall denote nor X and tan X respectively. We shall not distinguish between vector fields on M and $\bar{M}$ vector fields on M that are tangent to M. The *shape tensor* is defined by

$$II(V, W) = \mathrm{nor}\, \bar{D}_V W \tag{13.7}$$

for vector fields V, W on M, cf. p. 100 of [65]. Here $\bar{D}_V W$ should be interpreted as follows: first one extends V and W to become vector fields on $\bar{M}$, then one computes the covariant derivative and finally one restricts the result to M. That this operation is independent of the choice of extensions is a consequence of Lemma 1, p. 99 of [65].

Due to Lemma 4, p. 100 of [65], II is bilinear over the smooth functions on M and symmetric. Let $e_0 = N$ and let V and W be vector fields on M. Extending W to a vector field $\overline{W}$ which is always orthogonal to e_0, we have, at a point $p \in M$,

$$\langle II(V,W), e_0\rangle = \langle \bar{D}_V W, e_0\rangle = \bar{D}_V\langle \overline{W}, e_0\rangle - \langle \bar{D}_V e_0, W\rangle = -k(V,W). \tag{13.8}$$

Thus

$$II(V,W) = k(V,W)e_0.$$

We wish to compute $\bar{G}(e_0, e_0)$, where $\bar{G}$ is the Einstein tensor of $(\bar{M}, \bar{g})$. Let us, in a neigbourhood of a point of M, assume e_i, $i = 1, \dots, n$, is an orthonormal basis for the tangent space of M. Let $\overline{\mathrm{Ric}}$ denote the Ricci curvature of $(\bar{M}, \bar{g})$. Then one can compute that

$$\bar{G}(e_0, e_0) = \frac{1}{2}\sum_{\mu=0}^{n} \overline{\mathrm{Ric}}(e_\mu, e_\mu).$$

However, recalling Lemma 52, p. 87 of [65] and the difference in conventions concerning the Riemann curvature tensor, one can compute that

$$\sum_{\mu=0}^{n} \overline{\mathrm{Ric}}(e_\mu, e_\mu) = \sum_{i,j=1}^{n} \langle \bar{R}_{e_i e_j} e_j, e_i\rangle,$$

where $\bar{R}$ denotes the Riemann curvature tensor of $(\bar{M}, \bar{g})$. Due to the Gauß equations, cf. Theorem 5, p. 100 of [65], we have

$$\begin{aligned}\sum_{i,j=1}^{n} \langle \bar{R}_{e_i e_j} e_j, e_i\rangle = \sum_{i,j=1}^{n} \langle R_{e_i e_j} e_j, e_i\rangle + \sum_{i,j=1}^{n} \langle II(e_i, e_j), II(e_j, e_i)\rangle \\ - \sum_{i,j=1}^{n} \langle II(e_i, e_i), II(e_j, e_j)\rangle,\end{aligned}$$

where R is the Riemann curvature tensor of (M, g). Thus, due to the above observations, we obtain (13.5).

We also wish to compute $\bar{G}(e_0, e_i) = \overline{\mathrm{Ric}}(e_0, e_i)$. Again, recalling Lemma 52, p. 87 of [65] and the difference in conventions, we have

$$\overline{\mathrm{Ric}}(e_0, e_i) = \sum_{j=1}^{n} \langle \bar{R}_{e_i e_j} e_j, e_0\rangle. \tag{13.9}$$

In order to compute the right-hand side, we shall use the *Codazzi equation*, cf. Proposition 33, p. 115 of [65], which states that

$$\text{nor}\, \bar{R}_{VW} X = (\nabla_V II)(W, X) - (\nabla_W II)(V, X), \tag{13.10}$$

for vector fields V, W and X on M, where

$$(\nabla_V II)(X, Y) = D_V^{\perp}[II(X,Y)] - II(D_V X, Y) - II(X, D_V Y), \tag{13.11}$$

and

$$D^{\perp}_{V}Z = \text{nor}\,\bar{D}_V Z$$

for a vector field V on M and an $\bar{M}$ vector field Z on M such that nor $Z = Z$. Let us express $\nabla_V II$ in terms of k. We have

$$D^{\perp}_{V}[II(X,Y)] = D^{\perp}_{V}[k(X,Y)e_0] = V[k(X,Y)]e_0 + k(X,Y)D^{\perp}_{V}e_0.$$

Since

$$\langle D^{\perp}_{V}e_0, e_0\rangle = \langle \bar{D}_V e_0, e_0\rangle = \frac{1}{2}\bar{D}_V\langle e_0, e_0\rangle = 0,$$

we conclude that

$$\begin{aligned} D^{\perp}_{V}[II(X,Y)] &= D^{\perp}_{V}[k(X,Y)e_0] = V[k(X,Y)]e_0 \\ &= (D_V k)(X,Y)e_0 + k(D_V X, Y)e_0 + k(X, D_V Y)e_0. \end{aligned}$$

Combining this observation with (13.11), we obtain

$$(\nabla_V II)(X,Y) = (D_V k)(X,Y)e_0.$$

Thus, due to (13.9) and (13.10), we have

$$\overline{\text{Ric}}(e_0, e_i) = \sum_{j=1}^{n}\langle (D_{e_i}k)(e_j,e_j)e_0 - (D_{e_j}k)(e_i,e_j)e_0, e_0\rangle.$$

Thus (13.6) holds. □

13.3 Constraint equations, non-linear scalar field case

Let $(\bar{M}, \bar{g})$ be a time oriented Lorentz manifold, let ϕ be a smooth function on $\bar{M}$, let M be a smooth spacelike hypersurface, let g and k be the metric and second fundamental form induced on M by $\bar{g}$ and let N be the future directed unit normal to M. Finally, let D be the Levi-Civita connection on M induced by g. Assume that $\bar{g}$ and ϕ satisfy (13.1). Combining (13.5) with (13.1), we obtain

$$\frac{1}{2}[S - k_{ij}k^{ij} + (\text{tr}_g k)^2] = \rho, \tag{13.12}$$

where S is the scalar curvature of g and

$$\rho = \frac{1}{2}[(N\phi)^2 + D^i\phi D_i\phi] + V(\phi).$$

We refer to (13.12) as the *Hamiltonian constraint*. For any vector v tangent to M, we have (13.6). Combining this with (13.1), we obtain

$$D^j k_{ji} - D_i(\text{tr}_g k) = j_i, \tag{13.13}$$

the so-called *momentum constraint* in the non-linear scalar field case, where

$$j_i = N(\phi)D_i\phi.$$

14 Local existence

When proving local existence of solutions, it would be convenient if the Einstein non-linear scalar field system were a system of non-linear wave equations for the metric components and the scalar field. As we mentioned in the outline, this is unfortunately not the case. However, it is in practice possible to consider an associated (or gauge fixed) system, specified by choosing so-called gauge source functions, which is of such a type. In Section 14.1, we describe the associated system in detail. Furthermore, we construct gauge source functions that are convenient in proving local existence. The initial data for Einstein's equations do not uniquely determine the initial data for the gauge fixed system. However, in order for the solution to the associated system to yield a solution to the original equations, the data for the gauge fixed system have to satisfy certain conditions, which we specify at the end of Section 14.1. In Section 14.2, we then describe how to determine initial data for the associated system (given initial data for Einstein's equations) such that these conditions are satisfied. After these preliminary observations have been made, it is possible to proceed to the main result of the present chapter: the existence of a globally hyperbolic development of initial data to Einstein's equations. We present a proof of this fact in Section 14.3. It is of course also of interest to have a local uniqueness result. However, it is not as obvious how to state such a result in the present setting as in the case of the wave equation on Minkowski space. The reason for this is that a development of initial data consists of an abstract manifold, with a smooth Lorentz metric and a smooth function on it, together with a smooth embedding of the initial hypersurface into the manifold. How to compare two developments is thus not immediately obvious. One way to define local uniqueness is, however, to demand that, given two developments, there be a third development which embeds isometrically into the first two in such a way that the embeddings of the initial hypersurface are respected. We provide a proof of this fact in Section 14.4. It is interesting to note that the proof of local uniqueness (in the above sense of the word) is more involved than the proof of the existence of a development.

14.1 Gauge choice

There is of course a problem with (13.3)–(13.4) when expressed in local coordinates. These equations cannot yield unique solutions given initial data; $R_{\mu\nu}$, considered as a differential operator acting on the metric, is not hyperbolic. We shall use the ideas of [41] to overcome this problem. Let us consider the following modified system:

$$\widehat{R}_{\mu\nu} - \nabla_\mu\phi\nabla_\nu\phi - \frac{2}{n-1}V(\phi)g_{\mu\nu} = 0, \tag{14.1}$$

$$\nabla^\mu\nabla_\mu\phi - V'(\phi) = 0, \tag{14.2}$$

where we have modified the Ricci tensor according to the formula

$$\hat{R}_{\mu\nu} = -\frac{1}{2}g^{\alpha\beta}\partial_\alpha\partial_\beta g_{\mu\nu} + \nabla_{(\mu}F_{\nu)} + g^{\alpha\beta}g^{\gamma\delta}[\Gamma_{\alpha\gamma\mu}\Gamma_{\beta\delta\nu} + \Gamma_{\alpha\gamma\mu}\Gamma_{\beta\nu\delta} + \Gamma_{\alpha\gamma\nu}\Gamma_{\beta\mu\delta}], \tag{14.3}$$

or, in other words,

$$\hat{R}_{\mu\nu} = R_{\mu\nu} + \nabla_{(\mu}\mathcal{D}_{\nu)}, \tag{14.4}$$

where

$$\mathcal{D}_\mu = F_\mu - \Gamma_\mu$$

cf. (10.13). Note that Γ_μ are not necessarily the components of a covector. However, by a judicious choice of F_μ, we can ensure that $\mathcal{D}_\mu$ are the components of a covector. Let h be a fixed reference Lorentz metric on M. Let $\bar{\nabla}$ be the associated Levi-Civita connection. Define A by

$$A(X, Y, \eta) = \eta(\nabla_X Y - \bar{\nabla}_X Y), \tag{14.5}$$

for vector fields X, Y and a 1-form field η. We see that A is multilinear over the functions, so that it is a tensor field. Writing it out in components, we get

$$A^\mu_{\alpha\beta} = A(\partial_\alpha, \partial_\beta, dx^\mu) = \Gamma^\mu_{\alpha\beta} - \bar{\Gamma}^\mu_{\alpha\beta},$$

where

$$\bar{\Gamma}^\mu_{\alpha\beta} = \frac{1}{2}h^{\mu\nu}(\partial_\alpha h_{\beta\nu} + \partial_\beta h_{\alpha\nu} - \partial_\nu h_{\alpha\beta}).$$

Compute

$$g_{\mu\nu}g^{\alpha\beta}A^\mu_{\alpha\beta} = \Gamma_\nu - g_{\mu\nu}g^{\alpha\beta}\bar{\Gamma}^\mu_{\alpha\beta}.$$

The left-hand side are clearly the components of a covector, so if we let

$$F_\nu = g_{\mu\nu}g^{\alpha\beta}\bar{\Gamma}^\mu_{\alpha\beta}, \tag{14.6}$$

then

$$\mathcal{D}_\nu = -g_{\mu\nu}g^{\alpha\beta}A^\mu_{\alpha\beta}$$

are the components of a covector. There is one obvious objection to this choice of F; it depends on g, which is unknown. It is therefore not clear why the resulting equation should be better. However, F only depends on the metric and not on the first derivatives of the metric, and consequently $\hat{R}$, considered as a differential operator acting on g, is of a form that we can deal with using the methods described earlier in these notes.

Let us assume (M, g) is a globally hyperbolic Lorentz manifold, that Σ is a smooth spacelike Cauchy hypersurface, that there is a smooth function ϕ on M and that g and ϕ satisfy (14.1) and (14.2) where F_ν has been chosen as in (14.6) for some Lorentz metric h on M. We then wish to demonstrate that if $\mathcal{D}_\mu$ and $\nabla_\nu\mathcal{D}_\mu$ vanish on some subset Ω of Σ, then $\mathcal{D}$ vanishes on $D(\Omega)$. Due to (14.1), we have

$$G_{\mu\nu} - T_{\mu\nu} = -\nabla_{(\mu}\mathcal{D}_{\nu)} + \frac{1}{2}(\nabla^\gamma\mathcal{D}_\gamma)g_{\mu\nu}. \tag{14.7}$$

Furthermore, $G_{\mu\nu}$ is divergence free, cf. (10.12), and $T_{\mu\nu}$ is divergence free due to (14.2). Consequently,

$$\nabla_\mu \nabla^\mu \mathcal{D}_\nu + R_\nu{}^\mu \mathcal{D}_\mu = 0, \tag{14.8}$$

cf. (10.9). Using Corollary 12.12, we conclude that $\mathcal{D}_\nu = 0$ in $D(\Omega)$. Consequently, g and ϕ solve Einstein's equations (13.3) and (13.4) in $D(\Omega)$. In the case that $\Omega = \Sigma$, we get a solution in all of M. The problem of solving (13.3) and (13.4) is thus reduced to the problem of solving (14.1) and (14.2) and finding initial data for these equations such that $\mathcal{D}_\nu$ and $\nabla_\mu \mathcal{D}_\nu = 0$ initially.

14.2 Initial data

Definition 14.1. *Initial data* to (13.3)–(13.4) consist of an n dimensional manifold Σ, a Riemannian metric g_0, a covariant 2-tensor k and two functions ϕ_0 and ϕ_1 on Σ, all assumed to be smooth and to satisfy

$$r - k_{ij}k^{ij} + (\operatorname{tr} k)^2 = \phi_1^2 + D^i\phi_0 D_i\phi_0 + 2V(\phi_0), \tag{14.9}$$

$$D^j k_{ji} - D_i(\operatorname{tr} k) = \phi_1 D_i \phi_0, \tag{14.10}$$

where D is the Levi-Civita connection of g_0, r is the associated scalar curvature and indices are raised and lowered by g_0. Given initial data, the *initial value problem* is that of finding an $n+1$ dimensional manifold M with a Lorentz metric g and a $\phi \in C^\infty(M)$ such that (13.3) and (13.4) are satisfied, and an embedding $i : \Sigma \to M$ such that $i^*g = g_0$, $\phi \circ i = \phi_0$, and if N is the future directed unit normal and K is the second fundamental form of $i(\Sigma)$, then $i^*K = k$ and $(N\phi) \circ i = \phi_1$. Such a triple (M, g, ϕ) is referred to as a *development* of the initial data, the existence of an embedding i being tacit. If, in addition, $i(\Sigma)$ is a Cauchy hypersurface in (M, g), we say that (M, g, ϕ) is a *globally hyperbolic development*.

In the end, we shall mainly be interested in globally hyperbolic developments. In that case, M is diffeomorphic to $\mathbb{R} \times \Sigma$, and the constructions we shall carry out will always be on subsets of this topological space.

Let $(\Sigma, g_0, k, \phi_0, \phi_1)$ be initial data to (13.3)–(13.4) and let us set up initial data for (14.1) and (14.2). When setting up initial data for the latter equations we have an additional freedom, which we shall use to obtain $\mathcal{D}_\nu = 0$ initially. Since we wish to construct a globally hyperbolic development, let us take $M = \mathbb{R} \times \Sigma$ and let us specify the initial data on the hypersurface $\{0\} \times \Sigma$ (in the end we shall only get existence for a subset of $\mathbb{R} \times \Sigma$). Let

$$h = -dt^2 + g_0.$$

This is our fixed Lorentz metric on M. Let us fix a coordinate system $(x^1, \dots, x^n)$ on an open subset U of Σ. We then obtain coordinates $(x^0, \dots, x^n)$ on $\mathbb{R} \times U$ with $x^0 = t$. When we speak of $g_{\mu\nu}$, we shall mean $\langle \partial_\mu, \partial_\nu \rangle$, etc. Let us consider the initial data for $g_{\mu\nu}$. The spatial part of the metric, g_{ij}, is determined by g_0:

$$g_{ij}|_{t=0} = g_0(\partial_i, \partial_j). \tag{14.11}$$

However, we are free to choose g_{00} and g_{0i}. We make the choice

$$g_{00}|_{t=0} = -1, \quad g_{0i}|_{t=0} = 0. \tag{14.12}$$

Due to this choice, the future directed unit normal to the hypersurface $t = 0$ is ∂_t, so that if we had a metric g whose second fundamental form were K, then we would have

$$K_{ij} = \frac{1}{2}\partial_0 g_{ij}.$$

The natural definition to make is thus

$$\partial_0 g_{ij}|_{t=0} = 2k_{ij} = 2k(\partial_i, \partial_j). \tag{14.13}$$

The only objects that remain to be determined are $\partial_0 g_{00}$ and $\partial_0 g_{0i}$. We shall let the condition $\mathcal{D}_\mu|_{t=0} = 0$ determine these quantities. Assuming we had a metric g, we would obtain, for $t = 0$,

$$\Gamma_0 = -\frac{1}{2}\partial_0 g_{00} - \operatorname{tr} k,$$

where we have used (14.12) and (14.13). We thus require

$$\partial_0 g_{00}|_{t=0} = -2F_0|_{t=0} - 2\operatorname{tr} k \tag{14.14}$$

(note that F_μ is well defined for $t = 0$ since the metric has been specified for $t = 0$). We also have, assuming we had a metric g, at $t = 0$,

$$\Gamma_l = -\partial_0 g_{0l} + \frac{1}{2}g^{ij}(2\partial_i g_{jl} - \partial_l g_{ij}).$$

Consequently we require, for $t = 0$,

$$\partial_0 g_{0l}|_{t=0} = \Big[-F_l + \frac{1}{2}g^{ij}(2\partial_i g_{jl} - \partial_l g_{ij})\Big]\Big|_{t=0}. \tag{14.15}$$

Concerning ϕ, we require

$$\phi|_{t=0} = \phi_0, \quad (\partial_t\phi)|_{t=0} = \phi_1. \tag{14.16}$$

Due to (14.14) and (14.15), we know that $\mathcal{D}_\mu = 0$ for $t = 0$. However, in order to be allowed to conclude that $\mathcal{D}_\mu$ is zero in the development of the data, we need to know that $\partial_0\mathcal{D}_\mu$ is zero for $t = 0$. On the other hand, we have no more freedom left in specifying initial data. Let us disregard this and assume that we have a solution to (14.1) and (14.2) in some coordinate patch, where the initial data for these equations are given by (14.11)–(14.16). The original initial data are of course required to be solutions of the constraint equations. The solution we obtain solves (14.7). Let us contract this equation with $n^\mu X^\nu$ for $t = 0$, where X is orthogonal to n. Then the left-hand side is zero since the constraints are fulfilled and the right-hand side is

$$-\frac{1}{2}n^\mu X^\nu(\partial_\mu\mathcal{D}_\nu + \partial_\nu\mathcal{D}_\mu).$$

Note that the part of the covariant derivative involving Christoffel symbols vanishes due to the fact that $\mathcal{D}_\mu = 0$ originally. Since $X^\nu\partial_\nu\mathcal{D}_\mu = 0$ for $t = 0$, we obtain $\partial_0\mathcal{D}_i = 0$ for $t = 0$, $i = 1, \dots, n$. If we contract (14.7) with $n^\mu n^\nu$, we obtain $\partial_0\mathcal{D}_0 = 0$ by a similar argument.

14.3 Existence of a globally hyperbolic development

Theorem 14.2. *Let* $(\Sigma, g_0, k, \phi_0, \phi_1)$ *be initial data to* (13.3)–(13.4). *Then there is a globally hyperbolic development of the data.*

Proof. Define
$$h = -dt^2 + g_0 \tag{14.17}$$
on $\mathbb{R}\times\Sigma$. Let $p \in \Sigma$. Let $U \ni p$ be an open subset of Σ such that we have coordinates $x^1, \dots, x^n$ on U and define coordinates $x^0, \dots, x^n$ on $\mathbb{R} \times U$ by $x^0 = t$. Consider the equations (14.1)–(14.2) where the initial data are given by (14.11)–(14.16) and the F_ν are given by (14.6). We would like to apply the local existence result given in Corollary 9.16, but this result does not immediately apply to the present situation due to the global restrictions on g made and the fact that a Lorentz metric on $\mathbb{R}^{n+1}$ can never have compact support. Let $V \ni p$ be an open set such that its closure is compact and contained in U. Let $A_{\mu\nu}$ be the components of a Lorentz matrix-valued function depending smoothly on the components $g_{\alpha\beta}$ of g. Let $A_{00} = g_{00}$ for all $g_{00} \in [-3/2, -1/2]$ and have the property that the range of A_{00} is contained in $[-2, -1/4]$. Let $A_{0i} = g_{0i}$ for $g_{0i} \in [-1, 1]$ and have the property that the range of A_{0i} is contained in $[-2, 2]$. Finally, let $\mathcal{U}$ be an open subset of the set of symmetric $n \times n$ matrices such that the matrices with components $g_{ij}(q)$ for $q \in \bar{V}$ are contained in $\mathcal{U}$ and that the closure of $\mathcal{U}$ in the set of all $n \times n$ matrices is compact and contained in the set of positive definite ones. Let A_{ij} be such that $A_{ij} = g_{ij}$ for $\{g_{ij}\} \in \mathcal{U}$ and A_{ij} is everywhere positive definite with a positive lower bound and an upper bound. For convenience, we can assume the derivatives of $A_{\mu\nu}$ with respect to the metric components to have compact support. In considering (14.1)–(14.2), we replace $g^{\mu\nu}$, wherever it appears, with $A^{\mu\nu}$, the components of the inverse of A. Concerning F_μ, we modify it by multiplying it with a function $\psi_1 \in C_0^\infty[(-1, 1) \times U]$ such that $\psi_1(p) = 1$ for $p \in [-1/2, 1/2] \times \bar{V}$. As initial data we would ideally like to prescribe that (14.11)–(14.16) hold. However, that does not lead to an equation of the type considered in Corollary 9.16. Let $\psi \in C_0^\infty(U)$ be such that $\psi(q) = 1$ for all $q \in \bar{V}$. Modify all the initial data by multiplying them with ψ. Let u be the vector which collects ϕ and $g_{\mu\nu}$ for $\mu, \nu = 0, \dots, n$. We can consider the resulting equation as an equation on $\mathbb{R}^{n+1}$. Furthermore, it is of such a form that Corollary 9.16 is applicable. We thus get a smooth local solution. Due to the smoothness of the solution, there is an open neighbourhood W of p in $\mathbb{R} \times \Sigma$ with the property that $g_{\mu\nu}$ are such that $A_{\mu\nu} = g_{\mu\nu}$ and $\psi_1 = 1$ in W, and $\pi(W) \subseteq V$, where π is the projection to Σ so that $\Sigma_p := W \cap \{0\} \times \Sigma \subseteq \{0\} \times V$. Furthermore, we can assume that every inextendible causal curve in W intersects Σ, that W is contained in convex neighbourhood of p, that $\operatorname{grad} t$ is timelike on W (since $g^{00} < 0$) and that $J^-(q) \cap J^+(\Sigma_p)$ is contained in W for every $q \in W$ with positive t-coordinate and similarly for points of W with negative t-coordinate. Thus Σ_p is a Cauchy hypersurface in W and $J^-(q) \cap J^+(\Sigma_p)$ is compact for every $q \in W$, cf. Lemma 40, p. 423 of [65]. If we let $\mathcal{D}_\mu = F_\mu - \Gamma_\mu$, then $\mathcal{D}_\mu = \partial_0 \mathcal{D}_\mu = 0$ on Σ_p due to the argument presented at the end of Section 14.2. Furthermore, $\mathcal{D}_\mu$ satisfies (14.8). We are allowed to apply Lemma 12.10 in order to

conclude that $\mathcal{D}_\mu = 0$ in all of W. Let W_p be an open neighbourhood of p with the same properties as W and whose closure is contained in W.

We would like to define the manifold M to be the union of all the W_p. The first problem we are confronted with is that of constructing a metric on M. Say that $W_p \cap W_q \neq \emptyset$. The closures of W_p and W_q are compact and contained in open sets W_1, W_2, with properties as above, on which we have coordinates $x = (x^0, \dots, x^n)$ and $y = (y^0, \dots, y^n)$ respectively, where $x^0 = y^0 = t$. On W_1 and W_2, we have metrics g_1 and g_2 and smooth functions ϕ_a and ϕ_b respectively, both satisfying (14.1)–(14.2) when expressed with respect to the coordinates x and y respectively. Let us express both g_1 and g_2 with respect to the coordinates x in $W_1 \cap W_2$ and refer to the components as $g_{1\mu\nu}$ and $g_{2\mu\nu}$ respectively. Let us use the notation $\Sigma_1 = \Sigma_p$ and $\Sigma_2 = \Sigma_q$

Both are solutions. Since the equations (14.1)–(14.2) are geometric, both $g_{1\mu\nu}, \phi_a$ and $g_{2\mu\nu}, \phi_b$ satisfy them. Furthermore, since $\mathcal{D}_{i\mu}$ are the components of a covector and vanishes with respect to one of the coordinate systems, it vanishes with respect to the other coordinate system.

The initial data coincide. By the construction it is clear that $g_{1ij} = g_{2ij}, g_{100} = g_{200}$ and that $g_{20i} = g_{10i}$ for $t = 0$. Since $\mathcal{D}_{i\mu} = 0$ and the metrics coincide for $t = 0$, the contracted Christoffel symbols for g_1 and g_2 with respect to the x-coordinates have to coincide. Since we have (14.13) and the coordinates have the above special form, we conclude that $\partial_t g_{1\mu\nu} = \partial_t g_{2\mu\nu}$ for $t = 0$. Finally, it is clear that $\phi_a = \phi_b$ and $\partial_t \phi_a = \partial_t \phi_b$ for $t = 0$.

The solutions coincide. We wish to prove that the solutions coincide in $\overline{W}_p \cap \overline{W}_q$. For $t \geq 0$, let

$$S_t = [0, t] \times \Sigma \cap \overline{W}_p \cap \overline{W}_q.$$

Let $\mathcal{A}$ be the set of $t \in [0, \infty)$ such that $g_1 = g_2$ and $\phi_a = \phi_b$ in S_t and that for $r \in S_t$,

$$J_1^-(r) \cap J_1^+(\Sigma_1) = J_2^-(r) \cap J_2^+(\Sigma_2), \tag{14.18}$$

where $J_1^-(r)$ is the causal past of r with respect to the metric g_1 in W_1, etc. Note that $0 \in \mathcal{A}$, so that $\mathcal{A}$ is non-empty. Assume $t \in \mathcal{A}$ and $r \in S_t$ with $r = (t, \xi)$. Note that $J_i^-(r) \cap J_i^+(\Sigma_i) \subseteq W_1 \cap W_2$. If $\tau > t$ is close enough to t, the same is true with r replaced by (τ, ξ) due to Lemma 10.10. Taking the difference of (14.1)–(14.2) for the two solutions, keeping in mind that $sg_1 + (1 - s)g_2$ is a Lorentz metric for $s \in [0, 1]$ due to the fact that $g_{i00} < 0$ and g_{ib} is positive definite for $i = 1, 2$, we conclude that Lemma 12.8 is applicable with two choices for the coefficients of the highest order derivatives; either $g_1^{\mu\nu}$ or $g_2^{\mu\nu}$. We conclude that $g_1 = g_2$ and $\phi_a = \phi_b$ in

$$J_1^-[(\tau, \xi)] \cap J_1^+(\Sigma_1) \cup J_2^-[(\tau, \xi)] \cap J_2^+(\Sigma_2).$$

Consequently (14.18) holds with r replaced by (τ, ξ). This proves that $\mathcal{A}$ is open, due to the following argument. Assume there is no $\varepsilon > 0$ such that $[t, t + \varepsilon] \subseteq \mathcal{A}$. Then there is a sequence $r_i = (t_i, p_i)$ such that $t_i \to t+$ and either $g_1(r_i) \neq g_2(r_i)$, $\phi_a(r_i) \neq \phi_b(r_i)$ or (14.18) does not hold for $r = r_i$. Due to compactness, we can assume p_i to converge to, say, p. Applying the above argument with $\xi = p$, i.e., $r = (t, p)$, we arrive at a contradiction for i large enough. We conclude that $[t, t+\varepsilon] \subseteq \mathcal{A}$ for $\varepsilon > 0$ small enough. The closedness is less complicated to prove, keeping in mind that (14.18) follows from Lemma 10.10. We conclude that $\mathcal{A} = [0, \infty)$ so that $g_1 = g_2$ and $\phi_a = \phi_b$ in $\overline{W}_p \cap \overline{W}_q$. We conclude that we have a solution to (13.3)–(13.4) on M, defined to be the union of all the W_p. The embedding $i : \Sigma \to M$ is simply the inclusion $i(p) = (0, p)$. By construction, it is clear that $i^*g = g_0$, $i^*K = k$, $i^*\phi = \phi_0$ and $i^*(N\phi) = \phi_1$. Let γ be an inextendible causal curve in M. Then the image of γ has to intersect some W_p and $\gamma|_{\gamma^{-1}(W_p)}$ is an inextendible causal curve in W_p which by construction has to intersect Σ. Since grad t is timelike by construction, the t-coordinate of γ is strictly monotone, so that γ intersects Σ exactly once. □

14.4 Two developments are extensions of a common development

Theorem 14.3. *Let* $(\Sigma, g_0, k, \phi_0, \phi_1)$ *be initial data to* (13.3)–(13.4). *Assume that we have two developments* (M_a, g_a, ϕ_a) *and* (M_b, g_b, ϕ_b) *with corresponding embeddings* $i_a : \Sigma \to M_a$ *and* $i_b : \Sigma \to M_b$. *Then there is a globally hyperbolic development* (M, g, ϕ) *with corresponding embedding* $i : \Sigma \to M$ *and smooth time orientation preserving maps* $\psi_a : M \to M_a$ *and* $\psi_b : M \to M_b$ *which are diffeomorphisms onto their image such that* $\psi_a^* g_a = g$, $\psi_a^* \phi_a = \phi$, $\psi_b^* g_b = g$ *and* $\psi_b^* \phi_b = \phi$. *Finally,* $\psi_a \circ i = i_a$ *and* $\psi_b \circ i = i_b$.

Remark 14.4. A less technical statement of the same thing would be to say that every pair of developments of initial data are extensions of a common development.

Proof. Let us assume that we have constructed a solution (g, ϕ) as in the proof of Theorem 14.2 on an open subset D of $\mathbb{R} \times \Sigma$ containing $\{0\} \times \Sigma$. Let furthermore h be the reference metric (14.17). Then, in particular, if (V, y) are coordinates with $V \subseteq D$, we have

$$\Gamma_\mu = g_{\mu\nu} g^{\alpha\beta} \Xi^\nu_{\alpha\beta} \tag{14.19}$$

where Γ_μ are the contracted Christoffel symbols of g with respect to y and $\Xi^\nu_{\alpha\beta}$ are the Christoffel symbols of h with respect to y.

Let $p \in i_a(\Sigma)$. Since $i_a(\Sigma)$ is a spacelike hypersurface of M_a, there are coordinates (U, x) as in the statement of Lemma 12.5 and we can assume that $x(p) = 0$. Let $\hat{x}^i = x^i|_{U \cap i_a(\Sigma)}$. Then $\hat{x} = (\hat{x}^1, \dots, \hat{x}^n)$ are coordinates on $U \cap i_a(\Sigma)$. Define $\hat{y}^i = \hat{x}^i \circ i_a$. Then $(\hat{y}^1, \dots, \hat{y}^n)$ define coordinates on $U_\Sigma = i_a^{-1}(U)$. Let $y^0(t, q) = t$ for $(t, q) \in \mathbb{R} \times \Sigma$ and $y^i(t, q) = \hat{y}^i(q)$. Then $y = (y^0, \dots, y^n)$ are coordinates on $\mathbb{R} \times U_\Sigma$. If we consider (13.3)–(13.4) on $V = \mathbb{R} \times U_\Sigma \cap D$ with respect to the coordinates y, we see that we can replace Γ_μ by the right-hand side of (14.19). Note that with respect to the coordinates y, $g_{00} = -1$ and $g_{0i} = 0$ on $\{0\} \times U_\Sigma$. Furthermore g_{ij} and $\partial_0 g_{ij}$ are determined on $\{0\} \times U_\Sigma$ by g_0 and k. Finally, $\partial_0 g_{00}$ and $\partial_0 g_{0i}$ are determined by (14.19). Recall the convention that if f is a smooth function on U, then

$$\frac{\partial f}{\partial x^\mu} = D_\mu(f \circ x^{-1}) \circ x,$$

where D_μ denotes differentiation with respect to slot number μ of $f \circ x^{-1}$, which is a function on an open subset of $\mathbb{R}^{n+1}$.

General idea of the proof. Our wish is to set up a coordinate system $\tilde{x}$ on a subset of U, so that with respect to this coordinate system, g_a has contracted Christoffel symbols of the form

$$^{(a)}\tilde{\Gamma}^\mu = \tilde{g}_a^{\alpha\beta}\Theta^\mu_{\alpha\beta},$$

where $^{(a)}\tilde{\Gamma}^\mu_{\alpha\beta}$ are the Christoffel symbols of g_a with respect to the coordinates $\tilde{x}$, $\tilde{g}_a^{\alpha\beta}$ are the components of the inverse of the metric g_a with respect to the coordinates $\tilde{x}$ and

$$\Theta^\mu_{\alpha\beta} = \Xi^\mu_{\alpha\beta} \circ y^{-1} \circ \tilde{x}.$$

In the latter formula, one has to take care so that the right-hand side is defined. However, we shall write down the details below. The point is that with respect to these coordinates, (g_a, ϕ_a) would satisfy the same equation as (g, ϕ) with respect to the coordinates y. Assuming we can construct the coordinate system in such a fashion that the initial data coincide (computed with respect to the respective coordinates), then by uniqueness $\tilde{g}_{a\mu\nu}$, considered as a function of $\tilde{x}$, would have to coincide with $g_{\mu\nu}$, considered as a function of y, and similarly for ϕ_a and ϕ. But this would then define an isometry from some open subset of $i_a^{-1}(p)$ to an open subset of p by simply mapping a point q to its y-coordinates, identifying the y coordinates with the $\tilde{x}$ coordinates and then applying $\tilde{x}^{-1}$. The resulting map, say ψ_p, would have the property that $\psi_p^* g_a = g$ and $\psi_p^*\phi_a = \phi$. The idea is then to paste together such local isometries to get a global one.

Construction of coordinates. Let us proceed to the details. Note that

$$y(0, i_a^{-1}(p)) = 0.$$

Let $\eta \in C_0^\infty(\mathbb{R}^{n+1})$ have support in an open ball with center at the origin contained in $y(V) \cap x(U)$. Assume furthermore that $0 \leq \eta \leq 1$ and $\eta = 1$ in an open neighbourhood of 0. Define

$$\Theta^\mu_{\alpha\beta}(w) = \Xi^\mu_{\alpha\beta} \circ y^{-1}[\eta(w)w],$$

where $\Xi^\mu_{\alpha\beta}$ are the Christoffel symbols of h with respect to the coordinates y. Then $\Theta^\mu_{\alpha\beta}$ is a smooth function on $\mathbb{R}^{n+1}$. Let

$$\rho_{\mu\nu}(w) = g_{a\mu\nu} \circ x^{-1}[\eta(w)w],$$

where $g_{a\mu\nu}$ are the components of g_a with respect to the coordinates x. Note that then ρ defines a smooth Lorentz metric on $\mathbb{R}^{n+1}$. We can assume the support of η to be small enough that $\rho_{00} < 0$ everywhere, with a uniform negative upper bound, and

ρ_{ij} to be the components of a positive definite matrix with a uniform positive lower bound. Furthermore, all the components $\rho_{\mu\nu}$ are bounded. Let $\chi \in C_0^\infty(\mathbb{R}^n)$ be such that it has support in a ball with center at the origin contained in $\hat{x}[U \cap i_a(\Sigma)]$. Assume furthermore that $0 \leq \chi \leq 1$ and that $\chi = 1$ in a neighbourhood of the origin. Let us set up the following initial value problem on $\mathbb{R}^{n+1}$:

$$\begin{aligned} \Box_\rho \bar{x}^\gamma &= -\rho^{\alpha\beta} \frac{\partial \bar{x}^\mu}{\partial \xi^\alpha} \frac{\partial \bar{x}^\nu}{\partial \xi^\beta} \Theta^\gamma_{\mu\nu} \circ \bar{x} \\ \bar{x}(0, \xi^1, \dots, \xi^n) &= \chi(\xi^1, \dots, \xi^n)(0, \xi^1, \dots, \xi^n) \\ \frac{\partial \bar{x}}{\partial \xi^0}(0, \xi^1, \dots, \xi^n) &= \chi(\xi^1, \dots, \xi^n)(1, 0, \dots, 0). \end{aligned} \tag{14.20}$$

In these equations, ξ is taken to be the identity coordinate system on $\mathbb{R}^{n+1}$, so that, comparing with the above notation,

$$\frac{\partial \bar{x}^\nu}{\partial \xi^\beta} = D_\beta(\bar{x}^\nu \circ \mathrm{Id}^{-1}) \circ \mathrm{Id} = D_\beta \bar{x}^\nu.$$

Here $\Box_\rho$ should be interpreted as the scalar wave operator, i.e.,

$$\Box_\rho = \rho^{\alpha\beta}(\partial_{\xi^\alpha}\partial_{\xi^\beta} - \Lambda^\mu_{\alpha\beta}\partial_{\xi^\mu}),$$

where Λ are the Christoffel symbols of ρ. Note that Corollary 9.16 is applicable to this system of equations, so that we get a smooth local solution $\bar{x}$. Furthermore $\bar{x}$ satisfies $\partial\bar{x}^\mu/\partial\xi^\nu = \delta^\mu_\nu$ at the origin. In a neighbourhood of the origin, $\bar{x}$ are thus coordinates. To be more precise, $\bar{x} \circ x$ defines coordinates on some open subset on which it is defined. Let $\tilde{x}$ be coordinates defined by restricting $\bar{x} \circ x$ to some open subset W containing p. We can assume W to be small enough that

$$\partial\tilde{x}^\mu/\partial x^\nu(q) = \delta^\mu_\nu \tag{14.21}$$

for $q \in i_a(\Sigma) \cap W$. We can furthermore assume that $\rho_{\mu\nu} \circ x = g_{a\mu\nu}$ on W. Note that then

$$\frac{\partial \rho_{\mu\nu}}{\partial \xi^\alpha} \circ x = D_\alpha(g_{a\mu\nu} \circ x^{-1}) \circ x = \frac{\partial g_{a\mu\nu}}{\partial x^\alpha},$$

etc. Using these sorts of observations, one can conclude that

$$(\Box_{g_a}\tilde{x}^\gamma) \circ x^{-1} = \Box_\rho \bar{x}^\gamma = -\rho^{\alpha\beta} \frac{\partial \bar{x}^\mu}{\partial \xi^\alpha} \frac{\partial \bar{x}^\nu}{\partial \xi^\beta} \Theta^\gamma_{\mu\nu} \circ \bar{x} = -\tilde{g}_a^{\mu\nu} \circ x^{-1} \Xi^\gamma_{\mu\nu} \circ y^{-1} \circ \bar{x},$$

where $\tilde{g}_a^{\mu\nu}$ are the components of the inverse of the metric g_a with respect to the coordinates $\tilde{x}$, and we only consider these equations on a suitably small neighbourhood of 0. Thus

$$\Box_{g_a}\tilde{x}^\gamma = -\tilde{g}_a^{\mu\nu} \Xi^\gamma_{\mu\nu} \circ y^{-1} \circ \tilde{x}.$$

Expressing this equation with respect to $\tilde{x}$-coordinates, we get

$$-{}^{(a)}\tilde{\Gamma}^\gamma = -\tilde{g}_a^{\mu\nu} \Xi^\gamma_{\mu\nu} \circ y^{-1} \circ \tilde{x}, \tag{14.22}$$

where ${}^{(a)}\tilde{\Gamma}^{\mu}$ are the contracted Christoffel symbols of g_a computed with respect to the coordinates $\tilde{x}$.

Comparing the initial data. Let us analyze what $\tilde{g}_{a\mu\nu}$, ϕ_a and their derivatives are initially. Since (14.21) holds initially and due to the properties of the x-coordinate system, we have

$$\tilde{g}_{a00}(q) = -1, \quad \tilde{g}_{a0i}(q) = 0,$$

for $q \in i_a(\Sigma) \cap W$. Let us introduce the notation $W_\Sigma = i_a^{-1}[W \cap i_a(\Sigma)]$ and note that for $q \in \{0\} \times W_\Sigma$, we have $y(q) = x \circ i_a(q) = \tilde{x} \circ i_a(q)$ (by abuse of notation). Thus on $\tilde{x}[W \cap i_a(\Sigma)]$, $y^{-1} = i_a^{-1} \circ \tilde{x}^{-1}$ (by abuse of notation). Let us compute

$$\begin{aligned}
i_{a*}\partial_{y^i}|_q f &= \partial_{y^i}|_q(f \circ i_a) \\
&= D_i(f \circ i_a \circ y^{-1}) \circ y(q) \\
&= D_i(f \circ \tilde{x}^{-1}) \circ \tilde{x} \circ i_a(q) \\
&= \partial_{\tilde{x}^i}|_{i_a(q)} f.
\end{aligned}$$

Thus, for $q \in \{0\} \times W_\Sigma$,

$$i_{a*}\partial_{y^i}|_q = \partial_{\tilde{x}^i}|_{i_a(q)},$$

so that

$$\begin{aligned}
g_{ij}(q) &= g_0(\partial_{y^i}|_q, \partial_{y^j}|_q) \\
&= i_a^* g_a(\partial_{y^i}|_q, \partial_{y^j}|_q) \\
&= g_a(\partial_{\tilde{x}^i}|_{i_a(q)}, \partial_{\tilde{x}^j}|_{i_a(q)}) \\
&= \tilde{g}_{aij}[i_a(q)].
\end{aligned}$$

By the above discussions, we see that

$$\tilde{g}_{a\mu\nu} \circ \tilde{x}^{-1} = g_{\mu\nu} \circ y^{-1} \tag{14.23}$$

on $\tilde{x}[W \cap i_a(\Sigma)]$. Furthermore,

$$k_{ij}(q) = i_a^* k_a(\partial_{y^i}|_q, \partial_{y^j}|_q) = k_a(\partial_{\tilde{x}^i}|_{i_a(q)}, \partial_{\tilde{x}^j}|_{i_a(q)}) = \tilde{k}_{aij}[i_a(q)]$$

for $q \in \{0\} \times W_\Sigma$ where k_a is the second fundamental form of $i_a(\Sigma)$ in M_a. Since $\partial_{\tilde{x}^0}|_q$ is the future directed unit normal to $i_a(\Sigma)$ for $q \in i_a(\Sigma)$, we get

$$(\partial_t g_{ij})(q) = 2k_{ij}(q) = 2\tilde{k}_{aij}[i_a(q)] = (\partial_{\tilde{x}^0}\tilde{g}_{aij})[i_a(q)] \tag{14.24}$$

for $q \in \{0\} \times W_\Sigma$. Note that

$$\begin{aligned}
\Gamma^0 &= \frac{1}{2}\partial_t g_{00} + \operatorname{tr} k, \\
\Gamma^i &= -g^{ij}\partial_0 g_{0j} + g^{kl}g^{ij}\partial_k g_{lj} - \frac{1}{2}g^{ij}g^{kl}\partial_j g_{kl}
\end{aligned}$$

on $\{0\} \times W_\Sigma$ and similarly for g_a. Note furthermore that due to (14.19), (14.22) and (14.23), we have

$$^{(a)}\tilde{\Gamma}^\mu \circ \tilde{x}^{-1} = \Gamma^\mu \circ y^{-1}$$

on $\tilde{x}[W \cap i_a(\Sigma)]$. Since $k_{ij} \circ y^{-1} = \tilde{k}_{aij} \circ \tilde{x}^{-1}$, $g_{\mu\nu} \circ y^{-1} = g_{a\mu\nu} \circ \tilde{x}^{-1}$ and

$$(\partial_{y^k} g_{\mu\nu}) \circ y^{-1} = D_k[g_{\mu\nu} \circ y^{-1}] = D_k[\tilde{g}_{a\mu\nu} \circ \tilde{x}^{-1}] = (\partial_{\tilde{x}^k} \tilde{g}_{a\mu\nu}) \circ \tilde{x}^{-1},$$

on the same set, we can conclude that

$$(\partial_{\tilde{x}^0} \tilde{g}_{a\mu\nu}) \circ \tilde{x}^{-1} = (\partial_t g_{\mu\nu}) \circ y^{-1} \tag{14.25}$$

on $\tilde{x}[i_a(\Sigma) \cap W]$, where we have also used (14.24). Similarly

$$\phi_a \circ \tilde{x}^{-1} = \phi \circ y^{-1}, \quad (\partial_{\tilde{x}^0} \phi_a) \circ \tilde{x}^{-1} = (\partial_t \phi) \circ y^{-1} \tag{14.26}$$

on $\tilde{x}[i_a(\Sigma) \cap W]$, since

$$(\partial_t \phi) \circ y^{-1} = [i_a^*(\partial_{\tilde{x}^0} \phi_a)] \circ y^{-1} = (\partial_{\tilde{x}^0} \phi_a) \circ \tilde{x}^{-1}$$

on this set.

Existence of a local isometry. Note that $(g_{\mu\nu} \circ y^{-1}, \phi \circ y^{-1})$ and $(\tilde{g}_{a\mu\nu} \circ \tilde{x}^{-1}, \phi_a \circ \tilde{x}^{-1})$ satisfy the same equation on $\tilde{x}(W)$ due to the fact that

$$-{}^{(a)}\tilde{\Gamma}^\mu \circ \tilde{x}^{-1} = -\tilde{g}_a^{\mu\nu} \circ \tilde{x}^{-1} \Xi^\gamma_{\mu\nu} \circ y^{-1},$$

cf. (14.22). We can assume that W is such that $\tilde{x}(W)$ is globally hyperbolic with respect to the metric $g_{\mu\nu} \circ y^{-1}$ and that $\tilde{x}(W) \cap \{0\} \times \mathbb{R}^n$ is a Cauchy hypersurface. We can also assume that $g_{00} \circ y^{-1}$ and $\tilde{g}_{a00} \circ \tilde{x}^{-1}$ are bounded from above by a negative constant on $\tilde{x}(W)$, that $g_{ij} \circ y^{-1}$ and $\tilde{g}_{aij} \circ \tilde{x}^{-1}$ are the components of positive definite matrices on $\tilde{x}(W)$ with a uniform positive lower bound and that $g_{\mu\nu} \circ y^{-1}$ and $\tilde{g}_{a\mu\nu} \circ \tilde{x}^{-1}$ are bounded on $\tilde{x}(W)$. Note that as a consequence,

$$s g_{\mu\nu} \circ y^{-1} + (1-s)\tilde{g}_{a\mu\nu} \circ \tilde{x}^{-1}$$

are Lorentz metrics on $\tilde{x}(W)$ with a determinant uniformly bounded away from zero. Taking the difference of the equation for $(g_{\mu\nu} \circ y^{-1}, \phi \circ y^{-1})$ and $(\tilde{g}_{a\mu\nu} \circ \tilde{x}^{-1}, \phi_a \circ \tilde{x}^{-1})$ we get an equation for the differences. The resulting equation is one to which we can apply Lemma 12.8, since the initial data are zero, cf. (14.23), (14.25) and (14.26). We get the conclusion that

$$(g_{\mu\nu} \circ y^{-1}, \phi \circ y^{-1}) = (\tilde{g}_{a\mu\nu} \circ \tilde{x}^{-1}, \phi_a \circ \tilde{x}^{-1}) \tag{14.27}$$

on $\tilde{x}(W)$. On $y^{-1}[\tilde{x}(W)]$ we can define $\psi = \tilde{x}^{-1} \circ y$, which is a diffeomorphism onto its image. Note that

$$\psi_* \partial_{y^\mu}|_q f = D_\mu(f \circ \psi \circ y^{-1}) \circ y(q) = D_\mu(f \circ \tilde{x}^{-1}) \circ \tilde{x} \circ \psi(q) = \partial_{\tilde{x}^\mu}|_{\psi(q)} f.$$

In other words,

$$\psi_* \partial_{y^\mu}|_q = \partial_{\tilde{x}^\mu}|_{\psi(q)}.$$

Thus

$$\psi^* g_a(\partial_{y^\mu}|_q, \partial_{y^\nu}|_q) = g_a(\partial_{\tilde{x}^\mu}|_{\psi(q)}, \partial_{\tilde{x}^\nu}|_{\psi(q)}) = \tilde{g}_{a\mu\nu} \circ \tilde{x}^{-1} \circ y(q) = g_{\mu\nu}(q),$$

Where we have used (14.27) in the last step. Thus $\psi^* g_a = g$. The relation $\psi^* \phi_a = \phi$ follows immediately from (14.27) and finally $\psi \circ i = i_a$ where both sides are defined. By reducing the domain O of ψ suitably, we can assume that $O \cap \{0\} \times \Sigma$ is a Cauchy hypersurface in O.

Patching together. For each $p \in i_a(\Sigma)$, we can construct an open set O_p containing $i_a^{-1}(p)$ and a diffeomorphism ψ_p as above. We would like to define O to be the union of the O_p and ψ to be the map given by ψ_p on O_p. In order to be allowed to do that, we need to ensure that $\psi_p = \psi_q$ on $O_p \cap O_q$. Let $r \in O_p \cap O_q$ and let us say that r is to the future of $\{0\} \times \Sigma$. Let γ be a timelike geodesic with $\gamma(0) = r$. Let $s_0 < 0$ be such that $\gamma(s_0) \in \{0\} \times \Sigma$. Then $\gamma([s_0, 0]) \subseteq O_p \cap O_q$. Furthermore, $\psi_p \circ \gamma$ and $\psi_q \circ \gamma$ are both geodesics. Finally, $\psi_p \circ \gamma(s_0) = \psi_q \circ \gamma(s_0)$ and similarly for the tangents at s_0, since ψ_p and ψ_q maps a vector tangent to $\{0\} \times \Sigma$ to the same tangent vector and since both map the future directed unit normal to $\{0\} \times \Sigma$ to the future directed unit normal to $i_a(\Sigma)$. Consequently, $\psi_p \circ \gamma = \psi_q \circ \gamma$ so that $\psi_p(r) = \psi_q(r)$. We have thus constructed a globally hyperbolic open neighbourhood O_a of $\{0\} \times \Sigma$ and a diffeomorphism $\psi_a \colon O_a \to \psi_a(O_a)$ such that $\psi_a^* g_a = g$, $\psi_a^* \phi_a = \phi$ and $\psi \circ i = i_a$. Similarly, we get ψ_b and O_b and since $O_a \cap O_b$ is a globally hyperbolic Lorentz manifold when endowed with the metric g, we get the desired maps by restricting ψ_a and ψ_b to $O_a \cap O_b$. □

15 Cauchy stability

In the case of non-linear wave equations, we proved local stability of solutions in Chapter 9. We now wish to prove an analogous result in the Einstein non-linear scalar field setting. In order to be able to give a precise meaning to the concept of stability, we need to specify a topology on the set of initial data. In Section 15.1, we do so by defining Sobolev spaces on manifolds. There is no canonical choice of norm, but we demonstrate that any two choices are equivalent, cf. Lemma 15.3. The purpose of Cauchy stability is to compare two developments whose initial data are close. However, for reasons mentioned in the introduction to the previous chapter, in the present setting it is not completely obvious how to do so. In fact, it is necessary to choose a diffeomorphism of a subset of one of the developments to a subset of the other in order to make the comparison. In Section 15.2, we specify the structure of the background solutions we are going to consider, and this structure amounts to choosing a class of diffeomorphisms. In Section 15.3 we then prove Cauchy stability. Note that we shall use this result in Chapter 21 in order to prove that there is an open set of initial data such that the corresponding maximal globally hyperbolic developments are future and past timelike geodesically incomplete (i.e., they recollapse).

15.1 Sobolev spaces on manifolds

Definition 15.1. Let M be a compact n dimensional manifold, and assume ϕ_i, $i = 1, \dots, l$ is a finite partition of unity such that $\operatorname{supp}\phi_i \subset U_i$ for open sets U_i. Assume furthermore that (x_i, U_i) are coordinates. Given $T \in \mathcal{T}_s^r(M)$, define

$$\|T\|_{H^k} = \Big(\sum_{i=1}^{l} \sum_{j_1,\dots,j_r=1}^{n} \sum_{i_1,\dots,i_s=1}^{n} \sum_{|\alpha|\le k} \int_{x_i(U_i)} (\phi_i |\partial^\alpha T_{i_1\dots i_s}^{j_1\dots j_r}|^2) \circ x_i^{-1}\, dx_i^1 \dots dx_i^n\Big)^{1/2}, \tag{15.1}$$

where $T_{i_1\dots i_s}^{j_1\dots j_r}$ are the components of T relative to the coordinates x_i and ∂^α signifies differentiation with respect to x_i.

Remark 15.2. In order not to get too cumbersome notation we abuse notation by not clearly indicating with respect to which coordinates we compute components of tensors, etc. That a partition of unity of the desired form exists follows from Lemma 10.3 and the compactness of the manifold.

Note that in the context of the above definition, we can in fact define an inner product (S, T) by the expression

$$\sum_{i=1}^{l} \sum_{j_1,\dots,j_r=1}^{n} \sum_{i_1,\dots,i_s=1}^{n} \sum_{|\alpha|\le k} \int_{x_i(U_i)} (\phi_i \partial^\alpha S_{i_1\dots i_s}^{j_1\dots j_r} \partial^\alpha T_{i_1\dots i_s}^{j_1\dots j_r}) \circ x_i^{-1}\, dx_i^1 \dots dx_i^n.$$

Since (15.1) can be defined in terms of an inner product, it is clear that it defines a norm on the space of smooth tensorfields $\mathcal{T}_s^r(M)$. By taking the completion of the smooth tensorfields one obtains a real Hilbert space, but that will not be of any interest to us here. If one uses a different partition of unity one clearly obtains a different norm, but it is of some interest to note that different partitions of unity yield equivalent norms.

Lemma 15.3. *Let (ϕ_i, U_i), $i = 1, \dots, l$ and (ψ_i, V_i), $i = 1, \dots, m$ be partitions of unity satisfying the properties given in Definition 15.1. Let $\|\cdot\|_{H^k}$ and $|\cdot|_{H^k}$ be the corresponding norms defined in analogy with (15.1). Then there are constants $C_{i,k} > 0$, $i = 1, 2$, depending on k, r, s and on the partitions of unity, such that*

$$C_{1,k}|T|_{H^k} \leq \|T\|_{H^k} \leq C_{2,k}|T|_{H^k}$$

for all $T \in \mathcal{T}_s^r(M)$.

Remark 15.4. One consequence is that if $T_j \to T$ with respect to $|\cdot|_{H^k}$, then $T_j \to T$ with respect to $\|\cdot\|_{H^k}$ and vice versa. It thus makes sense to say that $T_j \to T$ with respect to H^k without any reference to a partition of unity.

Proof. In order to relate the different norms, consider

$$\begin{aligned}
\|T\|_{H^k}^2 & \\
&= \sum_{i=1}^{l} \sum_{j_1,\dots,j_r=1}^{n} \sum_{i_1,\dots,i_s=1}^{n} \sum_{|\alpha|\leq k} \int_{x_i(U_i)} (\phi_i |\partial^\alpha T_{i_1\dots i_s}^{j_1\dots j_r}|^2) \circ x_i^{-1}\, dx_i^1 \dots dx_i^n \\
&= \sum_{i=1}^{l} \sum_{j=1}^{m} \sum_{j_1,\dots,j_r=1}^{n} \sum_{i_1,\dots,i_s=1}^{n} \sum_{|\alpha|\leq k} \int_{x_i(U_i)} (\phi_i \psi_j |\partial^\alpha T_{i_1\dots i_s}^{j_1\dots j_r}|^2) \circ x_i^{-1}\, dx_i^1 \dots dx_i^n.
\end{aligned}$$

We need to change coordinates in

$$\int_{x_i(U_i)} (\phi_i \psi_j |\partial^\alpha T_{i_1\dots i_s}^{j_1\dots j_r}|^2) \circ x_i^{-1}\, dx_i^1 \dots dx_i^n,$$

which is not a problem, since $\phi_i \psi_j$ has compact support in $U_i \cap V_j$. All the terms that arise due to the change of coordinates are under control since we only need to estimate them on a compact subset of $U_i \cap V_j$. Thus

$$\begin{aligned}
&\int_{x_i(U_i)} (\phi_i \psi_j |\partial^\alpha T_{i_1\dots i_s}^{j_1\dots j_r}|^2) \circ x_i^{-1}\, dx_i^1 \dots dx_i^n \\
&\quad \leq C_k \sum_{\beta\leq\alpha} \sum_{i'_1,\dots,i'_s=1}^{n} \sum_{j'_1,\dots,j'_r=1}^{n} \int_{y_j(V_j)} (\psi_j |\partial^\beta T_{i'_1\dots i'_s}^{j'_1\dots j'_r}|^2) \circ y_j^{-1}\, dy_j^1 \dots dy_j^n,
\end{aligned}$$

where the components and derivatives of T in the right-hand side are computed with respect to the coordinates y_j. This proves one inequality, and since the only difference between the two partitions of unity are the labels of different objects, the other follows as well. □

Assume that $T_j \to T$ in H^k, let (U, x) be coordinates, where U is an open set, and let $V \subseteq U$ be such that the closure of V is contained in U (note that the closure is automatically compact). Then, by an argument similar to the proof of the above lemma,

$$\lim_{j\to\infty} \sum_{j_1,\dots,j_r=1}^{n} \sum_{i_1,\dots,i_s=1}^{n} \sum_{|\alpha|\leq k} \int_{x(V)} |\partial^\alpha [T^{j_1\dots j_r}_{i_1\dots i_s} - (T_j)^{j_1\dots j_r}_{i_1\dots i_s}]|^2 \circ x^{-1}\, dx^1 \dots dx^n = 0,$$

where components and derivatives are computed with respect to the coordinates x. In order to make a precise statement concerning stability, we need to be specific concerning the requirements of the background solution.

15.2 Background solutions

Definition 15.5. Assume $(\bar{M}, g)$ is an $n+1$ dimensional Lorentz manifold. A *canonical foliation* of $(\bar{M}, g)$ is a diffeomorphism $\chi\colon \mathbb{R} \times M \to \bar{M}$ for some smooth n-manifold M, such that ∂_t is timelike and the hypersurfaces $\{t\} \times M$ are spacelike with respect to $\chi^* g$.

Remark 15.6. Here ∂_t signifies differentiation with respect to the first coordinate in $\mathbb{R} \times M$. Not all Lorentz manifolds admit canonical foliations.

Proposition 15.7. *If $(\bar{M}, g)$ is an oriented, time oriented, connected and globally hyperbolic Lorentz manifold, then $(\bar{M}, g)$ admits a canonical foliation. Furthermore, if M_0 is a spacelike Cauchy hypersurface, the n-manifold appearing in the definition can be chosen to be M_0 and $\chi(0, p) = p$. On the other hand, if $(\bar{M}, g)$ allows a canonical foliation $\chi\colon \mathbb{R} \times M \to \bar{M}$ with M compact, then $(\bar{M}, g)$ is globally hyperbolic.*

Proof. Assuming $(\bar{M}, g)$ is an oriented, time oriented, connected and globally hyperbolic Lorentz manifold, let $\mathcal{T}$ be the function constructed in Theorem 11.27 where S is replaced by M_0. Define

$$T = \frac{\operatorname{grad} \mathcal{T}}{\langle \operatorname{grad} \mathcal{T}, \operatorname{grad} \mathcal{T} \rangle}.$$

Let $\gamma\colon (t_-, t_+) \to \bar{M}$ be an integral curve of T. We can assume that it is inextendible, because if it is extendible, either it is extendible as an integral curve, or $\gamma(s)$ converges to a point q as s tends to ∞ or $-\infty$. In the latter case, $T_q = 0$, a contradiction. For the details of these arguments, we refer the reader to the proof of Proposition 11.3. By definition, γ is timelike. Due to Theorem 11.27, we conclude that $\mathcal{T}[\gamma(s)] \to \pm\infty$ as $s \to t_\pm$. On the other hand,

$$\frac{d}{ds}(\mathcal{T} \circ \gamma)(s) = \dot{\gamma}(s)[\mathcal{T}] = d\mathcal{T}[\dot{\gamma}(s)] = \langle (\operatorname{grad} \mathcal{T})_{\gamma(s)}, \dot{\gamma}(s) \rangle = 1.$$

We conclude that $t_\pm = \pm\infty$. Thus T is a complete vector field. Due to the proof of Proposition 11.3, the map χ defined by $\chi(t,p) = \Phi(t,p)$ for $p \in M_0$, where Φ is the flow of T is a diffeomorphism onto $\bar{M}$. Since $\chi_* \partial_t = T$, ∂_t is timelike with respect to $\chi^* g$, and since $\{t\} \times M_0$ is mapped to a level set of $\mathcal{T}$, it is spacelike.

Assume now that $(\bar{M}, g)$ allows a canonical foliation with M compact. Let us use χ to identify $\bar{M}$ with $\mathbb{R} \times M$ and let t denote the function which maps (τ, p) to τ. Let x be coordinates on $U \subseteq M$. Then t and x together form coordinates on $\mathbb{R} \times U$. If we let $\partial_i = \partial/\partial x^i$, $\partial_0 = \partial/\partial t$ and $g_{\mu\nu} = g(\partial_\mu, \partial_\nu)$, then the definition of a canonical foliation implies that $g_{00} < 0$ and that g_{ij} are the components of a positive definite matrix. Consequently, Lemma 8.5 implies that g^{00} is negative. Since this is the inner product of the gradient of t with itself, we conclude that the gradient of t is timelike. Consequently, t is strictly increasing along causal curves. Every $\{\tau\} \times M$ is thus intersected at most once by every inextendible causal curve. Since $[T_1, T_2] \times M$ is compact, every inextendible causal curve must leave such an interval and this can be used to prove that every hypersurface $\{\tau\} \times M$ is a Cauchy hypersurface, which is also spacelike. The proposition follows. □

Definition 15.8. Let M be a compact n dimensional manifold and let g be a smooth Lorentz metric on $I \times M$ where I is an open interval. Let t be the function defined by $t(\tau, p) = \tau$ and assume that $g(\partial_t, \partial_t) = g_{00} < 0$ and that the hypersurfaces $\{\tau\} \times M$ are spacelike with respect to g for $\tau \in I$. Finally, assume that $\phi\colon I \times M \to \mathbb{R}$ is a smooth function which, together with g, satisfies

$$R_{\mu\nu} - \nabla_\mu \phi \nabla_\nu \phi - \frac{2}{n-1} V(\phi) g_{\mu\nu} = 0, \tag{15.2}$$

$$\nabla^\mu \nabla_\mu \phi - V'(\phi) = 0. \tag{15.3}$$

Then we shall call $(I \times M, g, \phi)$ a *background solution.*

Definition 15.9. Let g be a Lorentz metric and ϕ be a smooth function on $(T_-, T_+) \times M$ and let $\tau \in (T_-, T_+)$. Assume $\{\tau\} \times M$ is spacelike with respect to g and let $i\colon M \to \mathbb{R} \times M$ be defined by $i(p) = (\tau, p)$. Let g_0 be the Riemannian metric on M obtained by pulling back the Riemannian metric induced on $\{\tau\} \times M$ by g by i, let k_0 be the covariant 2-tensor obtained by pulling back the second fundamental form induced on $\{\tau\} \times M$ by g by i, let $\phi_0 = \phi \circ i$ and let $\phi_1 = (N\phi) \circ i$, where N is the future directed unit normal to $\{\tau\} \times M$ with respect to g. Then we shall refer to $(g_0, k_0, \phi_0, \phi_1)$ as the *initial data induced on* $\{\tau\} \times M$ by (g, ϕ), or simply the initial data induced on $\{\tau\} \times M$ if the solution is understood from the context.

15.3 Cauchy stability in general relativity

Theorem 15.10. *Let $(I \times M, g, \phi)$ be a background solution. Let $(g_0, k_0, \phi_0, \phi_1)$ be the initial data induced on $\{0\} \times M$ by (g, ϕ). Assume ρ_j is a sequence of Riemannian metrics on M, κ_j a sequence of covariant 2-tensors and $\psi_{0,j}$ and $\psi_{1,j}$ are a sequence of smooth functions such that ρ_j and $\psi_{0,j}$ converge to g_0 and ϕ_0 respectively in H^{l+1}*

and κ_j and $\psi_{1,j}$ converge to k_0 and ϕ_1 respectively in H^l, where $l > n/2+1$. Assume furthermore that $(\rho_j, \kappa_j, \psi_{0,j}, \psi_{1,j})$ satisfy the constraint equations (14.9)–(14.10) *with (g_0, k, ϕ_0, ϕ_1) replaced by $(\rho_j, \kappa_j, \psi_{0,j}, \psi_{1,j})$. Then there are $T_{j,+}, T_{j,-} \in \mathbb{R}$, a Lorentz metric h_j on $\bar{M}_j = (T_{j,-}, T_{j,+}) \times M$ and a smooth function ψ_j on $\bar{M}_j$ such that (h_j, ψ_j) satisfy* (15.2)–(15.3) *on $\bar{M}_j$. Furthermore, the initial data induced on $\{0\} \times M$ by (h_j, ψ_j) are $(\rho_j, \kappa_j, \psi_{0,j}, \psi_{1,j})$, ∂_t is timelike with respect to h_j and $\{\tau\} \times M$ is a spacelike Cauchy hypersurface with respect to h_j for all $\tau \in (T_{j,-}, T_{j,+})$. If $T \in I$, then $T \in (T_{j,-}, T_{j,+})$ for j large enough and the initial data induced on $\{T\} \times M$ by (h_j, ψ_j) converge to the corresponding initial data of (g, ϕ).*

Remark 15.11. When we speak of a time orientation, we take it for granted that ∂_t is future oriented.

Proof. Given the initial data $(\rho_j, \kappa_j, \psi_{0,j}, \psi_{1,j})$, we wish to find a Lorentz metric h and a smooth function ψ solving

$$\hat{R}_{\mu\nu} - \nabla_\mu \psi \nabla_\nu \psi - \frac{2}{n-1} V(\psi) h_{\mu\nu} = 0, \tag{15.4}$$

$$\nabla^\mu \nabla_\mu \psi - V'(\psi) = 0 \tag{15.5}$$

on $(T_{j,-}, T_{j,+}) \times M$, for some $T_{j,-} < 0$ and $T_{j,+} > 0$, such that the initial data induced on $\{0\} \times M$ by (h, ψ) are $(\rho_j, \kappa_j, \psi_{0,j}, \psi_{1,j})$. Here, the gauge source functions F_μ are given by

$$F_\mu = h_{\mu\nu} h^{\alpha\beta} \bar{\Gamma}^\nu_{\alpha\beta}$$

and $\bar{\Gamma}^\nu_{\alpha\beta}$ are the Christoffel symbols of g. Note that

$$\hat{R}_{\mu\nu} = R_{\mu\nu} + \nabla_{(\mu} \mathcal{D}_{\nu)},$$

where $R_{\mu\nu}$ is the Ricci tensor associated with h, $\mathcal{D}_\mu = F_\mu - \Gamma_\mu$ are the components of a covector and Γ_ν are the contracted Christoffel symbols of h. Concerning notation, we shall denote the solution we obtain by (h, ψ) even though it clearly depends on the label j of the initial data. The reason we do so is that we want to avoid cumbersome notation. In the end we shall distinguish between the different solutions by giving them a label j.

Construction of initial data. We wish to construct initial data for all the components of h and their time derivatives. Concerning the spatial components, we demand that

$$h_{im}|_{t=0} = \rho_j(\partial_i, \partial_m).$$

In order to specify the h_{00} and h_{0i} components, we use the criterion that N should be a unit normal to $\{0\} \times M$ for h as well as for g. Let

$$N = N^0 \partial_t + N^i \partial_i,$$

where $\partial_i = \partial/\partial x^i$ for coordinates x on some open subset U of M. Then N^i are the components of a smooth vector field on M and N^0 is a smooth function. In local coordinates, the condition that N be a unit normal to $\{0\} \times M$ translates into

$$h_{0m}|_{t=0} = -\frac{1}{N^0} N^i h_{im}|_{t=0},$$
$$h_{00}|_{t=0} = -\frac{1}{(N^0)^2}\{1 - N^i N^m h_{im}|_{t=0}\}.$$

Note that $h_{0i}|_{t=0}$ transforms as a covector on M under change of coordinates on M and that $h_{00}|_{t=0}$ is independent of choice of coordinates on M. Note that as $j \to \infty$, $h_{\mu\nu}|_{t=0}$ converges to $g_{\mu\nu}|_{t=0}$ in H^{l+1}. Some care is required in making this statement, since $g_{\mu\nu}$ and $h_{\mu\nu}$ are not intrinsic to $\{0\} \times M$. On the other hand, we can interpret $g_{\mu\nu}$ as a Riemannian metric with components g_{ij}, a one form with components g_{0i} and a function g_{00}. In this way, we can think of $g_{\mu\nu}|_{t=0}$ as consisting of three objects which are intrinsic to $\{0\} \times M$. In order to define $\partial_t h_{\mu\nu}|_{t=0}$, let us note that if we already had a metric h, the associated second fundamental form would have the property that

$$k(\partial_i, \partial_m) = \langle \nabla_{\partial_i} N, \partial_m \rangle = \partial_i(N^0) h_{0m} + N^0 \langle \nabla_{\partial_i} \partial_0, \partial_m \rangle + \langle \nabla_{\partial_i} X, \partial_m \rangle,$$

where X is defined by $N = N^0 \partial_0 + X$. Let us make some observations. First, $\partial_i(N^0) h_{0m}$ transforms as a covariant 2-tensor under change of coordinates on M and is already defined at $t = 0$, since $h_{0m}|_{t=0}$ is defined. The object $\langle \nabla_{\partial_i} X, \partial_m \rangle$ is intrinsic to M since all the vector fields appearing are tangent to M (and thus we can replace the Levi-Civita connection of the spacetime, i.e. that of h, by the Levi-Civita connection of M, i.e. that of ρ_j), so that it is already defined. It furthermore transforms as a covariant 2-tensor under change of coordinates on M. Finally, let us consider

$$\langle \nabla_{\partial_i} \partial_0, \partial_m \rangle = \frac{1}{2}(\partial_i h_{m0} - \partial_m h_{i0}) + \frac{1}{2}\partial_0 h_{im}.$$

Note that h_{i0} are the components of a covector and that $\partial_i h_{m0} - \partial_m h_{i0}$ are the components of the antisymmetrized covariant derivative of the corresponding 1-form with respect to the Levi-Civita connection of ρ_j. The expression $\partial_i h_{m0} - \partial_m h_{i0}$ can also be viewed as the components of $d\lambda$ where $\lambda = 2h_{0m} dx^m$. Consequently, the expression $\partial_i h_{m0} - \partial_m h_{i0}$ transforms as a covariant 2-tensor under changes of coordinates on M. Furthermore, this expression has already been defined. As a consequence of the above observations, we can define

$$\partial_t h_{im}|_{t=0} = \partial_m h_{i0} - \partial_i h_{m0} + \frac{2}{N^0}\left[\kappa_j(\partial_i, \partial_m) - \partial_i(N^0) h_{0m} - \langle \nabla_{\partial_i} X, \partial_m \rangle\right],$$

where ∇ in $\nabla_{\partial_i} X$ should be interpreted as the Levi-Civita connection of ρ_j. The right-hand side of this expression has already been defined and it transforms as a covariant 2-tensor under change of coordinates on M. Furthermore $\partial_t h_{im}|_{t=0}$ converges to $\partial_t g_{im}|_{t=0}$ in H^l as $j \to \infty$.

We shall let the condition that $\mathcal{D}_\mu|_{t=0} = 0$ determine $\partial_0 h_{00}|_{t=0}$ and $\partial_0 h_{0i}|_{t=0}$. Due to (8.6) and (8.8), h^{00} transforms as a function under changes of coordinates on M and h^{0i} transforms as a vector. Since

$$h^{kl}h_{lm} + h^{k0}h_{0m} = \delta^k_m,$$

and the inverse of the Riemannian metric with components h_{im} is a contravariant 2-tensor, we conclude that h^{im} transforms as a contravariant 2-tensor under changes of coordinates on M. Furthermore, h^{00}, h^{0i} and h^{im} converge to g^{00}, g^{0i} and g^{im} respectively in H^{l+1} as $j \to \infty$. In order to facilitate our seeing that the expressions $\partial_0 h_{0i}$ and $\partial_0 h_{00}$ the condition that $\mathcal{D}_\mu|_{t=0} = 0$ implies transform as a covector and scalar respectively under changes of coordinates on M, let us introduce a metric on $\mathbb{R} \times M$ by

$$\hat{h}_{\mu\nu}(\tau, p) = h_{\mu\nu}(0, p)$$

with respect to a coordinate system of the form (t, x) on $\mathbb{R} \times U$, where x are coordinates on $U \subseteq M$. Assuming we had a metric h, let us denote the corresponding Christoffel symbols by $\Gamma^\mu_{\alpha\beta}$. Let $\hat{\Gamma}^\mu_{\alpha\beta}$ and $\bar{\Gamma}^\mu_{\alpha\beta}$ denote the Christoffel symbols of $\hat{h}$ and g respectively. Note that due to the observations made in connection to (14.5), their difference are the components of a tensor. We can write

$$\hat{h}_{\mu\nu}\hat{h}^{\alpha\beta}\bar{\Gamma}^\nu_{\alpha\beta} = \hat{h}_{\mu\nu}\hat{h}^{\alpha\beta}(\bar{\Gamma}^\nu_{\alpha\beta} - \hat{\Gamma}^\nu_{\alpha\beta}) + \hat{\Gamma}_\mu.$$

The first term on the right-hand side are the components of a tensor. Consequently, we know that it transforms properly. At $t = 0$, $\hat{h}_{\mu\nu} = h_{\mu\nu}$, so that at $t = 0$, the equality we desire is

$$\Gamma_\mu = \hat{h}_{\mu\nu}\hat{h}^{\alpha\beta}(\bar{\Gamma}^\nu_{\alpha\beta} - \hat{\Gamma}^\nu_{\alpha\beta}) + \hat{\Gamma}_\mu. \tag{15.6}$$

Compute

$$\Gamma_0 = \frac{1}{2}h^{im}(\partial_i h_{m0} + \partial_m h_{i0} - \partial_0 h_{im}) + h^{0i}\partial_i h_{00} + \frac{1}{2}h^{00}\partial_0 h_{00}.$$

Note that all the terms appearing which involve a spatial derivative of a component of h cancel against the corresponding term in $\hat{\Gamma}_0$. In the corresponding expression for $\hat{h}$, the terms which involve time derivatives vanish, so that after the cancellation, no terms in $\hat{\Gamma}_0$ remain. The remaining terms in Γ_0 are

$$-\frac{1}{2}h^{im}\partial_0 h_{im} + \frac{1}{2}h^{00}\partial_0 h_{00}.$$

At $t = 0$, we have already defined the first term, and we know that it is independent of the choice of coordinates on M. To conclude, we see that (15.6) for $\mu = 0$ defines $\partial_0 h_{00}$ at $t = 0$ and that the definition is independent of choice of coordinates on M. Compute

$$\begin{aligned}\Gamma_i = {} & \frac{1}{2}h^{km}(\partial_k h_{mi} + \partial_m h_{ki} - \partial_i h_{km}) + h^{0m}(\partial_0 h_{mi} + \partial_m h_{0i} - \partial_i h_{0m}) \\ & + \frac{1}{2}h^{00}(2\partial_0 h_{0i} - \partial_i h_{00}).\end{aligned}$$

The terms that involve spatial derivatives of the components of h cancel against $\hat{\Gamma}_i$ and what remains is

$$h^{0m}\partial_0 h_{mi} + h^{00}\partial_0 h_{0i}.$$

The first term transforms as a covector under changes of coordinates on M, and consequently (15.6) defines $\partial_0 h_{0i}$ and demonstrates that this object transforms as a covector under change of coordinates on M. Due to the above definitions, $\partial_0 h_{00}$ and $\partial_0 h_{0i}$ converge to $\partial_0 g_{00}$ and $\partial_0 g_{0i}$ respectively in H^l as $j \to \infty$ in H^l. Note that these objects are a function and a covector respectively and that the convergence is most easily seen by not introducing the metric $\hat{h}$.

Finally, $\psi|_{t=0}$ is defined by $\psi_{0,j}$ and $\partial_t\psi$ is defined by

$$\partial_0\psi|_{t=0} = \frac{1}{N^0}(\psi_{1,j} - N^i\partial_i\psi_{0,j}).$$

Then ψ and $\partial_0\psi$ converge to ϕ and $\partial_t\phi$ respectively in H^{l+1} and H^l respectively as $j \to \infty$.

Geometric setup. In the end we shall be interested in proving that if $T \in I$, then it belongs to the existence interval of (h_j, ψ_j) for j large enough and that the initial data induced on $\{T\} \times M$ converges to the initial data of (g, ϕ). The problem is of course that we have a compact manifold so that we have to carry out the analysis in coordinate charts. Assume, without loss of generality, that $T > 0$ and let ρ be some fixed Riemannian metric on M. Let $\varepsilon > 0$ be such that $J_\varepsilon = [-\varepsilon, T+\varepsilon] \subset I$. For every $\tau \in J_\varepsilon$, g induces a Riemannian metric on $\{\tau\} \times M$. Let us denote the corresponding metric on M by $g_\flat(\tau)$. By for example looking at local coordinate systems of the form $I \times V$ for suitably chosen V, one can convince oneself that there is a constant $C_1 > 1$ such that

$$\frac{1}{C_1}\rho \leq g_\flat(\tau) \leq C_1\rho \tag{15.7}$$

on M for all $\tau \in J_\varepsilon$. Note that the components g_{0i} with respect to standard coordinates on $I \times V$ define a 1-form on each hypersurface $\{\tau\} \times M$. We shall denote the corresponding 1-form on M by $\eta(\tau)$. By arguments similar to the ones that led to (15.7), we can argue that there is a constant C_2 such that

$$|\eta(\tau)| \leq C_2 \tag{15.8}$$

on M for all $\tau \in J_\varepsilon$, where $|\eta(\tau)|$ denotes the length of $\eta(\tau)$ with respect to ρ. Finally, there is a $C_3 > 1$ such that

$$-C_3 \leq g_{00} \leq -\frac{1}{C_3} \tag{15.9}$$

on $J_\varepsilon \times M$. Let γ be a causal curve with respect to g. We shall denote the projection of this curve to M by $\gamma_\flat$, so that $\gamma = (\gamma^0, \gamma_\flat)$. The fact that the curve is causal can be expressed in terms of local coordinates as

$$g_{00}\dot{\gamma}^0\dot{\gamma}^0 + 2g_{0i}\dot{\gamma}^0\dot{\gamma}^i + g_{im}\dot{\gamma}^i\dot{\gamma}^m \leq 0.$$

We get, using (15.7),

$$\frac{1}{C_1}\rho(\dot{\gamma}_\flat, \dot{\gamma}_\flat) \le -g_{00}\dot{\gamma}^0\dot{\gamma}^0 - 2g_{0i}\dot{\gamma}^0\dot{\gamma}^i \le C_3\dot{\gamma}^0\dot{\gamma}^0 + 2|\eta||\dot{\gamma}_\flat||\dot{\gamma}^0|,$$

where $|\dot{\gamma}_\flat|$ denotes the length of $\dot{\gamma}_\flat$ with respect to ρ, $|\dot{\gamma}^0|$ is the ordinary absolute value of a real number and we have used (15.9). We conclude that

$$\rho(\dot{\gamma}_\flat, \dot{\gamma}_\flat) \le C_1C_3|\dot{\gamma}^0|^2 + \frac{1}{2}|\dot{\gamma}_\flat|^2 + 2C_1^2C_2^2|\dot{\gamma}^0|^2,$$

where we have used the Schwartz inequality. Consequently,

$$|\dot{\gamma}_\flat|^2 \le 2(C_1C_3 + 2C_1^2C_2^2)|\dot{\gamma}^0|^2.$$

Integrating this inequality, we conclude that there is a constant $C > 0$ such that for a curve $\gamma\colon [a,b] \to I \times M$, causal with respect to g, with the property that $t \circ \gamma$ takes values in J_ε,

$$l_\rho(\gamma_\flat) \le C\{t[\gamma(b)] - t[\gamma(a)]\}, \tag{15.10}$$

where $l_\rho(\gamma_\flat)$ is the length of $\gamma_\flat$ with respect to ρ. Furthermore, the constant C only depends on the bounds (15.7)–(15.9). Consequently, if we have a family of metrics for which bounds of this form holds, then we get an estimate of the form (15.10) with a constant independent of the choice of Lorentz metric in the family under consideration.

Let, for $p \in M$, $B_r(p)$ denote the set of points in M at a distance strictly less than r from p, where the distance is measured with respect to ρ. Let $p_i \in M, i = 1, \dots, N$ and $r_0 > 0$ be such that for each i, $B_{3r_0}(p_i)$ is a normal neighbourhood of p_i (with respect to ρ) and $B_{r_0}(p_i), i = 1, \dots, N$ is a covering of M.

Local existence. Let $V_i = I \times B_{3r_0}(p_i)$ and consider the equations (15.4)–(15.5) on V_i. Let $x^1, \dots, x^n$ be normal coordinates on $B_{3r_0}(p_i)$ with respect to ρ, so that we get coordinates $x^0, \dots, x^n$, where $x^0 = t$, on V_i. We shall call such coordinates standard coordinates. In order to be able to prove local existence, we need to modify the equations and the initial data similarly to the proof of Theorem 14.2. From now on, we shall restrict our attention to the set V_i on which we have the above coordinates. Consequently, we shall think of all the objects involved in the equations we are studying to be identical with the components of these objects with respect to the specified coordinates. We shall thus speak of $g_{\mu\nu}$, $h_{\mu\nu}$, etc., instead of the metrics. Let us define a smooth function A mapping Lorentz matrices to Lorentz matrices. We shall denote the components of A by $A_{\mu\nu}$. Let $I_0 = [a_0, b_0]$ be an interval with $b_0 < 0$ such that $g_{00}(\tau, p)$ and $h_{00}(0, p)$ are contained in I_0 for $p \in B_{2r_0}(p_i)$, $\tau \in [0, T]$ and $i = 1, \dots, N$. Let A_{00}, considered as a function of, say $c_{\alpha\beta}$, only depend on c_{00} and have the properties that $A_{00} = c_{00}$ for $c_{00} \in [a_0 - 1, b_0/2]$ and that the range of A_{00} is contained in $[a_0 - 2, b_0/4]$. Let $r_1 > 0$ be such that the Euclidean lengths of $g_{0m}(\tau, p)$ and $h_{0m}(0, p)$ are less than r_1 for $\tau \in [0, T]$, $p \in B_{2r_0}(p_i)$ and $i = 1, \dots, N$. Let A_{0m} be such that $A_{0m} = c_{0m}$ as long as the Euclidean length of c_{0m} is less than $2r_1$

and let A_{0m} have a range contained in the set of vectors of Euclidean length less than $3r_1$. Finally, let K_0 be a compact subset of the set of positive definite $n \times n$-matrices such that the matrices whose components are given by $g_{i_1 j_1}(\tau, p)$ and $h_{i_1 j_1}(0, p)$ are contained in K_0 for $\tau \in [0, T]$, $p \in B_{2r_0}(p_i)$ and $i = 1, \dots, N$. Let K_1 be a compact subset of the set of positive definite $n \times n$-matrices that contains K_0 in its interior and let $A_{km} = c_{km}$ as long as c_{km} are the components of a positive definite matrix contained in K_1. Assume finally that the range of the matrices with components A_{km} is contained in a compact subset of the positive definite $n \times n$-matrices. Note that since the constructed initial data for $h_{\mu\nu}$ converges to the initial data for $g_{\mu\nu}$, we can in fact use the same A for all the different initial data. When considering (15.4)–(15.5) we replace $h^{\mu\nu}$, where ever it appears, by $A^{\mu\nu}$, where $A^{\mu\nu}$ are the components of the inverse of A and we take it for granted that we evaluate it on $h_{\alpha\beta}$. Furthermore, we replace F_μ with $\chi_1 F_\mu$, where $\chi_1 \in C_0^\infty[I \times B_{3r}(p_i)]$ is such that $\chi_1 = 1$ on $J_\varepsilon \times B_{2r}(p_i)$. Thus we have modified the equation in such a way that our local existence results apply. However, we also need to modify the initial data. Let χ be a smooth function with compact support in $B_{3r_0}(p_i)$ such that $\chi = 1$ on $B_{2r_0}(p_i)$. Modify all the initial data by multiplying them by χ, pull them back to $\mathbb{R}^n$ using the normal coordinates and denote the collection of them u_0 (for the initial function) and u_1 (for the initial time derivative). The resulting equation on $\mathbb{R}^{n+1}$ is of such a form that Corollary 9.16 is applicable. Thus we get a local smooth solution u (in fact, it will be indexed by j). Before patching together these solutions, let us introduce some notation concerning the sets we shall be working with. Let

$$\mathcal{C}_{r,s,\tau_0,\tau_1}(p) = \{(u, q) \in \mathbb{R} \times M : q \in B_{r-s|u-\tau_0|}(p),\ |u - \tau_0| < \tau_1\}.$$

Consider standard coordinates on some region Ω of $\mathbb{R} \times B_{2r_0}(p_i)$. Assume that h is a metric such that with respect to these coordinates, it takes values in the range of A. Then h satisfies estimates of the form (15.7)–(15.9) on Ω. We conclude that there is a constant C, only depending on A, such that if γ is a causal curve with respect to h which is contained in Ω, then (15.10) holds. Note that if $\mathcal{C}_{r,s,\tau_0,\tau_1}(p)$ is contained in Ω and $s = C$, we can thus conclude that $\mathcal{C}_{r,s,\tau_0,\tau_1}(p)$ is globally hyperbolic with respect to h and a Cauchy hypersurface is given by $\{\tau_0\} \times B_r(p)$.

Patching together. In practice it will not be enough to start only at $\tau = 0$, we need to start at any $\tau \in [0, T]$. Let us assume we have initial data $h_{\mu\nu}|_{t=\tau}$, $\partial_t h_{\mu\nu}|_{t=\tau}$, $\psi|_{t=\tau}$ and $\partial_t \psi|_{t=\tau}$ which are indexed by j (which we suppress in order to avoid cumbersome notation) and converge to $g_{\mu\nu}|_{t=\tau}$, etc., respectively. Note that $\partial_t h_{\mu\nu}|_{t=\tau}$ should be interpreted as a Riemannian metric, a one form and a function, etc. Assume furthermore that the initial data are such that $\mathcal{D}_\mu|_{t=\tau} = 0$ and that the constraint equations are satisfied. Assume that we start at $\{\tau\} \times B_{3r_0}(p_i)$. Let A be constructed as above (note that it does not depend on the starting point τ, only on g and on the initial data at $\tau = 0$) and χ be a cutoff function as above. After cutting off the initial data, pulling them back to $\mathbb{R}^n$ and solving the modified equations, we get a solution u_j. After cutting off the initial data of $g_{\mu\nu}$ and ϕ, pulling them back to $\mathbb{R}^n$ and solving the modified equations, we get a solution u. Note that the cut off initial data for g and ϕ satisfy a bound in suitable Sobolev spaces after having been pulled back to $\mathbb{R}^n$ in the specific coordinates under consideration, and the bound is independent of $\tau \in [0, T]$ and p_i. Since the initial data of (h, ψ) converge to the initial data of (g, ϕ) where the norm used for the initial functions is H^{l+1} and the norm for the initial time derivative is H^l, we get an existence time for the solutions u_j and u, independent of τ, p_i and j (for j large enough). Let this existence time be $2\varepsilon_b > 0$. Our first goal is to prove that the solution u coincides with (g, ϕ) on some suitable set. Consider $\mathcal{C}_{i,\tau} = \mathcal{C}_{2r_0,C,\tau,\varepsilon_b}(p_i)$. Note that on this set (g, ϕ), expressed in terms of standard coordinates and pulled back to $\mathbb{R}^{n+1}$, solve the modified equations (i.e., the ones in which we replaced $h^{\mu\nu}$ by $A^{\mu\nu}$). Furthermore, $\mathcal{C}_{i,\tau}$ is globally hyperbolic with respect to both g and A (evaluated on the metric belonging to the solution u). Finally, u and (g, ϕ) have the same initial data on the Cauchy hypersurface $\{\tau\} \times B_{2r_0}(p_i)$. Deriving an equation for the difference of the two solutions along the lines of the proof of Lemma 9.7 and using Corollary 12.14, we conclude that u and (g, ϕ) have to equal on $\mathcal{C}_{i,\tau}$. For j large enough, u_j will be arbitrarily close to u when measured with respect to the $H^{l+1} \times H^l$ norm of the data induced on spatial hypersurfaces, cf. Proposition 9.17. As a consequence, A, when evaluated on the metric belonging to the solution u_j will coincide with the metric belonging to the solution u_j for j large enough, if we restrict our attention to $\mathcal{C}_{i,\tau}$. Thus on $\mathcal{C}_{i,\tau}$, we get a solution $h_{(j,i,\tau)}, \psi_{(j,i,\tau)}$ to the original equation. Since $\mathcal{D}_\mu|_{t=\tau} = 0$ and $\partial_t \mathcal{D}_\mu|_{t=\tau} = 0$ (since the constraints are satisfied, cf. the argument at the end of Section 14.2) and since $\mathcal{C}_{i,\tau}$ is globally hyperbolic with respect to $h_{(j,i,\tau)}$, (14.8) implies that $\mathcal{D}_\mu = 0$ in $\mathcal{C}_{i,\tau}$. Consequently, $h_{(j,i,\tau)}, \psi_{(i,j,\tau)}$ satisfy (15.2)–(15.3). Let us compare $h_{(j,i,\tau)}, \psi_{(i,j,\tau)}$ for two different i's, say i_1, i_2. For the sake of brevity, let us introduce the notation $h_k = h_{(j,i_k,\tau)}$ and similarly for ψ. If we change coordinates and express h_1 in terms of the standard coordinates on $\mathcal{C}_{i_2,\tau}$, say x, then the initial data of h_1, ψ_1 coincide with the initial data of h_2, ψ_2 when expressed in terms of the coordinates x (if $\tau = 0$, we have proved this statement, and if $\tau \neq 0$, it is part of the assumptions). Furthermore, since $\mathcal{D}_\mu$ are zero for both solutions and since they transform as covectors, $\mathcal{D}_\mu$ computed for h_1 with respect to x-coordinates vanishes. Thus h_k, ψ_k, both expressed with respect to x-coordinates satisfy (15.4)–(15.5). Finally, $\mathcal{C}_{i_1,\tau} \cap \mathcal{C}_{i_2,\tau}$ is globally hyperbolic with respect to both metrics h_k by the above construction. We conclude that h_k, ψ_k, when expressed in x-coordinates, coincide in this set. Consequently, we get a solution $h_{(j,\tau)}, \psi_{(j,\tau)}$ on

$$\bigcup_{i=1}^{N} \mathcal{C}_{i,\tau},$$

which is a globally hyperbolic manifold with respect to $h_{(j,\tau)}$. Let $\varepsilon_0 \leq \varepsilon_b$ be such that $\varepsilon_0 \leq r_0/C$. Then we get a solution on

$$\bigcup_{i=1}^{N} \mathcal{C}_{2r_0,C,\tau,\varepsilon_0}(p_i) = [\tau - \varepsilon_0, \tau + \varepsilon_0] \times M,$$

since $B_{r_0}(p_i) \subseteq B_{2r_0 - C\varepsilon_0}(p_i)$ by the definition of ε_0 and the union of the $B_{r_0}(p_i)$ cover M. Assuming the jth initial data at τ converge to the initial data around which we are perturbing, the initial data induced on $\tau' \in (\tau - \varepsilon_0, \tau + \varepsilon_0)$, converges to the initial data of the background solution, satisfy the constraint equations (since they are solutions to the Einstein scalar field system) and have the property that $\mathcal{D}_\mu|_{t=\tau'} = 0$. The reason for the convergence is that in each $\mathcal{C}_{i,\tau}$ we get convergence. Consequently, we get convergence on $\{\tau'\} \times B_{r_0}(p_i)$ for every i. Since $B_{r_0}(p_i)$, $i = 1, \dots, N$ is an open covering of M, we can choose a partition of unity subordinate to this covering and construct the corresponding H^l norms. Patching up the local convergence, we thus get convergence on $\{\tau'\} \times M$. Finally, starting at $\tau = 0$, we can progress a distance ε_0 into the future by choosing j large enough, and we get convergence for all hypersurfaces in between 0 and ε_0. Starting at $3\varepsilon_0/4$, we can then progress to $3\varepsilon_0/2$, etc. In the end, we get convergence at $\tau = T$. Finally, the global hyperbolicity of h_j on $(T_{j,-}, T_{j,+}) \times M$ follows from the geometric setup. □

16 Existence of a maximal globally hyperbolic development

In the present chapter, we prove the main result of these notes. Since the axiom of choice plays a central role in the argument, we begin by reminding the reader of the terminology in Section 16.1. In Section 16.2 we then define the concept of a maximal globally hyperbolic development and prove that it exists, given initial data. The arguments presented in this chapter are taken from the paper [10] by Yvonne Choquet-Bruhat and Robert Geroch.

16.1 Background from set theory

Definition 16.1. A *partial ordering* on a set X is a relation $\leq$ on X such that

- $a \leq a$ for all $a \in X$, i.e., the relation is reflexive;
- $a \leq b$ and $b \leq a$ implies $a = b$, i.e., the relation is antisymmetric;
- $a \leq b$ and $b \leq c$ implies $a \leq c$, i.e., the relation is transitive.

A set together with a partial ordering is called a *partially ordered set*.

Definition 16.2. A partially ordered set is said to be *totally ordered* if $a, b \in X$ implies $a \leq b$ or $b \leq a$.

Definition 16.3. If $(X, \leq)$ is a partially ordered set and $A \subseteq X$, then $x \in X$ is an *upper bound* for A if $a \in A$ implies $a \leq x$. A *maximal element* of X is an $x \in X$ such that $x' \geq x$ implies $x' = x$.

Let us state the theorem we shall need. The reader interested in a proof is referred to Theorem B.18, p. 526 of [8].

Theorem 16.4. *The following statements are equivalent:*

- (*Maximality principle*) *If X is a partially ordered set such that every totally ordered subset has an upper bound, then X has a maximal element.*
- (*Axiom of choice*) *If $\{S_\alpha : \alpha \in A\}$ is an indexed family of nonempty sets S_α, then there exists a function $f : A \to \bigcup S_\alpha$ such that $f(\alpha) \in S_\alpha$ for all $\alpha \in A$.*

Most people refer to the first statement as Zorn's lemma. We have not done so since, according to [8] p. 527, the result is due, independently, to R. L. Moore and Kuratowski a dozen years before Zorn. If we accept the axiom of choice, we are thus free to use the maximality principle.

16.2 Existence of a maximal globally hyperbolic development

Definition 16.5. Let $(\Sigma, g_0, k, \phi_0, \phi_1)$ be initial data to (13.3)–(13.4). Recall that a development (M, g, ϕ) of these initial data such that $i(\Sigma)$ is a Cauchy hypersurface in M is called a globally hyperbolic development of the initial data. We shall say that a globally hyperbolic development (M, g, ϕ), with embedding $i: \Sigma \to M$, is a *maximal globally hyperbolic development* of the initial data if for every other globally hyperbolic development (M', g', ϕ') with embedding $i': \Sigma \to M'$ there is a map $\psi: M' \to M$ which is a time orientation preserving diffeomorphism onto its image such that $\psi^* g = g'$, $\psi^*\phi = \phi'$ and $\psi \circ i' = i$.

Assume (M, g, ϕ) and (M', g', ϕ') are two maximal globally hyperbolic developments of the same data. Then there are maps $\psi: M \to M'$ and $\psi': M' \to M$ with properties as in the statement of the definition. In particular $(\psi' \circ \psi)^* g = g$ and $\psi' \circ \psi$ is the identity map on $i(\Sigma)$. Let $p \in J^+[i(\Sigma)]$ and let γ be a future directed inextendible timelike geodesic with $\gamma(0) = p$. Then there is an $s_0 < 0$ such that $\gamma(s_0) \in i(\Sigma)$. Thus $\psi' \circ \psi \circ \gamma(s_0) = \gamma(s_0)$. Furthermore, $\psi' \circ \psi$ sends a tangent vector to $i(\Sigma)$ to itself and since it is a time orientation preserving isometry, it sends the future directed unit normal to itself. In other words, $(\psi' \circ \psi)_*$ is the identity on $T_p M$ for $p \in i(\Sigma)$. We conclude that $(\psi' \circ \psi \circ \gamma)'(s_0) = \gamma'(s_0)$ so that $\psi' \circ \psi \circ \gamma = \gamma$. Thus $\psi' \circ \psi(p) = p$. The argument is the same for $p \in J^-[i(\Sigma)]$ and we conclude that $\psi' \circ \psi = \mathrm{Id}$. Similarly $\psi \circ \psi' = \mathrm{Id}$. As a consequence, we see that if (M, g, ϕ) and (M', g', ϕ') are two maximal globally hyperbolic developments of the same data, M and M' are diffeomorphic with a diffeomorphism $\psi: M \to M'$ such that $\psi^* g' = g$, $\psi^* \phi' = \phi$ and $\psi \circ i = i'$. If, given two developments of the same initial data, there is a diffeomorphism with the above properties, we shall say that the developments are *isometric*. Note that this means not only that the diffeomorphism is an isometry but also that it preserves the scalar field and respects the embedding. With this terminology, a maximal globally hyperbolic development is unique up to isometry.

Given initial data $(\Sigma, g_0, k, \phi_0, \phi_1)$ to (13.3)–(13.4), we shall below speak of the collection of all globally hyperbolic developments, denoted $\mathcal{M}$, of the initial data. For those readers who are uncomfortable about dealing with such a class, let us note that due to Proposition 11.3, a globally hyperbolic development (M, g, ϕ) has the property that $M \cong \mathbb{R} \times i(\Sigma)$, where $\cong$ signifies the existence of a diffeomorphism, since $i(\Sigma)$ is a spacelike Cauchy hypersurface. Thus $M \cong \mathbb{R} \times \Sigma$, since i is a diffeomorphism from Σ to $i(\Sigma)$. Consequently, we can consider $\mathcal{M}$ as a subset of the set of pairs (g, ϕ) on $\mathbb{R} \times \Sigma$, where g is a Lorentz metric and ϕ is a smooth function. Strictly speaking, the reason for this is that in practice, we shall only be interested in the collection of equivalence classes of globally hyperbolic developments where two developments are defined to be equivalent if they are isometric. By the above observation, this collection of equivalence classes can be considered to be a set.

Theorem 16.6. *Let $(\Sigma, g_0, k, \phi_0, \phi_1)$ be initial data to* (13.3)–(13.4). *Then there exists a maximal globally hyperbolic development of the data which is unique up to isometry.*

Proof. Let $\mathcal{M}$ denote the collection of all globally hyperbolic developments of the initial data. Due to Theorem 14.2, $\mathcal{M}$ is non-empty.

Let N and N' be elements of $\mathcal{M}$ (for the sake of brevity, we write N instead of (N, g, ϕ) and similarly for N'). Due to Theorem 14.3 there is an open subset U of N and a time orientation preserving isometry ψ from U onto an open subset of N' such that U is a globally hyperbolic development of the initial data and ψ respects the embedding. For technical reasons, it will be convenient to put additional conditions on U. In fact we shall demand that if $q \in \partial U$, then there is a spacelike hypersurface S in N such that $q \in S$ and $S - \{q\} \subset U$. If U has this property, we shall say that U has a *spacelike boundary*. Note that if U and U' are two globally hyperbolic developments contained in N, which both have spacelike boundary, then the union is a globally hyperbolic development with spacelike boundary. Let us prove that there is a globally hyperbolic development $U \subseteq N$ with a spacelike boundary, in addition to the properties listed above. Let $U \subseteq N$ be a globally hyperbolic development as above, whose boundary is not necessarily spacelike. Assume that $i : \Sigma \to N$ is the embedding corresponding to the development N. Then, for $q \in i(\Sigma)$, we can assume there is a neighbourhood $O_q \subseteq U$ of q such that O_q is globally hyperbolic with Cauchy hypersurface $O_q \cap i(\Sigma)$. By considering local coordinates of the form constructed in Lemma 12.5, one can also convince oneself that one can construct O_q in such a way that the boundary of O_q is spacelike (excepting the boundary points on $i(\Sigma)$; the shape to keep in mind is roughly that of a lens in $\mathbb{R}^{2+1}$). Taking the union of the O_q we get the desired development (which we shall also denote U). We shall denote the collection of pairs (U, ψ) by $C(N, N')$. Let (U, ψ) and $(\tilde{U}, \tilde{\psi})$ be two elements of $C(N, N')$ and let $p \in U \cap \tilde{U}$. Let γ be an inextendible timelike geodesic in $U \cap \tilde{U}$. Since $U \cap \tilde{U}$ is globally hyperbolic, $\gamma(s_0) \in i(\Sigma)$ for some s_0. By arguments similar to ones given earlier, $\psi \circ \gamma(s_0) = \tilde{\psi} \circ \gamma(s_0)$ and similarly for the first derivative. Thus $\psi \circ \gamma = \tilde{\psi} \circ \gamma$ so that $\psi(p) = \tilde{\psi}(p)$. Consequently, there exists an isometry from $\tilde{U} \cup U$ into N', and, as noted above, $\tilde{U} \cup U$ is globally hyperbolic and has a spacelike boundary.

By the above, $C(N, N')$ is partially ordered by inclusion of the U's, and every totally ordered subset of $C(N, N')$ has an upper bound. Due to Theorem 16.4, we conclude that $C(N, N')$ has a maximal element (U, ψ). By the above argument, the maximal element is unique.

Write $N \leq N'$ if $U = N$. Due to the uniqueness of maximal elements of $C(N, N')$, the relation $\leq$ introduces a partial ordering on $\mathcal{M}$, where two elements are taken to be identical if they are isometric (in other words, we are strictly speaking not considering the class of globally hyperbolic developments, but, as we mentioned above, the set of equivalence classes of globally hyperbolic developments, where two globally hyperbolic developments are defined to be equivalent if they are isometric). Let $\{N_\alpha\}$, $\alpha \in A$ be a totally ordered subset of $\mathcal{M}$. If $N_\alpha \leq N_\beta$, there is a map $\psi_{\beta\alpha} : N_\alpha \to N_\beta$ with properties listed above. By uniqueness, $\psi_{\gamma\beta} \circ \psi_{\beta\alpha} = \psi_{\gamma\alpha}$ assuming that $N_\alpha \leq N_\beta \leq N_\gamma$. Define K to be the disjoint union of the N_α

$$K = \coprod_{\alpha \in A} N_\alpha.$$

We define an equivalence relation on K by saying that $p_\alpha \in N_\alpha$ is equivalent to $p_\beta \in N_\beta$, written $p_\alpha \sim p_\beta$, if one of the following two statements are true:

$$N_\alpha \leq N_\beta,\ \psi_{\beta\alpha}(p_\alpha) = p_\beta, \quad \text{or} \quad N_\beta \leq N_\alpha,\ \psi_{\alpha\beta}(p_\beta) = p_\alpha.$$

Let Q be the quotient of K under this equivalence relation.

Q is a Hausdorff topological space. We shall write $q = [p_\alpha]$ for the equivalence class containing $p_\alpha \in N_\alpha$ and we define $\pi : K \to Q$ by $\pi(p_\alpha) = [p_\alpha]$. We define a topology on Q by declaring the open sets to be the union of finite intersections of sets of the form $\pi(U_\alpha)$ for open sets $U_\alpha \subseteq N_\alpha$. Thus Q is a topological space. To prove that it is Hausdorff, let $q_1, q_2 \in Q$ be such that $q_1 \neq q_2$. Let $p_\alpha \in q_1$ and $r_\beta \in q_2$. Either $N_\alpha \leq N_\beta$ or vice versa. Let us assume the former. Then $[\psi_{\beta\alpha}(p_\alpha)] = q_1$ and $\psi_{\beta\alpha}(p_\alpha) \neq r_\beta$. Thus there are open neighbourhoods U_β and V_β of $\psi_{\beta\alpha}(p_\alpha)$ and r_β respectively such that U_β and V_β are disjoint. Then $\pi(U_\beta)$ and $\pi(V_\beta)$ are disjoint open neighbourhoods containing q_1 and q_2 respectively. Thus Q is Hausdorff.

Q has a differentiable structure. Let us define a differentiable structure on Q. If $q \in Q$, then there is a $p_\alpha \in q$. Let U_α be an open neighbourhood of p_α and let x_α be coordinates on U_α. Since, given any $q \in \pi(U_\alpha)$, there is a unique $q_\alpha \in U_\alpha$ such that $\pi(q_\alpha) = q$, we can define an injective map y_α on $\pi(U_\alpha)$ by $y_\alpha([q_\alpha]) = x_\alpha(q_\alpha)$. We would like the y_α to constitute coordinate charts. First we need to prove that y_α is a homeomorphism. Since $y_\alpha^{-1}(W) = \pi[x_\alpha^{-1}(W)]$, we see that y_α is continuous. Since

$$\pi(U_\alpha) \cap \pi(U_\beta) = \pi[U_\alpha \cap \psi_{\alpha\beta}(U_\beta)] \quad \text{or} \quad \pi(U_\alpha) \cap \pi(U_\beta) = \pi[U_\alpha \cap \psi_{\beta\alpha}^{-1}(U_\beta)],$$

we see that any open subset of $\pi(U_\alpha)$ can be written $\pi(V_\alpha)$ for some open subset $V_\alpha \subseteq U_\alpha$. Consequently, y_α takes open sets to open sets, so that it is a homeomorphism. Let y_β be defined similarly on $\pi(U_\beta)$ and assume that $\pi(U_\alpha) \cap \pi(U_\beta)$ is non-empty. Assuming, without loss of generality, that $N_\alpha \leq N_\beta$, we get $y_\beta \circ y_\alpha^{-1} = x_\beta \circ \psi_{\beta\alpha} \circ x_\alpha^{-1}$ so that $y_\beta \circ y_\alpha^{-1}$ and $y_\alpha \circ y_\beta^{-1}$ are both smooth. Thus $[y_\alpha, \pi(U_\alpha)]$ constitutes an atlas which can be extended to become maximal.

Existence of a Lorentz metric on Q. In order to define a Lorentz metric on Q, let $\pi_\alpha : N_\alpha \to Q$ be defined by $\pi_\alpha(p_\alpha) = [p_\alpha]$. Note that π_α is injective, and since $y_\alpha \circ \pi_\alpha \circ x_\alpha^{-1}$ equals the identity when defined, π_α is a local diffeomorphism. We conclude that π_α is a diffeomorphism onto its image. Given $p \in Q$ and $X, Y \in T_p Q$, let p_α be such that $p = [p_\alpha]$ and $X_\alpha, Y_\alpha \in T_{p_\alpha} N_\alpha$ be such that $\pi_{\alpha *} X_\alpha = X$ and similarly for Y. Note that X_α and Y_α are unique since π_α is a diffeomorphism. Define

$$g(X, Y) = g_\alpha(X_\alpha, Y_\alpha).$$

We need to prove that this definition makes sense. Assume $[p_\beta] = p$ and define X_β, Y_β analogously to the definition of X_α, Y_α. We can, without loss of generality, assume that $N_\alpha \leq N_\beta$. By uniqueness, $\psi_{\beta\alpha *} X_\alpha = X_\beta$ and similarly for Y_β. Thus

$$g_\beta(X_\beta, Y_\beta) = g_\beta(\psi_{\beta\alpha *} X_\alpha, \psi_{\beta\alpha *} Y_\alpha) = \psi_{\beta\alpha}^* g_\beta(X_\alpha, Y_\alpha) = g_\alpha(X_\alpha, Y_\alpha).$$

The smoothness of g is immediate since

$$g(\partial_{y_\alpha^\mu}|_p, \partial_{y_\alpha^\nu}|_p) = g_\alpha(\partial_{x_\alpha^\mu}|_{\pi_\alpha^{-1}(p)}, \partial_{x_\alpha^\nu}|_{\pi_\alpha^{-1}(p)}).$$

To prove that we can define a smooth function ϕ on Q is less demanding. Furthermore we can define an embedding $i : \Sigma \to Q$ and all the desired relations will hold. Assuming all the maps $\psi_{\alpha\beta}$ are time orientation preserving and orientation preserving, we can assume Q to be time oriented and oriented. Furthermore, $i(\Sigma)$ is a spacelike Cauchy hypersurface in Q. Assuming Σ to be connected, we conclude that Q is connected.

Q is second countable. What remains to be proved is that Q is second countable. We shall do so using the geometric structures present. It is possible to define a vector field G on TQ such that the projection $\pi_1 : TQ \to Q$ yields a one to one correspondence between integral curves of G and geodesics of Q, cf. Proposition 28, p. 70 of [65]. At $v \in TQ$, G_v is given by the initial velocity of the curve $\gamma_v'(s)$, where γ_v is the geodesic satisfying $\gamma_v(0) = \pi_1(v)$ and $\gamma_v'(0) = v$. Due to Theorem 4.26, p. 54 of [25], there is an open subset $\mathcal{D}$ of $\mathbb{R} \times TQ$ containing $\{0\} \times TQ$ and a smooth flow $\alpha : \mathcal{D} \to TQ$, i.e., $\alpha'(t, v) = G_{\alpha(t,v)}$. In fact, $\pi_1 \circ \alpha(t, v) = \gamma_v(t)$ and $\alpha(t, v) = \gamma_v'(t)$ with the above notation. Let

$$E = \bigcup_{p \in i(\Sigma)} \mathbb{R} \times C_p Q \cap \mathcal{D},$$

where $C_p Q$ is the set of timelike vectors in $T_p Q$. Then E is second countable since Σ is second countable. If Σ is n dimensional, E is a $2n + 2$ dimensional manifold. Let W_j be a countable basis for the topology of E and let $O_j = \pi_1 \circ \alpha(W_j)$. We claim that the O_j form a basis for the topology of Q. Let $q \in Q$ be contained in an open neighbourhood U. Due to the fact that $i(\Sigma)$ is a Cauchy hypersurface, there is a $(t, v) \in E$ such that $\pi_1 \circ \alpha(t, v) = q$. By continuity of $\pi_1 \circ \alpha$, we conclude that there is a W_j such that $(t, v) \in W_j$ and $\pi_1 \circ \alpha(W_j) \subseteq U$. All that remains to be proved is that the O_j are open. Before we do so, let us note the following fact. If $v \in TQ$ is a timelike vector, let γ_v be the geodesic with initial velocity v. Since $i(\Sigma)$ is a Cauchy hypersurface, there is a unique $s_v \in \mathbb{R}$ such that $\gamma_v(s_v) \in i(\Sigma)$. In this way we get a map $\sigma : \mathcal{T} \to \mathbb{R}$ given by $\sigma(v) = s_v$, where $\mathcal{T} \subset TQ$ is the set of timelike vectors. We wish to prove that σ is continuous. Let $v_i \to v$ and assume that v is future directed. Assume $\sigma(v_i) \le \sigma(v) - \varepsilon$ for some $\varepsilon > 0$. Then

$$\gamma_{v_i}[\sigma(v_i)] \le \gamma_{v_i}[\sigma(v) - \varepsilon] \ll \gamma_v[\sigma(v)]$$

for i large enough, and we can choose ε to be small enough that $\gamma_{v_i}[\sigma(v) - \varepsilon]$ is well defined. The reason for the last, strict, inequality is the fact that $\gamma_{v_i}[\sigma(v) - \varepsilon]$ converges to $\gamma_v[\sigma(v) - \varepsilon]$. Since $\gamma_{v_i}[\sigma(v_i)]$ and $\gamma_v[\sigma(v)]$ both belong to $i(\Sigma)$ we get a timelike curve with endpoints in $i(\Sigma)$, contradicting the fact that $i(\Sigma)$ is a Cauchy hypersurface. Similarly we cannot have $\sigma(v_i) \ge \sigma(v) + \varepsilon$. We conclude that σ is continuous. Let $q \in O_j$. Then there is a $(t, v) \in E$ such that $q = \pi_1 \circ \alpha(t, v)$. Let $w = \alpha(t, v)$, let O be an open neighbourhood of w in TQ and assume O to be small enough that it only consists of timelike vectors. In particular, $\pi_1(O)$ is an open neighbourhood of q. Note that as O shrinks, $\alpha(\sigma(u), u)$ converges to $\alpha(\sigma(w), w) = v$ and $\sigma(u)$ converges to $-t$ (for $u \in O$). Thus

$$[-\sigma(u), \alpha(\sigma(u), u)] \tag{16.1}$$

converges to (t, v). For O small enough (16.1) is thus contained in W_j, so that $\pi_1(O)$ is contained in O_j and O_j is open. We conclude that Q is second countable.

Existence of a maximal element. Since Q is a development and $N_\alpha \leq Q$ for all $\alpha \in A$, we conclude that Q is an upper bound for $\{N_\alpha\}$, $\alpha \in A$. In other words, each totally ordered subset of $\mathcal{M}$ has an upper bound. We conclude that $\mathcal{M}$ has a maximal element M. We know that the only extension of M is M, up to isometry, but we need to prove that it is an extension of every development.

The maximal element is a maximal globally hyperbolic development. Let M' be another development of the initial data. Let (U, ψ) be the maximal element of $C(M', M)$. We can construct an equivalence relation on the disjoint union $M \amalg M'$ by requiring $q \sim q$ for all q, $q \sim \psi(q)$ for $q \in U$ and $\psi(q) \sim q$ for $q \in U$. Taking the quotient of $M \amalg M'$ under this equivalence relation, we get a topological space $\widetilde{M}$ and we shall denote the projection from $M \amalg M'$ to $\widetilde{M}$ by π_2. This space has a differentiable structure and it is second countable. If we were able to prove that it is Hausdorff, it would thus be a manifold. That it is a globally hyperbolic development of the initial data would then be immediate. Since it is an extension of M, we get the conclusion that $U = M'$, since the only extension of M is M itself.

Elementary consequences of non-Hausdorffness. In order to prove that $\widetilde{M}$ is Hausdorff, let us assume the opposite. Then there are $p, p' \in \widetilde{M}$ such that for every pair V, V' of open neighbourhoods of p and p' respectively, the intersection of V and V' is non-empty. We can then, without loss of generality, assume $p' \in \partial U \cap J^+(i'(\Sigma), M')$ and $p \in \partial\psi(U) \cap J^+(i(\Sigma), M)$. Note that

$$I^-(p, M) \cap J^+(i(\Sigma), M) \subseteq \psi(U), \quad I^-(p', M') \cap J^+(i'(\Sigma), M') \subseteq U.$$

The reason is that if

$$q' \in I^-(p', M') \cap J^+(i'(\Sigma), M')$$

for example, and $q' \notin U$, then $I^+(q', M')$ and its projection to $\widetilde{M}$ are open neighbourhoods of p' that belong to the complement of U, since otherwise one could construct an inextendible timelike geodesic in U which does not intersect $i'(\Sigma')$. We obtain a contradiction to the fact that $p' \in \partial U$. The argument to prove the first inclusion is identical. Let $q' \ll p'$. Then $I^+(q', M')$ and its projection to $\widetilde{M}$ are open neighbourhoods of p' in M' and $\widetilde{M}$ respectively. Let $q \in M$ be such that $p \ll q$. Then $I^-(q, M)$ and its projection to $\widetilde{M}$ is an open neighbourhood of p. Since the intersection of these neighbourhoods has to be non-zero, we conclude that $\psi(q') \ll q$. Since M is globally hyperbolic, the relation $\leq$ is closed on M, so that $\psi(q') \leq p$. However, if $q' \ll p'$, there is a q'' such that $q' \ll q'' \ll p'$ so that $\psi(q'') \leq p$ and thus $\psi(q') \ll p$. Analogously, if $q \ll p$ and $q \in \psi(U)$, then $\psi^{-1}(q) \ll p'$. Let γ be a future directed timelike curve in M' with the property that $\gamma(0) = p'$. Let $r \ll p$. Then $\psi^{-1}(r) \ll p'$ so that $\psi^{-1}(r) \ll \gamma(t)$ for t close enough to 0. Thus there is an $\varepsilon > 0$ such that $r \ll \psi \circ \gamma(t) \ll p$ for $t \in (-\varepsilon, 0)$. Since, for every neighbourhood V of p there is an $r \ll p$ such that $J^+(r, M) \cap J^-(p, M)$ is contained in V (note that this statement is based on the fact that M is globally hyperbolic), we conclude that $\psi \circ \gamma(t) \to p$ as $t \to 0-$.

Definition of φ. Since the boundary of U is spacelike, there is a smooth spacelike hypersurface S such that $p' \in S$ and $S - \{p'\} \subseteq U$. Let us define $\varphi \colon S \to M$ by $\varphi(q) = \psi(q)$ for $q \in S - \{p'\}$ and $\varphi(p') = p$.

The map φ is smooth. The only problem occurs at the point p'. Let n be the future directed unit normal to S. There is a neighbourhood V of p' in S and an $\varepsilon > 0$ such that geodesics starting at $q \in V$ with initial velocity n_q are defined on the interval $(-\varepsilon, \varepsilon)$. For reasons mentioned above the map taking q to $\gamma'_{n_q}(-\varepsilon/2)$ is a smooth map from V into TM'. Note that if ε is small enough, $\gamma_{n_q}(-\varepsilon/2)$ belongs $J^+[i'(\Sigma)]$ and thus to U, since one could otherwise construct an inextendible timelike geodesic in U which never intersects $i'(\Sigma)$. Thus $\eta(q) = \psi_* \gamma'_{n_q}(-\varepsilon/2)$ defines a smooth map from V to TM. We would now like to compute $\gamma_{\eta(q)}(\varepsilon/2)$. However, it is not clear that this is defined. For $q \neq p'$ it is defined due to the existence of the isometry, since

$$\gamma_{\eta(q)}(s) = \psi \circ \gamma_{n_q}(s - \varepsilon/2)$$

for $q \neq p'$ and for $q = p'$, the statement made at the end of the paragraph on the elementary consequences of non-Hausdorffness ensures that it is defined. Since $\gamma_v(\varepsilon/2)$ is a smooth function of v on the open set on which it is defined, we conclude that $\gamma_{\eta(q)}(\varepsilon/2)$ is a smooth function of $q \in V$. Since this map coincides with φ on V, we conclude that φ is smooth.

The map φ is an embedding. Let e_i, $i = 1, \dots, n$ be an orthonormal basis of TS in a neighbourhood of p'. Then, for $q \neq p'$ in this neighbourhood,

$$\delta_{ij} = g'(e_i|_q, e_j|_q) = \psi^* g(e_i|_q, e_j|_q) = g(\varphi_* e_i|_q, \varphi_* e_j|_q).$$

By continuity, we conclude that the same equality holds for $q = p'$. Thus φ_* is injective when restricted to $T_{p'}S$. In other words, φ is an immersion. By restricting S if necessary, we can ensure that φ is an embedding, cf. p. 19 of [65]. Furthermore, the pull back of the initial data induced on $\varphi(S)$ by (g, ϕ) under φ coincides with the initial data induced on S by (g', ϕ') due to the fact that ψ is an isometry and the smoothness of φ.

As a consequence of the existence of the embedding φ, M and M' are both developments of S. Due to Theorem 14.3 we conclude that there is an open neighbourhood D of S such that D is globally hyperbolic with respect to the metric g' and S is a Cauchy hypersurface, and an isometry χ from D into M such that $\chi = \varphi$ when restricted to S. We can furthermore assume D has a spacelike boundary. By a uniqueness argument of a type we have already gone through several times, $\chi = \psi$ on $D \cap U$. Furthermore

$D \cup U$ is a globally hyperbolic development of Σ with spacelike boundary. Thus we can extend ψ to a larger domain, contradicting the maximality of (U, ψ). As a consequence, $\widetilde{M}$ is Hausdorff so that $\widetilde{M}$ is an extension of M. By the maximality of M, we conclude that $U = M'$ so that $M' \leq M$. □

The following two corollaries are of some interest. The proofs are similar to the proof of the fact that the maximal globally hyperbolic development is unique up to isometry.

Corollary 16.7. *Let (M, g, ϕ) be the maximal globally hyperbolic development of initial data $(\Sigma, g_0, k, \phi_0, \phi_1)$ to* (13.3)–(13.4) *and let Σ' be a Cauchy hypersurface in M. Then (M, g, ϕ) is the maximal globally hyperbolic development of the initial data induced on Σ' by (g, ϕ).*

Corollary 16.8. *Let (M, g, ϕ) be a globally hyperbolic Lorentz manifold satisfying* (13.3)–(13.4). *Let Σ and Σ' be two Cauchy hypersurfaces in M. Then the maximal globally hyperbolic developments of the initial data induced on Σ and Σ' by (g, ϕ) are isometric.*

Finally, let us remark that an isometry of the initial data leads to an isometry of the maximal globally hyperbolic development (MGHD) due to the abstract properties of the MGHD. However, it is also of interest to consider the case that there is a smooth Lie group action on the initial hypersurface under which the initial data are invariant. Does such a smooth group action induce a smooth group action on the MGHD? We shall not attempt to answer this question here, but rather refer to [69], pp. 176–177. See also [22].

Part IV

Pathologies, strong cosmic censorship

17 Preliminaries

17.1 Purpose

In the previous chapter, we demonstrated that, given suitable initial data, there is a unique maximal globally hyperbolic development (MGHD). It is then natural to ask if the MGHD is extendible, and, if so, if the extension is unique. The purpose of the current part of these notes is to demonstrate that there are examples of initial data for which the MGHD is extendible in inequivalent ways. The examples we shall give are certain initial data for Einstein's vacuum equations on SU(2) which are invariant under left translations. It is important to point out that the extensions we shall construct are in themselves inextendible, and in this sense maximal; given one extension, removing a suitable subset from it typically leads to a topologically inequivalent extension. If there were always one unique maximal extension of the MGHD, the issue of predictability would still have a satisfactory answer; given initial data there would then be a unique maximal development. For this reason, the existence of inequivalent maximal extensions of the MGHD is important, since it can be interpreted as a lack of determinism of the general theory of relativity. However, one is naturally led to the so-called strong cosmic censorship conjecture, which we shall briefly discuss in Section 17.2 below.

To prove the existence of the above mentioned extensions, it would suffice to describe the corresponding spacetimes and their properties. However, we here wish to describe a somewhat more general situation: the case of initial data for Einstein's vacuum equations on 3-dimensional unimodular Lie groups, cf. Definition 19.1, that are invariant under left translations. The reasons for wanting to consider this more general situation are as follows. First of all, at some stage, it is necessary to verify that Einstein's vacuum equations are satisfied, and in order to do so, it is necessary to compute the Ricci tensor. There is a perspective on solutions to Einstein's equations corresponding to initial data on unimodular Lie groups, due to Ellis and MacCallum [37], that covers all the unimodular Lie groups at once. Consequently, considering all the unimodular Lie groups at once is not much more complicated that considering only SU(2). Secondly, in all the cases where the initial data are such that the MGHD allows an extension, the construction of the extensions is, from an abstract Lie group perspective, independent of the unimodular Lie group under consideration, cf. Section 17.3 for a more detailed explanation. Finally, it is of course of interest to ask if the MGHD is inextendible for initial data that are generic in the class of left invariant initial data (on a specific unimodular Lie group). To answer this question in case the Lie group is SU(2) is non-trivial. It is easier to do so in the case of some other unimodular Lie groups. Consequently, we shall analyze the situation in the simpler cases, and only quote the results in the case of SU(2), the objective being to illustrate the type of ideas involved in proving generic inextendibility in as simple a setting as possible.

17.2 Strong cosmic censorship

The question asked at the end of the preceding section can of course be asked in greater generality, and it naturally leads to the *strong cosmic censorship* (SCC) conjecture, which we shall phrase as follows.

Conjecture 17.1 (Strong Cosmic Censorship). *For generic initial data to Einstein's equations, the MGHD is inextendible.*

Note that in order to make a precise statement, one must restrict one's attention to a specific matter model, and it is also necessary to specify what is meant by inextendibility and genericity. Furthermore, it might be necessary to qualify the conjecture to state that it only holds for matter models which do not develop singularities in the absence of gravity. The above form of the statement is to be found in the work of Chruściel, cf. [22], Section 1.3, based on ideas due to Eardley and Moncrief [32], and Penrose [67]. At this stage, it is not realistic to hope to prove this conjecture in all generality, but one can consider the same conjecture in classes of spacetimes with symmetries, as was done at the end of the preceding section. To be more precise: given a manifold Σ and a Lie group G acting smoothly on Σ, cf. p. 207 of [55], the conjecture would be that for initial data that are generic in the class of initial data invariant under the Lie group action, the MGHD is inextendible. In this way, one obtains a sequence of problems corresponding to different Lie group actions, the difficulty, reasonably, increasing as the dimension of G decreases, assuming the group action to be *effective*, i.e., the only group element that fixes all points of Σ is the identity element. As we shall see, this conjecture is not true for all Lie group actions; the left invariant vacuum initial data on SU(2) whose MGHD's allow inequivalent extensions are exactly given by initial data allowing an extra rotational isometry. Nevertheless, if the conjecture does not hold in the case of a smooth action of a Lie group G on Σ, but it does hold if one restricts the action to some subgroup H of G, then the failure of the conjecture to hold in the case of G is not of essential importance. This is what happens in the case of SU(2). Let us give a brief, imprecise and incomplete description of results that have been obtained concerning strong cosmic censorship, the goal being to illustrate some different techniques and notions of genericity and inextendibility that have been used. Before doing so, it is, however, necessary to explain the division into *asymptotically flat* and *cosmological* spacetimes.

For the purposes of the present discussion, we shall take asymptotically flat initial data to mean initial data that are suitably close to those of a $t = 0$ hypersurface in Minkowski space in the components of an asymptotic region. It is assumed that the asymptotic region consists of a finite number of disjoint sets, each diffeomorphic to $\mathbb{R}^3$ minus a ball, and that it is obtained by subtracting a compact subset from the initial manifold. The exact norm measuring the "closeness" to Minkowski initial data depends on the situation under consideration. For a more precise definition in a particular context, cf. [81]. Asymptotically flat spacetimes are taken to be the MGHD's of asymptotically flat initial data. The asymptotically flat setting is intended to describe isolated systems such as stars, black holes, etc., and the basic example of a (non-trivial) solution is Schwarzschild, cf. Chapter 13 of [65].

Cosmological spacetimes are, for the purposes of the present discussion, defined to be the MGHD's corresponding to initial data specified on compact manifolds. The basic examples of solutions are suitable quotients of the Friedmann–Robertson–Walker solutions, cf. Chapter 12 of [65]. In case the initial data are homogeneous or locally homogeneous (i.e., the initial data induced on the universal covering space are homogeneous), the compactness/non-compactness is of no importance when analyzing the asymptotics (in the expanding direction and in the direction of the big bang/big crunch), and therefore, the restriction to compact initial manifolds may seem artificial. However, in the non-compact case, the matter content, energies (to the extent they are defined), etc., should be expected to be infinite. The reason for this is that the initial data of interest are such that, as opposed to the asymptotically flat case, there are no preferred regions in which the energy/matter density could be less than in other regions. It should also be mentioned that, if one is interested in studying the case of initial data that are close to, but not exactly, homogeneous, but do not admit compact quotients, one first needs to develop a suitable theory for the existence of solutions to the constraint equations in that setting, something which has, to the best of our knowledge, not been achieved at this time.

17.2.1 The asymptotically flat case. Since homogeneity of the initial data is incompatible with asymptotic flatness, with the exception of Minkowski initial data, the first natural type of symmetry to consider is that of spherical symmetry. In the case of vacuum, the only spherically symmetric solution is the Schwarzschild solution. This is a consequence of *Birkhoff's theorem*, cf. Appendix B of [48]. To obtain a non-trivial evolution problem, it is thus necessary to couple Einstein's equation to some matter field. This was done in the work of Christodoulou, who, in a series of papers [11]–[18], studied Einstein's equations coupled to a scalar field, i.e., the stress energy tensor is assumed to be of the form (13.2) with $V = 0$. In [18], Christodoulou considered initial data of bounded variation and demonstrated that solutions corresponding to initial data that do not belong to an exceptional set $\mathcal{E}$ are C^0-inextendible. Furthermore, he proved that $\mathcal{E}$ has the property that if $\vartheta \in \mathcal{E}$, then there is a 2-dimensional linear subspace Π_ϑ of the set of initial data such that the intersection of $\Pi_\vartheta - \{\vartheta\}$ and $\mathcal{E}$ is empty. This property defines the concept of genericity in [18]. However, it should be noted that [18] is based on a characteristic initial value formulation of the equations, so that the statement is not exactly of the form given in Conjecture 17.1.

In [26], [27], Dafermos considered the Einstein–Maxwell scalar field equations in spherical symmetry and proved that, given certain conditions concerning the fall-off of the data in a characteristic initial value problem setting, the corresponding solution is C^0-*extendible* but C^1-inextendible. Finally, let us draw the attention of the reader to the paper [28] containing related results.

17.2.2 The cosmological case. The natural Lie group actions to begin with in the cosmological case are the transitive ones. In other words, it is natural to begin by a study of homogeneous initial data. In the paper [24], Chruściel and Rendall studied the spatially compact, spatially locally homogeneous vacuum case. They proved that the assumption that there is a smooth extension of the MGHD leads to the conclusion that the local Killing vector algebra is at least four-dimensional. In other words, one obtains C^∞-inextendibility in the generic locally homogeneous case, i.e., if the Killing vector algebra is at most three-dimensional. Here, we shall prove Theorem 24.12 concerning the case of left invariant vacuum initial data specified on a unimodular Lie group. The statement is that either one can construct a smooth extension of the MGHD, or the Kretschmann scalar, $R_{\alpha\beta\gamma\delta}R^{\alpha\beta\gamma\delta}$, is unbounded in the incomplete directions of causal geodesics. Furthermore, the initial data leading to extendible MGHD's will be demonstrated to be non-generic in the sense that they constitute a set of positive codimension in the set of initial data. As a consequence of the unboundedness of the Kretschmann scalar, one obtains C^2-inextendibility of the MGHD, cf. Lemma 18.18. This constitutes a small improvement of some of the results of [24]. However, the main advantage of Theorem 24.12 is the conclusion concerning curvature blow up; the singularity theorems of Hawking and Penrose demonstrate that Lorentz manifolds have singularities in the sense of causal geodesic incompleteness under quite general circumstances, but it is of interest to show that the gravitational fields are unbounded in the approach to the singularities. In view of the above, it is natural to generalize Conjecture 17.1 to the following statement.

Conjecture 17.2 (Curvature blow up). *For generic initial data to Einstein's equations, the Kretschmann scalar, $R_{\alpha\beta\gamma\delta}R^{\alpha\beta\gamma\delta}$, is unbounded in the incomplete directions of causal geodesics in the MGHD.*

Due to Lemma 18.18, Conjecture 17.2 implies Conjecture 17.1, assuming that one is content with the conclusion of C^2-inextendibility.

Turning to the inhomogeneous case, the natural starting point is to consider the so-called polarized Gowdy spacetimes. For a definition of the Gowdy and polarized Gowdy spacetimes, see [21] and the references cited therein. In [21], Chruściel, Isenberg and Moncrief proved that there is an open and dense subset of the space of smooth initial data (for Einstein's vacuum equations in the polarized Gowdy case) such that the corresponding MGHD's are inextendible. In fact, they also proved that the timelike observers, in the generic case, experience curvature blow up, cf. [21] for the details. In [72]–[76], the general Gowdy metrics were considered in the case of $\mathbb{T}^3$-topology. The result was that there is a set of initial data for Einstein's vacuum equations in the case of Gowdy symmetry which is open with respect to the $C^1 \times C^0$-topology and dense with respect to the C^∞-topology such that all causal geodesics in the corresponding development are future complete and past incomplete (given a suitable time orientation) and the Kretschmann scalar is unbounded along all past inextendible causal curves. In other words, the results in the case of Gowdy are in the spirit of Conjecture 17.2.

Finally, let us mention the work of Dafermos and Rendall, [29]–[31]. In [29], the authors considered Einstein's equations coupled to matter of Vlasov type (in particular, vacuum is included) in the case of a $\mathbb{T}^2$ isometry group, proving that expanding solutions cannot be extended to the future. Readers interested in the definition of Vlasov, i.e. collisionless matter, are referred to [29] and the references cited therein. Results concerning future inextendibility in the case of surface symmetry are also contained in [29]. In [30], the authors proved generic C^2-inextendibility of solutions to the Einstein–Vlasov equations in the case of $\mathbb{T}^2$-symmetry (this includes Gowdy as a special case). However, the genericity assumption includes an assumption that the distribution function be non-zero for arbitrarily high momenta. In particular, the vacuum case is not contained as a subcase of the results of [30]. Furthermore, the results do not address the question posed by Conjecture 17.2, since the proof consists of a contradiction argument starting with the assumption that there is an extension. Finally, one does not obtain detailed information concerning the behaviour in the approach to the singularity. If one is interested in proving curvature blow up along incomplete causal geodesics, it is to be expected that one has to analyze the asymptotics. However, the fact that one can obtain strong cosmic censorship without such an analysis is an advantage; analyzing the asymptotics in detail is often quite complicated, and therefore, it is of great interest to develop methods that yield the desired conclusion of inextendibility without such an analysis. Further results concerning the surface symmetric situation are contained in [31].

17.2.3 Genericity. Let us make some general comments concerning the concept of genericity. In the spatially homogeneous case, the set of initial data can be given the structure of a finite dimensional manifold. Consequently, there are in that case many different concepts of genericity available: a subset can be said to be generic if, e.g.,

- the complement is of measure zero with respect to the measure induced by a Riemannian metric on the manifold of initial data;
- the complement is a countable union of submanifolds of positive codimension;
- the set is open and dense;
- the set is a G_δ set, i.e., a countable intersection of open and dense sets with respect to some complete topological metric.

These concepts are of course quite different, but there is a minimal requirement: if a subset satisfies the conditions of the definition of genericity, then the complement must not satisfy these conditions. According to this requirement, to define a set to be generic if it is dense is not appropriate since the rational and irrational real numbers would then both be generic. That the complement of a G_δ set (in a complete metric space) is not a G_δ set is a consequence of Baire's theorem, cf. pp. 97–98 of [79].

In the inhomogeneous case, the set of initial data is infinite dimensional, so that the measure theoretic notion of genericity becomes somewhat less clear. The notion of genericity used by Christodoulou illustrates that it is possible to generalize the definition based on the condition that the complement have positive codimension, and the definition that the generic sets are the G_δ sets of course generalizes trivially (assuming that one has a natural complete metric on the set of initial data).

17.3 Construction of extensions in the unimodular case

Let us describe the construction of the extensions in the unimodular vacuum case. All the MGHD's (corresponding to left invariant vacuum initial data on a 3-dimensional unimodular Lie group) that allow extensions can be written in the form

$$g_{\mathrm{MGHD}} = -\frac{L^2}{X}du^2 + X\xi^1 \otimes \xi^1 + Y^2(\xi^2 \otimes \xi^2 + \xi^3 \otimes \xi^3) \tag{17.1}$$

on $M = I \times G$, where I is an interval and G is a 3-dimensional unimodular Lie group. Here, the ξ^i are the duals of a basis e_i of the Lie algebra with suitable properties, see Chapter 24 for the details. Furthermore, $L > 0$ is a constant and Y^2 is a strictly positive polynomial in u of degree at most 2. Finally, XY^2 is a polynomial in u of degree at most 2, and the interval I is a proper subset of $\mathbb{R}$ on which this polynomial is strictly positive. In order to construct an extension, one defines

$$h_\pm(u) = \pm \int_{u_a}^{u} \frac{L}{X(s)}ds$$

for some $u_a \in I$. Let $\gamma\colon \mathbb{R} \to G$ be the smooth homomorphism which is the integral curve of e_1 which passes through the identity element of G. Then there are two diffeomorphisms of M defined by

$$\phi_\pm(u, g) = (u, g\gamma[h_\pm(u)]),$$

and one can compute, cf. Corollary 24.4, that

$$\phi_\pm^* g_{\mathrm{MGHD}} = \pm L du \otimes \xi^1 \pm L\xi^1 \otimes du + X\xi^1 \otimes \xi^1 + Y^2(\xi^2 \otimes \xi^2 + \xi^3 \otimes \xi^3).$$

The metric on the right-hand side, which we shall denote $g_{\mathrm{EXT},\pm}$, is a Lorentz metric on all of $M_\pm = \mathbb{R} \times G$, and consequently, the MGHD is extendible. One can also demonstrate that the extension is a solution to Einstein's vacuum equations and, if the Lie group G is compact, that it is C^2-inextendible.

Let us illustrate the above constructions in the simplest setting possible. Consider the metric

$$g_K = -\frac{1}{u}du^2 + 4udx^2 + dy^2 + dz^2$$

on $\mathbb{R}_+ \times \mathbb{T}^3$, where $\mathbb{R}_+ = (0, \infty)$ and $\mathbb{T}^3$ is the 3-torus, cf. (23.5). Viewed as a metric on $\mathbb{R}_+ \times \mathbb{R}^3$, it is isometric to an open subset of Minkowski space so that the Riemann curvature tensor is identically zero, cf. the discussion in connection with (23.5). The extension can then be written

$$g_{EK,\pm} = \pm 2du \otimes dx \pm 2dx \otimes du + 4udx^2 + dy^2 + dz^2$$

on $\mathbb{R} \times \mathbb{T}^3$. Observe that the circle corresponding to the ∂_x direction is spacelike for $u > 0$, null for $u = 0$ and timelike for $u < 0$. In other words, the extension has closed timelike curves. Note that if one removes one point in $(-\infty, 0] \times \mathbb{T}^3$, one obtains an inequivalent extension, but the resulting extension is extendible and in this sense not maximal. By removing more points, one obtains an infinite collection of inequivalent extensions. For this reason we shall below be interested in the existence of inequivalent *maximal* extensions. The precise definition of the concept maximal is given in Definition 18.17. When considering the metric g_K, the null geodesics corresponding to an initial velocity in the $\partial_u\partial_x$-plane for $u > 0$ are of special interest. After changing the parameter to u, they take the form

$$\gamma(u) = \left(u, x_0 \pm \frac{1}{2}\ln u, y_0, z_0\right),$$

where x_0, y_0 and z_0 are fixed and $x_0 \pm \ln u/2$ in the second slot should be taken mod 1. In other words, both geodesics wind an infinite number of turns around the circle corresponding to the x direction as u tends to zero. The effect of $\phi_\pm$ on these geodesics is to map one of them to a straight line $\gamma(u) = (u, x_0, y_0, z_0)$, whereas the other still winds around the circle corresponding to the x-direction. Which geodesic is straightened out depends on the choice of $+$ and $-$ in $\phi_\pm$.

17.4 Sketch of the proof of existence of inequivalent extensions

Let us sketch the proof of the statement that there are inequivalent maximal extensions in the SU(2)-case. The argument, which is taken from [23], see also [62], is based on considerations of the properties of null geodesics corresponding to an initial velocity in the $\partial_u e_1$-plane. First of all, if a null geodesic starts in this plane, it remains in it. At a given point in the MGHD, there are two future oriented null vectors in this plane. One of the corresponding null geodesics is mapped to part of a complete null geodesic in the extensions $(M_\pm, g_{\mathrm{EXT},\pm})$ and the other is mapped to a geodesic which is future and past incomplete. In all the cases except SU(2), the MGHD is future causally geodesically complete (assuming that one has chosen a suitable time orientation). However, in the SU(2) case, the Cauchy horizon of the MGHD, considered as a subset of the extension, consists of two components, and this makes it possible to construct additional extensions. If, in this case, the MGHD is $(I \times G, g_{\mathrm{MGHD}})$, where $I = (u_-, u_+)$, then u_- and u_+ are finite. Furthermore, we can construct an extension as follows: consider the disjoint union of $(-\infty, u_+) \times G$ and $(u_-, \infty) \times G$ with metrics $g_{\mathrm{EXT},-}$ and $g_{\mathrm{EXT},+}$ respectively. Using $\phi_\pm$ one can then identify the solutions on $(u_-, u_+) \times G$. This leads to an extension which solves the Einstein vacuum equations and is C^2-inextendible. Let us denote the manifold M_{+-} and the metric $g_{\mathrm{EXT},+-}$. Considering a null vector in the $\partial_u e_1$-plane in the MGHD, there are two possibilities for the corresponding null geodesic in $(M_{+-}, g_{\mathrm{EXT},+-})$: either it is future complete and past incomplete or it is future incomplete and past complete. In other words, the situation is different from the case $(M_+, g_{\mathrm{EXT},+})$. In order to obtain a contradiction to the assumption that there is an isometry from $(M_+, g_{\mathrm{EXT},+})$ to $(M_{+-}, g_{\mathrm{EXT},+-})$, it is thus enough to prove that such an isometry would have to preserve the $\partial_u e_1$-plane of the globally hyperbolic part of the respective Lorentz manifolds. In order to prove the latter statement, one first verifies that an isometry would have to preserve the globally hyperbolic part. Thus an isometry would restrict to an isometry, say χ, of (M, g_{MGHD}). In order to prove that χ would respect the $\partial_u e_1$-plane, one observes that M is foliated by constant mean curvature (CMC) hypersurfaces, cf. the next chapter for a definition. In the spacetimes of interest, hypersurfaces with non-zero constant mean curvature are unique. Since χ would have to preserve CMC hypersurfaces, and since ∂_u is orthogonal to the CMC hypersurfaces, χ would have to map ∂_u to a function times ∂_u. In order to prove that χ would have to map e_1 to a multiple of itself, one notes that χ would have to restrict to an isometry from any given CMC hypersurface to some other CMC hypersurface. For a generic CMC hypersurface, the vector space generated by e_1 can be characterized as a particular eigenspace of the Ricci tensor of the CMC hypersurface, viewed as a Riemannian manifold, if we consider the Ricci tensor to be a linear operator from the tangent space to itself. Thus χ would have to map the $\partial_u e_1$-plane to itself, and we have a contradiction.

17.5 Outline

Let us give an outline of the present part of these notes. In Chapter 18 we discuss compact spacelike hypersurfaces of constant, non-zero mean curvature. The main result is that such hypersurfaces are unique in globally hyperbolic Lorentz manifolds satisfying the timelike convergence condition (i.e., $\text{Ric}(v, v) \geq 0$ for all timelike tangent vectors v). As was noted above, this will be used in the proof of the fact that there are MGHD's with inequivalent maximal extensions. In Chapter 19 we analyze the constraint equations in the unimodular case. We also compute the Ricci tensor of the initial metric, something which is of use in the proof of the existence of inequivalent extensions, cf. the above discussion. The perspective we take in this chapter and the following is based on the one taken in [37]. In Chapter 20 we compute the Ricci tensor for the spacetimes of interest, construct developments of initial data and establish the basic properties of the developments, e.g. proving that they are the maximal globally hyperbolic developments of the initial data and analyzing to what extent they are causally geodesically complete. Among the 3-dimensional unimodular Lie groups, SU(2) is special in that the MGHD's corresponding to left invariant vacuum initial data on SU(2) are both future and past causally geodesically incomplete. Chapter 21 contains a proof of this fact, which is due to Lin and Wald, see [56], as well as a proof of the fact that for any such initial data, there is an open neighbourhood of them such that all timelike geodesics are past and future incomplete in the corresponding MGHD's. In Chapter 22 we introduce the Wainwright–Hsu variables, which are suited for analyzing the asymptotics, and we carry out the analysis in case of the simpler unimodular Lie groups. Concerning the more complicated cases, we simply state the results.

All the developments that allow extensions have an additional symmetry, referred to as a local rotational symmetry. In Chapter 23 we demonstrate that, in all the cases for which it was not already demonstrated in Chapter 22 that the MGHD is inextendible, the metric can be written in the form (17.1). In Chapter 24 we then construct extensions of such metrics. We also establish the fact that given left invariant vacuum initial data on a unimodular Lie group, there are two possibilities: either there is an extension of the MGHD which solves Einstein's vacuum equations or the Kretschmann scalar, $R_{\alpha\beta\gamma\delta}R^{\alpha\beta\gamma\delta}$, is unbounded in the incomplete directions of causal geodesics in the MGHD, in which case the MGHD is C^2-inextendible. Since the MGHD is only extendible for initial data that have an extra symmetry, it is clear that extendibility of the MGHD is non-generic within the class of left invariant vacuum initial data on unimodular Lie groups. Finally, in Chapter 25, we prove that in the cases of left invariant vacuum initial data on SU(2) for which the MGHD is extendible, there are two non-isometric extensions which solve the Einstein vacuum equations and are C^2-inextendible.

18 Constant mean curvature

As was pointed out in the introduction, we shall need to know that constant mean curvature foliations are unique in the classes of spacetimes of interest. The main tool for proving uniqueness is Proposition 18.8, and this is a consequence of the calculus of variations, a subject to which we shall devote the first section of this chapter. There is one situation in which it is not so clear that this proposition is of use; if the Ricci tensor equals zero, one does not obtain a contradiction to the assumption that there are different hypersurfaces with the same constant mean curvature, the main purpose of Proposition 18.8. In the beginning of Section 18.2, we shall therefore prove a lemma to the effect that one in that case can perturb one of the hypersurfaces in order to obtain a contradiction. The remainder of Section 18.2 is concerned with writing down the consequences of this. In order to prove that the MGHD is extendible, it is of course necessary to provide conditions that ensure that a globally hyperbolic development is the maximal globally hyperbolic development. In Section 18.3 we state such conditions. In Section 18.4 we state conditions that ensure the inextendibility of the MGHD.

18.1 Calculus of variations

Let us recall and slightly generalize some results from [65] concerning the calculus of variations. Let (M, g) be a Lorentz manifold and let $\alpha\colon [a,b] \to M$ be a piecewise smooth curve. We shall call a map $\boldsymbol{x} : [a,b] \times (-\delta, \delta) \to M$, where $\delta > 0$, a *piecewise smooth variation* of α if the following holds:

- $\boldsymbol{x}(u, 0) = \alpha(u)$ for all $u \in [a, b]$;
- $\boldsymbol{x}$ is continuous;
- there is a $0 \leq k \in \mathbb{Z}$ and *breaks* $a < u_1 < \cdots < u_k < b$ such that the restriction of $\boldsymbol{x}$ to $[u_{i-1}, u_i] \times (-\delta, \delta), i = 1, \dots, k+1$ is smooth, where we use the notation $u_0 = a$ and $u_{k+1} = b$.

Note that we can assume α and $\boldsymbol{x}$ to have the same breaks u_i by inserting trivial breaks. We denote the length of the curve $u \mapsto \boldsymbol{x}(u, v)$ by $L_{\boldsymbol{x}}(v)$. The *variation vector field* of a variation $\boldsymbol{x}$ is the vector field V on α given by $V(u) = \boldsymbol{x}_v(u, 0)$; cf. p. 216 and pp. 122–123 of [65] for the notation. The discontinuity of α' at a break is measured by

$$\Delta\alpha'(u_i) := \alpha'(u_i+) - \alpha'(u_i-) \in T_{\alpha(u_i)}M.$$

The *sign* ε of a curve α is defined to be the sign of $g(\alpha', \alpha')$ (we shall assume the sign of this quantity to be constant for the curves under consideration) and the *speed* is defined to be $c = |g(\alpha', \alpha')|^{1/2}$. Finally, we shall use the notation $\langle v, w\rangle := g(v, w)$. Let us recall Proposition 2, p. 264 of [65].

Proposition 18.1. *Let $\alpha\colon [a,b] \to M$ be a piecewise smooth curve segment with constant speed $c > 0$ and sign ε. If $\boldsymbol{x}$ is a piecewise smooth variation of α, then*

$$L'_{\boldsymbol{x}}(0) = -\frac{\varepsilon}{c}\int_a^b \langle \alpha'', V\rangle\, du - \frac{\varepsilon}{c}\sum_{i=1}^{k}\langle \Delta\alpha'(u_i), V(u_i)\rangle + \frac{\varepsilon}{c}\langle \alpha', V\rangle\Big|_a^b,$$

where $u_1 < \cdots < u_k$ are the breaks of α and $\boldsymbol{x}$.

For $L_{\boldsymbol{x}}$ to be zero for all fixed endpoint variations, i.e., variations $\boldsymbol{x}$ such that $\boldsymbol{x}(a,\,\cdot) = \alpha(a)$ and $\boldsymbol{x}(b,\,\cdot) = \alpha(b)$, α has to be a geodesic, cf. Corollary 3, p. 265 of [65], and in that case, it is of interest to consider the second variation. Note that if α is a geodesic with $|\alpha'| > 0$ and Y is a vector field on α, then

$$(Y^{\perp})' = (Y')^{\perp} =: \overset{\perp}{Y'},$$

where $Y^{\perp}$ is the component of Y perpendicular to α and the last equality is a definition. We shall call $A(u) = \boldsymbol{x}_{vv}(u,0)$ the *transverse acceleration vector field* of $\boldsymbol{x}$. Let us quote Theorem 4, p. 266 of [65], keeping in mind that our definition of the Riemann curvature tensor differs from that of [65].

Theorem 18.2. *Let $\sigma\colon [a,b] \to M$ be a geodesic segment of speed $c > 0$ and sign ε. If $\boldsymbol{x}$ is a variation of σ, then*

$$L''_{\boldsymbol{x}}(0) = \frac{\varepsilon}{c}\int_a^b \{\langle \overset{\perp}{V'}, \overset{\perp}{V'}\rangle - \langle R_{\sigma' V}V, \sigma'\rangle\}\, du + \frac{\varepsilon}{c}\langle \sigma', A\rangle\Big|_a^b,$$

where V is the variation vector field and A the transverse acceleration vector field of $\boldsymbol{x}$.

Let us describe the situation in which we shall be interested. Let (M,g) be a Lorentz manifold, let Σ_i, $i = 1,2$, be two spacelike hypersurfaces of (M,g) and let $[0,b]$ be an interval with $b > 0$. Let $\Omega(\Sigma_1,\Sigma_2)$ denote the set of all piecewise smooth curves $\alpha\colon [0,b] \to M$ such that $\alpha(0) \in \Sigma_1$ and $\alpha(b) \in \Sigma_2$. We call a piecewise smooth variation $\boldsymbol{x}\colon [0,b] \times (-\delta,\delta) \to M$ of α such that the ranges of $\boldsymbol{x}(0,\,\cdot)$ and $\boldsymbol{x}(b,\,\cdot)$ are contained in Σ_1 and Σ_2 respectively a (Σ_1,Σ_2)-variation of α.

Definition 18.3. The *tangent space* $T_\alpha(\Omega)$ to $\Omega(\Sigma_1,\Sigma_2)$ at α consists of all piecewise smooth vector fields V on α such that $V(0) \in T_{\alpha(0)}\Sigma_1$ and $V(b) \in T_{\alpha(b)}\Sigma_2$.

It will be of interest to know that given a $V \in T_\alpha(\Omega)$, there is a (Σ_1,Σ_2)-variation $\boldsymbol{x}$ of α such that the variation vector field of $\boldsymbol{x}$ coincides with V.

Lemma 18.4. *Let $\alpha \in \Omega(\Sigma_1,\Sigma_2)$ and let $V \in T_\alpha(\Omega)$. Then there is a (Σ_1,Σ_2)-variation $\boldsymbol{x}$ of α such that the variation vector field of $\boldsymbol{x}$ is V.*

Proof. Let E_μ, $\mu = 0,\dots,n$, be a parallel transported orthonormal frame along α and let $\gamma_i\colon (-\delta,\delta) \to \Sigma_i$, $i = 1,2$ be smooth curves such that $\gamma_1'(0) = V(0)$ and

$\gamma_2'(0) = V(b)$ and such that they are contained in normal neighbourhoods of $\alpha(0)$ and $\alpha(b)$ respectively. Let W_i, $i = 1, 2$, be the curves in $T_{\alpha(0)}M$ and $T_{\alpha(b)}M$ respectively such that $\exp[W_i(s)] = \gamma_i(s)$ for $s \in (-\delta, \delta)$. Note that

$$W_1(s) = W_1^{\mu}(s)E_{\mu}(0), \quad W_2(s) = W_2^{\mu}(s)E_{\mu}(b), \quad V(u) = V^{\mu}(u)E_{\mu}(u)$$

for smooth functions W_i^{μ} and piecewise smooth functions V^{μ}. Observe that $W_i^{\mu}(0) = 0$, that $(W_1^{\mu})'(0) = V^{\mu}(0)$ and that $(W_2^{\mu})'(0) = V^{\mu}(b)$. Let $\chi \in C^{\infty}(\mathbb{R})$ be such that $\chi(t) = 1$ for $t \leq b/4$ and $\chi(t) = 0$ for $t \geq 3b/4$. Define

$$\begin{aligned} Z(u, v) = \{&\chi(u)[v(V^{\mu}(u) - V^{\mu}(0)) + W_1^{\mu}(v)] \\ &+ [1 - \chi(u)][v(V^{\mu}(u) - V^{\mu}(b)) + W_2^{\mu}(v)]\}E_{\mu}(u). \end{aligned}$$

Note that $Z(0, v) = W_1(v)$, that $Z(b, v) = W_2(v)$ and that if we write $Z(u, v) = Z^{\mu}(u, v)E_{\mu}(u)$, then for every fixed $u \in [0, b]$ we have $Z^{\mu}(u, v) = vV^{\mu}(u) + O(v^2)$. Consequently, $\boldsymbol{x}(u, v) = \exp[Z(u, v)]$ has the desired properties. □

Lemma 18.5. *Let $\alpha \in \Omega(\Sigma_1, \Sigma_2)$ have constant speed $c > 0$. Then $L_{\boldsymbol{x}}'(0) = 0$ for every (Σ_1, Σ_2)-variation $\boldsymbol{x}$ of α if and only if α is a geodesic normal to Σ_1 and Σ_2.*

Proof. By Proposition 18.1, we see that if α is a geodesic normal to Σ_1 and Σ_2, then $L_{\boldsymbol{x}}'(0) = 0$ for every (Σ_1, Σ_2)-variation $\boldsymbol{x}$ of α. To prove the converse, note that a fixed endpoint variation of α, i.e., a variation $\boldsymbol{x}$ such that $\boldsymbol{x}(0, \cdot)$ and $\boldsymbol{x}(b, \cdot)$ are constant and equal to $\alpha(0)$ and $\alpha(b)$ respectively, is a (Σ_1, Σ_2)-variation of α. As a consequence, we can apply Corollary 3, p. 265 of [65] in order to conclude that α has to be an unbroken geodesic. Thus

$$L_{\boldsymbol{x}}'(0) = \frac{\varepsilon}{c}\langle \alpha', V\rangle\Big|_a^b.$$

Since, for every $v \in T_{\alpha(0)}\Sigma_1$, we can choose a (Σ_1, Σ_2)-variation $\boldsymbol{x}$ such that the corresponding variation vector field V has the property that $V(0) = v$ and $V(b) = 0$, cf. Lemma 18.4, we conclude that α has to be normal to Σ_1. For similar reasons, α has to be normal to Σ_2. □

Before we discuss the second variation for (Σ_1, Σ_2)-variations, let us introduce some terminology.

Definition 18.6. Let (M, g) be a time oriented Lorentz manifold, let Σ be a spacelike hypersurface in (M, g) and let k be the second fundamental form of Σ, cf. Definition 13.1. Then the mean curvature κ of Σ is defined by

$$\kappa(p) = \sum_{i=1}^{n} k(e_i, e_i),$$

for $p \in \Sigma$, assuming that Σ is n-dimensional and $\{e_i\}$ is an orthonormal basis of $T_p\Sigma$ with respect to the induced metric. Finally, Σ is said to be a constant mean curvature (CMC) hypersurface if κ is constant on Σ.

Let Σ be a spacelike hypersurface of a time oriented Lorentz manifold (M, g), let γ be a curve in Σ, let $\ddot{\gamma}$ denote the acceleration of γ in M and let γ'' denote the acceleration of γ in Σ. Then, due to Corollary 9, p. 103 of [65], we have

$$\ddot{\gamma} = \gamma'' + II(\gamma', \gamma'),$$

where II is the shape tensor defined in (13.7), cf. also p. 100 of [65]. If $T_{\gamma(t)}$ is the future oriented unit timelike normal at $\gamma(t)$, we thus get

$$\langle \ddot{\gamma}, T_{\gamma(t)} \rangle = \langle II[\gamma'(t), \gamma'(t)], T_{\gamma(t)} \rangle = -k[\gamma'(t), \gamma'(t)], \tag{18.1}$$

cf. (13.8).

Lemma 18.7. *Let (M, g) be a time oriented Lorentz manifold, let Σ_i, $i = 1, 2$, be spacelike hypersurfaces and assume there is a future directed timelike geodesic $\sigma \in \Omega(\Sigma_1, \Sigma_2)$ with speed $c > 0$ which is normal to both Σ_1 and Σ_2. Let $\boldsymbol{x}$ be a (Σ_1, Σ_2)-variation of σ with associated variation vector field V. Then*

$$L''_{\boldsymbol{x}}(0) = -\frac{1}{c}\int_0^b \{\langle \overset{\perp}{V'}, \overset{\perp}{V'} \rangle - \langle R_{\sigma' V} V, \sigma' \rangle\}\, du + k_2[V(b), V(b)] - k_1[V(0), V(0)],$$

where k_i is the second fundamental form of Σ_i.

Proof. The lemma follows from Theorem 18.2, once we have made the following observations. First of all, $\varepsilon = -1$ in the current context. Since $\sigma'(0)/c$ is the future directed unit normal to Σ_1 and $\sigma'(b)/c$ is the future directed unit normal to Σ_2 at $\sigma(0)$ and $\sigma(b)$ respectively, we can use (18.1) in order to conclude that

$$\frac{\varepsilon}{c}\langle \sigma', A \rangle(0) = k_1[V(0), V(0)], \quad \frac{\varepsilon}{c}\langle \sigma', A \rangle(b) = k_2[V(b), V(b)].$$

The lemma follows. □

Proposition 18.8. *Let (M, g) be a time oriented Lorentz manifold, let Σ_i, $i = 1, 2$, be spacelike hypersurfaces and assume there is a future directed timelike curve $\sigma \in \Omega(\Sigma_1, \Sigma_2)$ with constant speed $c > 0$ which has maximal length among the future oriented timelike curves in $\Omega(\Sigma_1, \Sigma_2)$. Then σ is a timelike geodesic which is normal to both Σ_1 and Σ_2. Furthermore,*

$$\frac{1}{c}\int_0^b \mathrm{Ric}[\sigma'(s), \sigma'(s)]\, ds + \kappa_2[\sigma(b)] - \kappa_1[\sigma(0)] \leq 0,$$

where κ_i is the mean curvature of Σ_i.

Proof. That σ is a timelike geodesic normal to Σ_1 and Σ_2 follows from Lemma 18.5. Let E_i, $i = 1, \dots, n$, be parallel propagated vector fields along σ such that $\{E_i(0)\}$ is an orthonormal basis for $T_{\sigma(0)}\Sigma_1$. Since σ is a geodesic, we conclude that $E_i(s)$, $i = 1, \dots, n$, constitutes an orthonormal basis for $\sigma'(s)^\perp$ for all $s \in [0, b]$. Let $\boldsymbol{x}_i$ be a (Σ_1, Σ_2)-variation of σ such that its associated variation vector field is E_i. Such a variation exists due to Lemma 18.4. By definition, $E_i' = 0$, so the fact that σ is length maximizing and Lemma 18.7 lead to the conclusion that

$$0 \geq L''_{\boldsymbol{x}_i}(0) = \frac{1}{c}\int_a^b \{\langle R_{\sigma' E_i} E_i, \sigma'\rangle\}\, du + k_2[E_i(b), E_i(b)] - k_1[E_i(0), E_i(0)].$$

Summing over i, we obtain the desired conclusion, cf. Lemma 52, p. 87 of [65], keeping the differences in conventions concerning the Riemann curvature tensor in mind. □

18.2 Constant mean curvature hypersurfaces

As has already been mentioned, it will be of interest to know that hypersurfaces of constant mean curvature are unique in spacetimes satisfying a certain condition, the timelike convergence condition, which we shall now define.

Definition 18.9. A Lorentz manifold (M, g) is said to satisfy the timelike convergence condition if $\mathrm{Ric}(v, v) \geq 0$ for all timelike vectors $v \in TM$.

In preparation for the proof of uniqueness, let us make the following observation.

Lemma 18.10. *Let (M, g) be a time oriented Lorentz manifold satisfying the timelike convergence condition and assume that there is a compact spacelike hypersurface Σ in M with constant mean curvature $\kappa \neq 0$. Given an open set U containing Σ, there are two compact spacelike hypersurfaces Σ_i, $i = 1, 2$, such that $\Sigma_1 \subset U \cap I^+(\Sigma)$, $\Sigma_2 \subset U \cap I^-(\Sigma)$ and if κ_i is the mean curvature of Σ_i, then*

$$\sup_{p \in \Sigma_1} \kappa_1(p) < \kappa < \inf_{p \in \Sigma_2} \kappa_2(p).$$

Remark 18.11. This argument is to be found in [60], p. 119.

Proof. Due to Proposition 26, p. 200 of [65], there is a normal neighbourhood of Σ. Since Σ is compact, there is thus an $\varepsilon > 0$ and a map $\phi : (-\varepsilon, \varepsilon) \times \Sigma \to M$ defined by

$$\phi(t, p) = \exp(tN_p),$$

where exp is the exponential map and N_p is the future directed unit timelike normal to Σ at p, such that ϕ is a diffeomorphism onto its image. It will be of interest to know that tangent vectors to $\{t\} \times \Sigma$ are orthogonal to ∂_t with respect to $\phi^* g$. To this end, let $\gamma : (-\delta, \delta) \to \Sigma$ be a smooth curve for some $\delta > 0$. Let $\boldsymbol{x} : (-\varepsilon, \varepsilon) \times (-\delta, \delta) \to M$ be defined by $\boldsymbol{x}(t, s) = \phi[t, \gamma(s)]$. Note that $g(\boldsymbol{x}_t, \boldsymbol{x}_t) = -1$. Using the notation on p. 122–123 of [65], we then have $g(\boldsymbol{x}_{ts}, \boldsymbol{x}_t) = 0$. Due to Proposition 44, p. 123 of [65], we have $\boldsymbol{x}_{st} = \boldsymbol{x}_{ts}$ so that $g(\boldsymbol{x}_{st}, \boldsymbol{x}_t) = 0$. Due to the fact that the curves of constant s are geodesics, we have $\boldsymbol{x}_{tt} = 0$. Combining this observation with the above, we conclude that the derivative with respect to t of $g(\boldsymbol{x}_s, \boldsymbol{x}_t)$ equals zero. Since

$g(\boldsymbol{x}_s, \boldsymbol{x}_t)$ is zero for $t = 0$ it is thus always zero, so that the desired conclusion holds. Let ξ be the smooth vector field defined on the image of ϕ by

$$\xi|_p = \phi_* \partial_t|_{\phi^{-1}(p)}.$$

Note that the integral curves of ξ are timelike geodesics and that $\xi_\alpha \xi^\alpha = -1$. For a fixed $p \in \Sigma$, let $\gamma(t) = \phi(t, p)$. Then γ is a geodesic. Furthermore, $\gamma'(t) = \xi|_{\gamma(t)}$. Due to Proposition 18, p. 65 of [65] and the fact that γ is a geodesic, we have

$$0 = \gamma''(t) = \nabla_{\gamma'(t)} \xi.$$

As a conclusion, $\nabla_\xi \xi = 0$. In index notation, we thus have

$$\xi^\alpha \nabla_\alpha \xi_\beta = 0, \quad \xi^\alpha \nabla_\beta \xi_\alpha = 0, \tag{18.2}$$

where the latter equality is due to the fact that $\xi_\alpha \xi^\alpha = -1$. Define $K_{\alpha\beta} = \nabla_\alpha \xi_\beta$. Then $K_{\alpha\beta}$ are the components of a covariant 2-tensorfield on the image of ϕ. Note that if i_t is the inclusion map from $\Sigma_t := \phi(\{t\} \times \Sigma)$ to M, then $i_t^* K$ is the second fundamental form of Σ_t. The reason why this is true is that the future directed unit normal to Σ_t is ξ. As a consequence of this and the fact that (18.2) holds, $(g^{\alpha\beta} K_{\alpha\beta})(q)$ is the mean curvature of Σ_t at q, assuming $q \in \Sigma_t$. Let us define $\mathcal{K} = g^{\alpha\beta} K_{\alpha\beta}$. In order to compute the change of the mean curvature, let us compute

$$\begin{aligned}
\xi^\gamma \nabla_\gamma K_{\alpha\beta} &= \xi^\gamma \nabla_\gamma \nabla_\alpha \xi_\beta = \xi^\gamma \nabla_\alpha \nabla_\gamma \xi_\beta + \xi^\gamma R_{\gamma\alpha\beta}{}^\delta \xi_\delta \\
&= \nabla_\alpha (\xi^\gamma \nabla_\gamma \xi_\beta) - (\nabla_\alpha \xi^\gamma) \nabla_\gamma \xi_\beta + R_{\gamma\alpha\beta\delta} \xi^\delta \xi^\gamma \\
&= -K_\alpha{}^\gamma K_{\gamma\beta} - R_{\alpha\gamma\beta\delta} \xi^\delta \xi^\gamma,
\end{aligned}$$

where we have used (10.9), (18.2) and (10.8). Note that this is the same computation as (9.2.10), p. 218 of [87]. Contracting this identity with $g^{\alpha\beta}$, keeping (10.11) and (10.8) in mind, we obtain

$$\xi \mathcal{K} = -K^{\alpha\beta} K_{\alpha\beta} - R_{\alpha\beta} \xi^\alpha \xi^\beta.$$

Since (M, g) satisfies the timelike convergence condition and since $\mathcal{K}(p) = \kappa \neq 0$ for $p \in \Sigma$, so that $(K^{\alpha\beta} K_{\alpha\beta})(p) \neq 0$ for $p \in \Sigma$, we obtain the conclusion of the lemma by choosing $\Sigma_i = \phi(t_i, \Sigma)$ for suitable t_i, $i = 1, 2$, such that $t_1 > 0$ and $t_2 < 0$. □

Proposition 18.12. *Let (M, g) be a time oriented, globally hyperbolic Lorentz manifold satisfying the timelike convergence condition. Assume (M, g) has two compact spacelike hypersurfaces Σ_i, $i = 1, 2$, and let κ_i be the associated mean curvatures. If*

$$\sup_{p \in \Sigma_1} \kappa_1(p) < \inf_{p \in \Sigma_2} \kappa_2(p),$$

then there is no future directed timelike curve from Σ_1 to Σ_2.

Proof. Assume there is a future directed timelike curve from Σ_1 to Σ_2. Note that the time separation function τ, cf. Definition 15, p. 409 of [65], is a continuous map from $M \times M$ to $[0, \infty)$, cf. Lemma 21, p. 412 of [65]. As a consequence, τ, restricted to $\Sigma_1 \times \Sigma_2$, attains its maximum on $\Sigma_1 \times \Sigma_2$. Let us denote one such point by (p, q). Due to our assumption, the maximum is positive and by Proposition 19 (recalling the different characterizations of global hyperbolicity), p. 411 of [65], there is a future oriented timelike geodesic σ from p to q of length $\tau(p, q)$. The conclusions of Proposition 18.8 then contradict the assumptions of the present proposition. The conclusion follows. □

Proposition 18.13. *Let (M, g) be a time oriented, globally hyperbolic Lorentz manifold satisfying the timelike convergence condition. Assume (M, g) has two compact spacelike hypersurfaces Σ_i, $i = 1, 2$, with the same constant mean curvature $\kappa \neq 0$. Then there is no future directed timelike curve from Σ_1 to Σ_2.*

Proof. Assume there is a future directed timelike curve from Σ_1 to Σ_2. Due to Lemma 18.10 there is a compact spacelike hypersurface Σ_3 in M such that

$$\sup_{p \in \Sigma_3} \kappa_3(p) < \kappa,$$

where κ_3 is the mean curvature of Σ_3. Furthermore, there is a future directed timelike curve from Σ_3 to Σ_2. We get a contradiction to Proposition 18.12. □

Corollary 18.14. *Let (M, g) be a time oriented, globally hyperbolic Lorentz manifold satisfying the timelike convergence condition. Assume (M, g) has two compact spacelike Cauchy hypersurfaces Σ_i, $i = 1, 2$, with the same constant mean curvature $\kappa \neq 0$. Then $\Sigma_1 = \Sigma_2$.*

Proof. Assume $\Sigma_1 \neq \Sigma_2$. Then there is a $p \in \Sigma_1$ such that $p \notin \Sigma_2$. Let γ be an inextendible timelike geodesic passing through p. Since Σ_2 is a Cauchy hypersurface, γ has to intersect Σ_2. Thus, without loss of generality, we can assume there is a future directed timelike curve from Σ_1 to Σ_2. This contradicts the conclusion of Proposition 18.13. □

The above corollary is all we shall need in connection with the discussion of the pathological example. Nevertheless, it is of interest to note that one can drop the assumption that Σ_i be Cauchy hypersurfaces; this follows from the remaining assumptions.

Corollary 18.15. *Let (M, g) be a connected, oriented, time oriented, globally hyperbolic Lorentz manifold satisfying the timelike convergence condition. Assume (M, g) has two compact spacelike hypersurfaces Σ_i, $i = 1, 2$, with the same constant mean curvature $\kappa \neq 0$. Then $\Sigma_1 = \Sigma_2$.*

Proof. Due to Proposition 18.13, Σ_1 and Σ_2 are achronal. By Proposition 11.29, they are thus Cauchy hypersurfaces so that Corollary 18.14 yields the desired result. □

18.3 Conditions ensuring maximality

We are interested in the extendibility properties of maximal globally hyperbolic developments of initial data. In order to be able to discuss this issue, we need to have criteria that ensure that a globally hyperbolic development is the maximal globally hyperbolic development. The purpose of this section is to provide such criteria.

Proposition 18.16. *Let (M, g) be a connected and time oriented Lorentz manifold and assume that one of the following conditions is satisfied:*

- *(M, g) is future causally geodesically complete and there are real numbers κ_j such that $\kappa_j \to \infty$ as $j \to \infty$ and smooth spacelike Cauchy hypersurfaces Σ_j in (M, g) with constant mean curvature κ_j.*
- *(M, g) is past causally geodesically complete and there are real numbers κ_j such that $\kappa_j \to -\infty$ as $j \to \infty$ and smooth spacelike Cauchy hypersurfaces Σ_j in (M, g) with constant mean curvature κ_j.*
- *There are real numbers $\kappa_{\pm,j}$ such that $\kappa_{\pm,j} \to \pm\infty$ as $j \to \infty$ and smooth spacelike Cauchy hypersurfaces $\Sigma_{\pm,j}$ in (M, g) with constant mean curvature $\kappa_{\pm,j}$.*

Assume $(\bar{M}, \bar{g})$ to be a time oriented and connected Lorentz manifold satisfying the timelike convergence condition and $i : M \to \bar{M}$ to be a smooth embedding such that $i(M)$ is an open set. Furthermore, assume that there is a Cauchy hypersurface S in (M, g) such that $i(S)$ is a Cauchy hypersurface in $(\bar{M}, \bar{g})$. Then $i(M) = \bar{M}$.

Proof. Let us first note that for any Cauchy hypersurface Σ in M, $i(\Sigma)$ is a Cauchy hypersurface in $\bar{M}$. The reason is as follows. Let γ be an inextendible timelike curve in $\bar{M}$. Let ξ be γ restricted to $\gamma^{-1}[i(M)]$. Note that since $i(S)$ is a Cauchy hypersurface in $(\bar{M}, \bar{g})$ and $i(M)$ is open, $\gamma^{-1}[i(M)]$ is a non-empty open subset of $\mathbb{R}$. Furthermore, we can consider ξ to be a map into M. Restricting ξ to any connected component of its domain, we get an inextendible timelike curve in M. Consequently ξ has to intersect Σ, and thus γ has to intersect $i(\Sigma)$. In order for γ to intersect $i(\Sigma)$ twice, the domain of ξ has to have at least two components; since ξ restricted to one component is an inextendible timelike curve in M, it can only intersect Σ once. But then ξ has to intersect S twice, which implies that γ has to intersect $i(S)$ twice, contradicting the fact that $i(S)$ is a Cauchy hypersurface.

Assume $i(M) \neq \bar{M}$. Then there is a point $p \in \bar{M} - i(M)$. Let γ be an inextendible timelike geodesic passing through p and ξ be as above. By the above argument, we know that the domain of ξ only consists of one component, and that ξ is an inextendible timelike curve in M. Thus, the image of ξ is either to the future or the past of p. By changing the time orientation, we can assume it to be to the past. Then $I^+(p)$, which is an open set, cannot intersect $i(M)$ for reasons similar to ones given above. Thus we might as well assume $p \notin \overline{i(M)}$. Let q be the point of intersection between ξ and S. By the above assumption, $q \ll p$. If M is future causally geodesically complete, we get a contradiction. We can thus assume there is a sequence of CMC Cauchy hypersurfaces Σ_j such that the CMC of Σ_j tends to $-\infty$ as $j \to \infty$. By the above observations, ξ intersects all the hypersurfaces Σ_j. Let p_j be the point of intersection between ξ and Σ_j and let t_j and t be such that $\gamma(t_j) = i(p_j)$ and $\gamma(t) = p$. Note that the length of the curve $\gamma|_{[t_j,t]}$ has a positive lower bound independent of j, namely the length of the curve outside $i(M)$. We then get a contradiction to Hawking's singularity theorem, Theorem 55A, p. 431 of [65]. In the application of this theorem, note that the concept *future convergence* used in the statement is the same as the constant mean curvature, except for a factor $1/n$, where $n+1$ is the dimension of M, and *a minus sign*. □

18.4 Conditions ensuring inextendibility

We shall prove the inextendibility of certain spacetimes in what follows, but in order to be able to do so, we shall need criteria ensuring inextendibility.

Definition 18.17. Let (M, g) be a connected Lorentz manifold which is at least C^2. Assume there is a connected C^2 Lorentz manifold $(\widehat{M}, \hat{g})$ of the same dimension as M and an isometric embedding $i : M \to \widehat{M}$ such that $i(M) \neq \widehat{M}$. Then M is said to be *C^2-extendible*. If (M, g) is not C^2-extendible, it is said to be *C^2-inextendible* or *maximal*.

Lemma 18.18. *Let (M, g) be a connected Lorentz manifold which is at least C^2. Assume that the Kretschmann scalar,*

$$R_{\alpha\beta\gamma\delta} R^{\alpha\beta\gamma\delta},$$

i.e., the Riemann curvature tensor of (M, g) contracted with itself, is unbounded in the incomplete directions of timelike geodesics. Then (M, g) is C^2-inextendible.

Remark 18.19. When we say that the Kretschmann scalar is unbounded in the incomplete directions of timelike geodesics, we mean the following: for any unit timelike geodesic ξ (i.e., curve satisfying $\xi'' = 0$ and $\langle \xi', \xi' \rangle = -1$) with maximal existence interval (s_-, s_+), we have

$$\limsup_{s \to s_+ -} |(R_{\alpha\beta\gamma\delta} R^{\alpha\beta\gamma\delta}) \circ \xi(s)| = \infty$$

if $s_+ < \infty$ and

$$\limsup_{s \to s_- +} |(R_{\alpha\beta\gamma\delta} R^{\alpha\beta\gamma\delta}) \circ \xi(s)| = \infty$$

if $s_- > -\infty$.

Proof. Assume (M, g) to be C^2-extendible and let $(\widehat{M}, \hat{g})$ be an extension with embedding $i : M \to \widehat{M}$. In order to obtain a contradiction, all we need to prove is that there is a timelike geodesic γ which intersects both $i(M)$ and $\widehat{M} - i(M)$. The reason is as follows. Consider γ on an open interval $I = (s_-, s_+)$ such that $\gamma[(s_-, s_0)]$, where $s_0 \in I$, is contained in $i(M)$ but $\gamma(s_0) \notin i(M)$ (by reversing the orientation of γ if necessary, we can assume ourselves to be in this situation). Then $\gamma|_{(s_-,s_0)}$ is inextendible in one time direction, corresponding to increasing s, if we consider it to be a causal geodesic in M. Therefore, according to the assumptions of the lemma, it is either complete in this direction or such that the Kretschmann scalar blows up along the curve as $s \to s_0-$. Neither of these possibilities is compatible with the fact that the geodesic can be continued to s_0 and beyond. In order to prove that there is a timelike geodesic as described above, let $p \in \hat{M} - i(M)$ be on the boundary of $i(M)$ and let γ be a timelikegeodesic through p. If γ intersects $i(M)$, we are done. If γ is contained in $\hat{M} - i(M)$, let U be a convex neighbourhood of p and let $q = \gamma(s_0) \in U$ be such that there is a timelike geodesic from q to p. By the standard properties of convex neighbourhoods and the fact that p is on the boundary of $i(M)$, we conclude the existence of the desired geodesic. □

19 Initial data

In the present chapter, we discuss left invariant metrics on unimodular Lie groups and, for a given left invariant metric, preferred bases for the Lie algebra. Furthermore, we compute the Ricci and scalar curvature. The latter of these quantities is needed when expressing the constraints in terms of the variables we shall be using, and the former is needed in the proof of the existence of initial data such that the corresponding MGHD has inequivalent extensions. Finally, we express the constraint equations for left invariant vacuum initial data on a 3-dimensional unimodular Lie group in terms of the variables we shall be using. The motivation for the perspective on initial data taken here is that it should fit naturally together with the formulation of Einstein's equations, due to Ellis and MacCallum [37], presented in the following chapter. For a general discussion of the curvature of left invariant metrics on Lie groups, the reader is referred to the work of Milnor, cf. [61].

19.1 Unimodular Lie groups

Let G be an n dimensional connected Lie group and let $\mathfrak{g}$ be the associated Lie algebra, i.e., the vector space of left invariant vector fields on G. Define, given $X \in \mathfrak{g}$, the linear transformation $\mathrm{ad}_X : \mathfrak{g} \to \mathfrak{g}$ by $\mathrm{ad}_X Y = [X, Y]$.

Definition 19.1. A connected Lie group G is said to be *unimodular* if ad_X has trace zero for every $X \in \mathfrak{g}$.

Remark 19.2. Of course, the unimodularity is a property of the Lie algebra, so it is also meaningful to speak of a unimodular Lie algebra. In [61], p. 316, a Lie group is defined to be unimodular if the left invariant Haar measure is also right invariant. Nevertheless, for connected Lie groups, this condition is shown to be equivalent to the above in Lemma 6.3, p. 317 of [61]. However, we shall not need any characterization of unimodularity other than the one given in the definition above.

Let e_i, $i = 1, \dots, n$, be a basis for $\mathfrak{g}$. The *structure constants* γ^i_{jk} associated with this basis are defined by

$$[e_j, e_k] = \gamma^i_{jk} e_i.$$

Note that the condition of unimodularity is equivalent to the condition that $\gamma^i_{ji} = 0$, where we use the Einstein summation convention. Here, we shall be interested in 3-dimensional Lie groups, and in that case, the information contained in the structure constants can be reduced to a symmetric matrix.

Lemma 19.3. *If G is a 3-dimensional unimodular Lie group, e_i is a basis of $\mathfrak{g}$ and γ^i_{jk} are the associated structure constants, there is a unique symmetric matrix with components ν^{ij} such that $\gamma^i_{jk} = \varepsilon_{jkl}\nu^{li}$.*

Remark 19.4. The definition and elementary properties of the permutation symbols ε_{ijk} are discussed in Section B.1.

Proof. Define

$$\nu^{ij} = \frac{1}{2}\gamma^{(i}_{kl}\varepsilon^{j)kl}, \tag{19.1}$$

where the parenthesis denotes symmetrization, i.e.,

$$\nu^{ij} = \frac{1}{4}(\gamma^{i}_{kl}\varepsilon^{jkl} + \gamma^{j}_{kl}\varepsilon^{ikl}),$$

then, after some computations using the identities (B.1) and (B.2) and the fact that the group is unimodular, we get

$$\varepsilon_{nmi}\nu^{ij} = \gamma^{j}_{nm}.$$

Furthermore, if there are two matrices ν and κ such that

$$\varepsilon_{nmi}\nu^{ij} = \varepsilon_{nmi}\kappa^{ij},$$

then $\kappa = \nu$ due to (B.3). □

Definition 19.5. If G is a 3-dimensional unimodular Lie group and e_i is a basis of $\mathfrak{g}$, then the matrix ν associated with the corresponding structure constants as described in Lemma 19.3 will be referred to as the associated *commutator matrix*.

It will be of interest to know what happens to the commutator matrix under a change of basis of the Lie algebra.

Lemma 19.6. *Let G be a 3-dimensional unimodular Lie group, let e_i and e'_i be two bases of $\mathfrak{g}$ and let ν and $\tilde{\nu}$ denote the associated commutator matrices respectively. Then there is a non-degenerate constant matrix A such that*

$$e'_i = A_i{}^j e_j \tag{19.2}$$

and

$$\nu = (\det A)^{-1} A^t \tilde{\nu} A. \tag{19.3}$$

Proof. Let $\tilde{\gamma}^i_{jk}$ be the structure constants associated with e'_i and γ^i_{jk} the structure constants associated with e_i. Then

$$\tilde{\gamma}^i_{lk} = A_l{}^p A_k{}^m \gamma^n_{pm} B_n{}^i = A_l{}^p A_k{}^m \varepsilon_{pmq}\nu^{qn} B_n{}^i = A_l{}^p A_k{}^m A_r{}^q B_s{}^r \varepsilon_{pmq}\nu^{sn} B_n{}^i,$$

where $B = A^{-1}$. In this expression,

$$A_l{}^p A_k{}^m A_r{}^q \varepsilon_{pmq} = (\det A)\varepsilon_{lkr},$$

so that

$$\tilde{\gamma}^i_{lk} = (\det A)\varepsilon_{lkr}(B_s{}^r \nu^{sn} B_n{}^i).$$

We conclude that (19.3) holds. □

Let us introduce the following terminology in order to conform with that used in the relativity community.

Definition 19.7. A *Bianchi class A* Lie group (algebra) is a 3-dimensional unimodular Lie group (algebra).

The transformation rule (19.3) can be used to classify the Bianchi class A Lie algebras; one can use it to diagonalize ν and then put the diagonal elements in a canonical form. When ν is diagonal, we shall denote the diagonal elements by ν_1, ν_2 and ν_3.

Lemma 19.8. *Table* 19.1 *constitutes a classification of the Bianchi class A Lie algebras, i.e., any Bianchi class A Lie algebra admits a basis such that the associated commutator matrix is diagonal with diagonal components of one of the forms given in Table* 19.1, *and two Bianchi class A Lie algebras are isomorphic if and only if they allow bases with associated commutator matrices which are of the same form* (*as given in Table* 19.1). *Consider an arbitrary basis e_i of $\mathfrak{g}$. Then there is an orthogonal matrix A such that if e_i' is defined by* (19.2), *the associated commutator matrix $\tilde{\nu}$ is diagonal, with diagonal elements of one of the types given in Table* 19.1.

Table 19.1. Bianchi class A.

Type	ν_1	ν_2	ν_3
I	0	0	0
II	+	0	0
VI_0	0	+	−
VII_0	0	+	+
VIII	−	+	+
IX	+	+	+

Remark 19.9. The above lemma should be compared with Lemma 21.1, p. 488 of [71]. Recall that simply connected Lie groups are determined up to isomorphism by their Lie algebras, cf. Theorem 3.27, p. 101 of [88] or Theorem 20.15, p. 532 of [55].

Proof. Let e_i be a basis for $\mathfrak{g}$ and ν be the associated commutator matrix. If we change the basis according to $e_i' = (A^{-1})_i{}^j e_j$ (i.e., A^{-1} of the proof equals A of the statement), then ν transforms to

$$\tilde{\nu} = (\det A)^{-1} A^t \nu A \tag{19.4}$$

due to (19.3). Since ν is symmetric, we assume from now on that the basis is such that it is diagonal. The matrix $A = \mathrm{diag}(1\ 1\ -1)$ changes the sign of ν. Suitable orthogonal matrices perform permutations of the diagonal. The number of non-zero elements on the diagonal is invariant under transformations (19.4) taking one diagonal matrix to another. If the components of A are $A_i{}^j$ and the diagonal matrix $\tilde{\nu}$ is constructed as in (19.4), we have $\tilde{\nu}^{kk} = (\det A)^{-1} \sum_{i=1}^{3} (A_i{}^k)^2 \nu^{ii}$, so that if all the diagonal elements of ν have the same sign, the same is true for $\tilde{\nu}$. The statements of the lemma follow, except for the classification statement. However, using a suitably chosen (typically not orthogonal) matrix $A = \mathrm{diag}(a_1, a_2, a_3)$, it is possible to transform every $+$ sign in Table 19.1 to 1 and every minus sign to -1, and the desired conclusion follows. □

19.2 Curvature

Definition 19.10. Let G be a Lie group. A left invariant Riemannian metric on G is a Riemannian metric g on G such that for every $h \in G$, left translation by h is an isometry of g. If G is a Bianchi class A Lie group and g is a left invariant Riemannian metric on G, we shall call a basis e_i of $\mathfrak{g}$ such that

- $g(e_i, e_j) = \delta_{ij}$,
- if ν is the commutator matrix associated with e_i, then ν is diagonal and of one of the forms given in Table 19.1

a canonical basis for the Lie algebra.

Due to Lemma 19.8, there is, given a Bianchi class A Lie group and a left invariant Riemannian metric on it, a canonical basis for the Lie algebra. We wish to express the Ricci and scalar curvature of g in terms of the matrix ν corresponding to an orthonormal basis of $\mathfrak{g}$.

Lemma 19.11. *Let G be a Bianchi class A Lie group, let g be a left invariant metric on G and let e_i be an orthonormal basis of $\mathfrak{g}$. Then*

$$\mathrm{Ric}(e_i, e_m) = 2\nu_{il}\nu^l{}_m - (\mathrm{tr}\,\nu)\nu_{im} - \nu_{kl}\nu^{kl}\delta_{im} + \frac{1}{2}(\mathrm{tr}\,\nu)^2\delta_{im}, \tag{19.5}$$

$$r = -\nu_{ij}\nu^{ij} + \frac{1}{2}(\mathrm{tr}\,\nu)^2, \tag{19.6}$$

where ν is the commutator matrix associated with e_i and Ric *and r denote the Ricci and scalar curvature of g respectively.*

For completeness, we have included the necessary computations in Section B.1, cf. also [61].

Corollary 19.12. *If g is a left invariant metric on a Bianchi class A Lie group G, the scalar curvature of g can only be positive if G is of Bianchi type* IX.

Proof. Let e_i be a canonical basis for the Lie algebra and let ν be the associated commutator matrix. Expressing the scalar curvature in terms of the diagonal components of ν, we conclude that

$$r = -\frac{1}{2}(\nu_1^2 + \nu_2^2 + \nu_3^2 - 2\nu_1\nu_2 - 2\nu_2\nu_3 - 2\nu_3\nu_1).$$

Consider the Bianchi classification given in Table 19.1. In the case of Bianchi class I and II, we see immediately that $r \leq 0$. In the case of class VI_0 and VII_0, we can assume $\nu_1 = 0$, and then we obtain

$$r = -\frac{1}{2}(\nu_2 - \nu_3)^2 \leq 0.$$

In the case of Bianchi VIII, we can assume $\nu_1 < 0$ and $\nu_2, \nu_3 > 0$ and rewrite

$$r = -\frac{1}{2}[\nu_1^2 + (\nu_2 - \nu_3)^2 - 2\nu_1(\nu_2 + \nu_3)] \leq 0.$$

The statement follows. □

19.3 The constraint equations

Lemma 19.13. *Let G be a Bianchi class A Lie group, let g and k be a left invariant metric and symmetric covariant two tensor on G respectively and let ρ be a constant. Finally, let e_i be an orthonormal basis of $\mathfrak{g}$ and let ν be the associated commutator matrix. Then the equations*

$$\begin{aligned} \frac{1}{2}[r - k_{ij}k^{ij} + (\mathrm{tr}_g\, k)^2] &= \rho, \\ D_i k^i{}_j - D_j \,\mathrm{tr}_g\, k &= 0, \end{aligned} \tag{19.7}$$

where D is the Levi-Civita connection associated with g and r is the associated scalar curvature and indices are raised and lowered with g, are equivalent to

$$-\nu_{ij}\nu^{ij} + \frac{1}{2}(\mathrm{tr}\,\nu)^2 - k_{ij}k^{ij} + (\mathrm{tr}_g\, k)^2 = 2\rho, \tag{19.8}$$

$$K\nu - \nu K = 0, \tag{19.9}$$

respectively, where the components of the matrix K in the last equation are given by $k(e_i, e_j)$.

Proof. Let $k_{ij} = k(e_i, e_j)$ and let us express (19.7) in terms of ν_{ij} and k_{ij} (note that k_{ij} are constants due to the left invariance of k). Due to the fact that $\mathrm{tr}_g\, k$ is constant, the second term on the left-hand side of (19.7) is zero. In order to compute the first term, note that since the Levi-Civita connection commutes with contraction,

$$(Dk)(e_i, e_j, e_l) = D_{e_i}[k(e_j, e_l)] - k(D_{e_i}e_j, e_l) - k(e_j, D_{e_i}e_l) = -\Gamma^m_{ij}k_{ml} - \Gamma^m_{il}k_{jm},$$

where the relation $\Gamma^i_{jk}e_i = D_{e_j}e_k$ defines Γ^i_{jk}. Contracting this identity with δ^{ij}, we obtain

$$D_i k^i{}_l = -\Gamma^m_{ii}k_{ml} - \Gamma^m_{il}k_{im},$$

where it is understood that we sum over repeated indices and that the indices are with respect to the orthonormal basis e_i. Since the first term on the right-hand side is zero by the observations made in Section B.1, we conclude that

$$D_i k^i{}_l = -\Gamma^m_{il} k_{im} = -\frac{1}{2}(-\gamma^i_{lm} + \gamma^l_{mi} + \gamma^m_{il})k_{im} = \gamma^i_{lm} k_{im},$$

where we have used (B.4) and the fact that k is symmetric. Summing up, we conclude that (19.7) is equivalent to

$$\gamma^i_{lm} k_{im} = 0.$$

Contracting with ε^{lpq}, this is equivalent to

$$0 = \varepsilon^{lpq}\varepsilon_{lmr}\nu^{ri}k_{im} = (\delta^p_m\delta^q_r - \delta^p_r\delta^q_m)\nu^{ri}k_{im} = \nu^{qi}k_{ip} - k^{qi}\nu_{ip}, \qquad (19.10)$$

where we have used (B.2). We conclude that (19.7) is equivalent to (19.9). The remaining statement is a direct consequence of (19.6). □

Corollary 19.14. *Let G be a Bianchi class A Lie group and let g and k be a left invariant metric and symmetric covariant two tensor on G respectively. Then, if* (19.7) *is satisfied, there is a canonical basis e_i of the Lie algebra such that $k(e_i, e_j)$ are the components of a diagonal matrix.*

Proof. This is a direct consequence of (19.9). □

Definition 19.15. Bianchi class A *initial data* for Einstein's vacuum equations are given by (G, g, k), where G is a class A Lie group and g and k are a left invariant metric and symmetric covariant two tensor on G respectively satisfying the so-called *constraint equations*

$$\frac{1}{2}[r - k_{ij}k^{ij} + (\mathrm{tr}_g\, k)^2] = 0, \qquad (19.11)$$

$$D_i k^i{}_j - D_j\, \mathrm{tr}_g\, k = 0, \qquad (19.12)$$

where indices are raised and lowered with the Riemannian metric g, D is the Levi-Civita connection associated with g and r is the associated scalar curvature.

Let us define a special class of initial data, which will be of great importance in what follows.

Definition 19.16. Let (G, g, k) be Bianchi class A initial data for Einstein's vacuum equations, let e_i be a canonical basis for the Lie algebra such that k is diagonal with respect to this basis, cf. Corollary 19.14, and let ν be the associated commutator matrix. Let ν_i and k_i denote the diagonal components of ν and K, where K is the diagonal matrix with components $k(e_i, e_j)$. Then, if $\nu_2 = \nu_3$ and $k_2 = k_3$ (or one of the permuted conditions hold), we shall say that the initial data are *locally rotationally symmetric* or of *Taub type*.

Remark 19.17. The concept local rotational symmetry also makes sense in the non-vacuum case. For the sake of brevity, we shall sometimes speak of LRS initial data. Note that a rotation in the plane defined by e_2 and e_3 defines an isomorphism of the Lie algebra if $\nu_2 = \nu_3$. To be more precise, let $\phi \in \mathbb{R}$ and let L_ϕ be defined by

$$L_\phi e_1 = e_1, \quad L_\phi e_2 = \cos\phi e_2 - \sin\phi e_3, \quad L_\phi e_3 = \sin\phi e_2 + \cos\phi e_3.$$

Then L_ϕ is a Lie algebra isomorphism. As was mentioned above, this Lie algebra isomorphism arises from a Lie group isomorphism, assuming that the group is simply connected, cf. [88], Theorem 3.27, p. 101 or [55], Theorem 20.15, p. 532. Furthermore, we of course have $L_{\phi_1} L_{\phi_2} = L_{\phi_1+\phi_2}$. Consequently, we have an action of S^1 on G corresponding to rotations in the e_2, e_3 directions. If the initial data are LRS with $\nu_2 = \nu_3$ and $k_2 = k_3$, then the isomorphisms L_ϕ constructed above all yield isometries of the initial data.

20 Einstein's vacuum equations

In the previous chapter, we introduced variables suitable for treating the case of left invariant vacuum initial data on a 3-dimensional unimodular Lie group. Furthermore, we expressed the constraint equations in terms of them. The purpose of the present chapter is to construct, and establish the basic properties of, the maximal globally hyperbolic development of such data in the vacuum case.

In Section 20.1, we compute the Ricci curvature of a class of model metrics of the form (20.1). The result is expressed in terms of a certain choice of variables due to Ellis and MacCallum, cf. [37]. This choice is very convenient in that it covers all the unimodular Lie groups at once.

In Section 20.2, we then use these computations in order to construct a development, given initial data. The manifold on which the development is defined is of the form $I \times G$ and the metric of the form (20.1). Here G is the Lie group on which the initial data are specified and I is an interval determined as the maximal existence interval of a solution to a certain system of ODE's. The development so constructed will be referred to as the Bianchi class A development. In order for this construction to be useful, it is necessary to prove that the Bianchi class A development is the maximal globally hyperbolic development. The proof of this fact proceeds in several steps. First of all, it is proved that the Bianchi class A development is a globally hyperbolic development such that every constant t hypersurface is a Cauchy hypersurface. Each constant t hypersurface is of course also a CMC hypersurface, and it is of interest to compute the interval exhausted by the mean curvature as t exhausts the interval I. This computation is fairly straightforward for most of the unimodular Lie groups with one exception: SU(2). In the case of SU(2), the mean curvature exhausts $\mathbb{R}$ and the development recollapses; i.e., all timelike geodesics are both future and past incomplete, and there is a moment of maximal expansion at which the mean curvature is zero. Note that SU(2) is the only 3-dimensional unimodular Lie group which allows a left invariant metric of positive scalar curvature, cf. the previous chapter. In fact, there is a general conjecture, called the recollapse conjecture, relating the existence of positive scalar curvature metrics on the initial manifold with the issue of recollapse, cf. the following chapter. The proof of the fact that there is recollapse in the case of left invariant vacuum initial data on SU(2), due to Lin and Wald, cf. [56], is rather lengthy, and we devote a separate chapter to it.

Finally, we analyze the question of causal geodesic completeness/incompleteness and conclude that the Bianchi class A development is the maximal globally hyperbolic development. The presentation of the material given here should be compared with that of [71].

20.1 Model metrics

Let G be a Bianchi class A Lie group and let e_i' be a basis of the Lie algebra such that the associated commutator matrix ν is diagonal and of one of the forms given in

Table 19.1. Let ξ^i be the basis dual to e_i' and consider a metric of the form

$$g = -dt^2 + \sum_{i=1}^{3} a_i^2(t)\xi^i \otimes \xi^i, \tag{20.1}$$

on $I \times G$, for some interval I, where the a_i are positive functions of t. Then $e_i = a_i^{-1} e_i'$ together with $e_0 = \partial_t$ form an orthonormal frame. Due to (19.3), we conclude that if n is the commutator matrix associated with e_i, then it is diagonal with diagonal components

$$n_1 = \frac{a_1}{a_2 a_3}\nu_1, \quad n_2 = \frac{a_2}{a_1 a_3}\nu_2, \quad n_3 = \frac{a_3}{a_1 a_2}\nu_3. \tag{20.2}$$

We shall write n_{ij} when convenient, an object we define to be the components of the diagonal matrix whose diagonal elements are given by n_1, n_2 and n_3. Finally, define

$$\theta_{ij} = \langle \nabla_{e_i} \partial_t, e_j \rangle = \langle \nabla_{e_i} e_0, e_j \rangle, \quad \theta = \delta^{ij}\theta_{ij},$$

where $g = \langle \cdot\,, \cdot \rangle$ and ∇ is the Levi-Civita connection associated with g. One can easily compute that θ_{ij} are the components of a diagonal matrix, cf. Section B.2.

Lemma 20.1. *With the above notation,*

$$\mathrm{Ric}(e_0, e_0) = -\dot{\theta} - \theta^{ij}\theta_{ij}, \tag{20.3}$$

$$\mathrm{Ric}(e_0, e_m) = \varepsilon_{mjl} n^{li}\theta_{ij}, \tag{20.4}$$

$$\mathrm{Ric}(e_l, e_m) = \dot{\theta}_{lm} + \theta\theta_{lm} + 2n_m{}^i n_{il} - n^{ij}n_{ij}\delta_{lm} + \frac{1}{2}(\mathrm{tr}\, n)^2\delta_{lm} - (\mathrm{tr}\, n)n_{lm}, \tag{20.5}$$

where Ric *is the Ricci tensor of the metric g, given by* (20.1), *and the dot signifies differentiation with respect to t.*

Due to the fact that θ_{ij} and n_{ij} are the components of diagonal matrices, (20.4) implies that $\mathrm{Ric}(e_0, e_m) = 0$. The necessary computations are presented in Section B.2.

Lemma 20.2. *With the above notation,*

$$\dot{n}^{ik} - \theta^k{}_l n^{li} - \theta^i{}_j n^{jk} + \theta n^{ki} = 0. \tag{20.6}$$

This is a consequence of the Jacobi identity for the frame e_α. The proof is to be found in Section B.3. Given the above computations, one can write down the equations in the case that the matter model is a perfect fluid, a scalar field, etc.

20.2 Constructing a spacetime

In the previous section we computed the Ricci tensor assuming that we had a metric of the form (20.1). In this section we wish to start with initial data and construct a development. Let (G, g, k) be Bianchi class A initial data for Einstein's vacuum equations. Due to Corollary 19.14, there is there is a canonical basis e_i' of the Lie algebra such that $k_{ij} = k(e_i', e_j')$ are the components of a diagonal matrix. Define $n(0) = \nu$, $\theta(0) = \operatorname{tr}_g k$ and $\sigma_{ij}(0) = k_{ij} - \theta(0)\delta_{ij}/3$. Define n, θ, σ to be the solution to

$$\dot{\theta} = -\frac{3}{2}\sigma_{ij}\sigma^{ij} - \frac{1}{2}n_{ij}n^{ij} + \frac{1}{4}[\operatorname{tr} n]^2, \tag{20.7}$$

$$\dot{\sigma}_{lm} = -\theta\sigma_{lm} - s_{lm}, \tag{20.8}$$

$$\dot{n}_{ij} = 2\sigma^k{}_{(i}n_{j)k} - \frac{1}{3}\theta n_{ij} \tag{20.9}$$

with these initial data, where the parenthesis denotes symmetrization and

$$s_{lm} = b_{lm} - \frac{1}{3}(\operatorname{tr} b)\delta_{lm}, \tag{20.10}$$

$$b_{lm} = 2n_m{}^i n_{il} - (\operatorname{tr} n)n_{lm}. \tag{20.11}$$

Let $I = (t_-, t_+)$ be the maximal existence interval. Define a_i by the conditions

$$\dot{a}_i = \left(\sigma_{ii} + \frac{1}{3}\theta\right)a_i, \quad a_i(0) = 1, \tag{20.12}$$

(no summation) and define a metric on $M = I \times G$ by

$$\bar{g} = -dt^2 + \sum_{i=1}^{3} a_i^2(t)\xi^i \otimes \xi^i, \tag{20.13}$$

where ξ^i are the duals of the e_i'.

Proposition 20.3. *Let (G, g, k) be Bianchi class A initial data for Einstein's vacuum equations. Then the Lorentz manifold $(M, \bar{g})$, where $M = I \times G$, constructed above is a solution to Einstein's vacuum equations such that if $i : G \to \{0\} \times G$ is the natural embedding and κ is the second fundamental form induced on $\{0\} \times G$ by $\bar{g}$, then $i^*\bar{g} = g$ and $i^*\kappa = k$. Furthermore, $\{t\} \times G$ is a Cauchy hypersurface for every $t \in I$.*

Proof. Note that (19.11) is equivalent to

$$\frac{2}{3}\theta^2 - \sigma_{ij}\sigma^{ij} - n_{ij}n^{ij} + \frac{1}{2}(\operatorname{tr} n)^2 = 0 \tag{20.14}$$

for $t = 0$, cf. Lemma 19.13. Considering (20.8) and (20.9), it is clear that if one collects the off-diagonal elements of n_{ij} and σ_{ij} into one vector, say v, then v satisfies an equation of the form $\dot{v} = Cv$ for some matrix C depending on the unknowns. Since $v(0) = 0$, we conclude that n_{ij} and σ_{ij} are the components of diagonal matrices as long as the solution exists. Note also that σ is trace free for all $t \in I$. Finally, note that if we denote the left-hand side of (20.14) by f, then $\dot{f} = 0$, due to (20.7)–(20.11). To conclude: for all $t \in I$, (20.7)–(20.11) and (20.14) are satisfied, n_{ij} and σ_{ij} are the components of diagonal matrices and σ_{ij} are the components of a trace free matrix.

Below, we shall denote the diagonal components of n_{ij} and σ_{ij} by n_i and σ_i respectively. Define $e_i = a_i^{-1} e_i'$. Then $e_0 = \partial_t$ and e_i together constitute an orthonormal frame for $\bar{g}$. Note that

$$\frac{a_1}{a_2 a_3}\nu_1 = \exp\left[\int_0^t \left(2\sigma_1 - \frac{1}{3}\theta\right) ds\right] n_1(0) = n_1(t),$$

where we have used (20.9) and the fact that σ is trace free, and similarly for the other expressions appearing on the right-hand sides of the equations in (20.2). As a consequence, if we define γ^i_{jk} by

$$[e_j, e_k] = \gamma^i_{jk} e_i,$$

then

$$\gamma^i_{jk} = \varepsilon_{jkl} n^{li},$$

where n^{li} are the same objects as the ones appearing in (20.7)–(20.11) and (20.14). Let ∇ be the Levi-Civita connection associated with $\bar{g}$ and define

$$\hat{\theta}(X, Y) = \langle \nabla_X e_0, Y \rangle, \quad \hat{\theta}_{\mu\nu} = \hat{\theta}(e_\mu, e_\nu).$$

Then, according to the arguments presented in the beginning of Section B.2, $\hat{\theta}_{ij}$ is diagonal with diagonal components given by

$$\hat{\theta}_{ii} = a_i^{-1} \dot{a}_i = \sigma_i + \frac{1}{3}\theta$$

(no summation). In other words,

$$\hat{\theta}_{ij} = \sigma_{ij} + \frac{1}{3}\theta\delta_{ij}.$$

Thus, we are exactly in the situation studied in Section 20.1, and the objects θ_{ij} and n_{ij} appearing there coincide with the variables appearing in the equations (20.7)–(20.11) and (20.14) if we define $\theta_{ij} = \sigma_{ij} + \theta\delta_{ij}/3$.

Let us check that $(M, \bar{g})$ satisfy the Einstein vacuum equations. Due to (20.4) and the fact that n_{ij} and θ_{ij} are the components of diagonal matrices, we conclude that $\mathrm{Ric}(e_0, e_m) = 0$. Due to (20.8) and (20.5) we conclude that the trace free part of $\mathrm{Ric}(e_l, e_m)$ is zero. The only components of Ric that could still potentially be non-zero are thus

$$\mathrm{Ric}(e_0, e_0) = -\dot{\theta} - \theta^{ij}\theta_{ij},$$

$$\sum_l \mathrm{Ric}(e_l, e_l) = \dot{\theta} + \theta^2 - n^{ij} n_{ij} + \frac{1}{2}(\mathrm{tr}\, n)^2.$$

Due to (20.14), we see that the sum of the right-hand sides vanishes. Due to (20.7) we see that the sum of the first right-hand side plus a third times the second right-hand side is zero. Consequently, Ric is zero, i.e., Einstein's vacuum equations are satisfied.

Before turning to the question of global hyperbolicity, let us note that a left invariant Riemannian metric, say ρ, on a Lie group G is geodesically complete. To prove this, assume $\gamma\colon (t_-,t_+) \to G$ is a geodesic satisfying $\rho(\gamma',\gamma') = 1$, with $t_+ < \infty$. There is a $\delta > 0$ such that every geodesic λ satisfying $\lambda(0) = e$, where e is the identity element of G, and $\lambda'(0) = v$ with $\rho(v,v) \leq 1$ is defined on $(-\delta,\delta)$. If $L_h\colon G \to G$ is defined by $L_h(h_1) = hh_1$, then L_h is by definition an isometry. Let $t_0 \in (t_-,t_+)$ satisfy $t_+ - t_0 \leq \delta/2$. Let $v \in T_eG$ be the vector corresponding to $\gamma'(t_0)$ under the isometry $L_{\gamma(t_0)}$. Let λ be a geodesic with $\lambda(0) = e$ and $\lambda'(0) = v$. Then $L_{\gamma(t_0)} \circ \lambda$ is a geodesic extending γ.

Let us introduce the notation $M_v = \{v\} \times G$ and let $\pi_1\colon M \to I$ be projection to the first factor. We wish to prove that for any $v \in I$, M_v is a Cauchy surface. First of all, note that a causal curve cannot intersect M_v twice since the t-component of such a curve must be strictly monotone; in fact, we have

$$\frac{d\pi_1 \circ \gamma}{ds}(s) = -\langle \partial_t|_{\gamma(s)}, \gamma'(s)\rangle$$

and the object on the right-hand side is either strictly positive or strictly negative depending on the time orientation of γ. Assume that $\gamma\colon (s_-,s_+) \to M$ is an inextendible causal curve that never intersects M_v. Let $s_0 \in (s_-,s_+)$ and assume that $\pi_1[\gamma(s_0)] = t_1 < v$ and that $\langle \gamma',\partial_t\rangle < 0$ where it is defined. Thus $\pi_1[\gamma(s)]$ increases with s and $\pi_1[\gamma([s_0,s_+))] \subseteq [t_1,v]$. Since we have uniform bounds on a_i from below and above on $[t_1,v]$ and the curve is causal, we get

$$\Big[\sum_{i=1}^{3} \xi^i(\gamma')^2\Big]^{1/2} \leq -C\langle \gamma', e_0\rangle \tag{20.15}$$

on that interval, where C is a positive constant. Since

$$\int_{s_0}^{s_+} -\langle \gamma', e_0\rangle\, ds = \int_{s_0}^{s_+} \frac{d\pi_1 \circ \gamma}{ds}\, ds \leq v - t_1, \tag{20.16}$$

the curve $\gamma|_{[s_0,s_+)}$, projected to G, will have finite length with respect to the metric ρ on G defined by making e_i' an orthonormal basis. Since ρ is a left invariant metric on a Lie group, it is complete by the above observations, and sets closed and bounded with respect to the corresponding topological metric must be compact by the Hopf–Rinow theorem, Theorem 21, p. 138 of [65]. Adding the above observations, we conclude that $\gamma([s_0,s_+))$ is contained in a compact set, and thus there is a sequence $s_k \in [s_0,s_+)$ with $s_k \to s_+$ such that $\gamma(s_k)$ converges. Since $\pi_1[\gamma(s)]$ is monotone and bounded it converges. Using (20.15) and an analogue of (20.16), we conclude that γ has to converge as $s \to s_+$. Consequently, γ is extendible contradicting our assumption. By this and similar arguments covering the other cases, we conclude that M_v is a Cauchy surface for each $v \in (t_-,t_+)$. The proposition follows. □

Definition 20.4. Let (G, g, k) be Bianchi class A initial data for Einstein's vacuum equations. Then the Lorentz manifold $(M, \bar{g})$ constructed above is referred to as the Bianchi class A development of (G, g, k).

Remark 20.5. We shall also speak of Bianchi type I developments, etc., if we wish to specify the Bianchi type of the initial data. If the initial data are locally rotationally symmetric (LRS, of Taub type), we shall also speak of LRS or Taub type Bianchi class A developments.

20.3 Elementary properties of developments

Let us sort out the elementary properties of Bianchi class A developments. It will be of interest to know that the Raychaudhuri equation,

$$e_0(\theta) + \theta_{ij}\theta^{ij} = 0, \tag{20.17}$$

holds. This is simply a consequence of (20.3) and the fact that $\bar{g}$ is a solution to the vacuum equations. Before stating any results, let us observe that if $I = (t_-, t_+)$ is the maximal existence interval of a solution to an ODE of the form $x' = f \circ x$ for some $f \in C^\infty(\mathbb{R}^n, \mathbb{R}^n)$, then the only possibility for t_+ to be finite is if the solution blows up to the future, i.e., if $|x(t)| \to \infty$ as $t \to t_+$, cf. Lemma 56, p. 30 of [65]. The statement concerning t_- is similar.

Lemma 20.6. *Consider a Bianchi class A development which is not of type* IX. *Let the existence interval be $I = (t_-, t_+)$. Then there are two possibilities:*

(1) $\theta \neq 0$ *for the entire development. We then time orient the manifold so that $\theta > 0$. With this time orientation, $t_- > -\infty$ and $t_+ = \infty$. Furthermore, θ is a monotone function and $\theta(t) \to \infty$ as $t \to t_-$ and $\theta(t) \to 0$ as $t \to \infty$.*

(2) *The initial data are a quotient of* $\mathbb{R}^3$ *with Riemannian metric given by the standard Euclidean metric and symmetric covariant* 2*-tensor given by* 0*, so that the development is a quotient of* 4*-dimensional Minkowski space.*

Remark 20.7. In particular, in case the Bianchi class A development is not a quotient of Minkowski space, there is a CMC foliation of the spacetime, the constant mean curvature exhausting the range $(0, \infty)$.

Proof. Since n_{ij} is diagonal we can formulate the constraint, (20.14), as

$$\sigma_{ij}\sigma^{ij} + \frac{1}{2}[n_1^2 + (n_2 - n_3)^2 - 2n_1(n_2 + n_3)] = \frac{2}{3}\theta^2, \tag{20.18}$$

where the n_i are the diagonal components of n_{ij}. Considering Table 19.1, we see that, excepting type IX, the expression in the n_i is always non-negative (this is of course the same observation that led to Corollary 19.12). Thus we deduce the inequality

$$\sigma_{ij}\sigma^{ij} \leq \frac{2}{3}\theta^2. \tag{20.19}$$

Combining this observation with (20.17), we conclude that $|e_0(\theta)| \leq \theta^2$. Consequently, if θ is zero once, it is always zero; this is an immediate consequence of Grönwall's lemma. Time orient the developments with $\theta \neq 0$ so that $\theta > 0$. Due to (20.17), we have

$$\partial_t \theta \leq -\frac{1}{3}\theta^2. \tag{20.20}$$

This inequality proves that θ has to blow up in finite time to the past. Note also that the solution exists until θ blows up; by the constraint, σ_{ij} is bounded as long as θ is bounded and (20.9) then implies that n_{ij} remains bounded as long as θ is bounded (since θ is bounded to the future, this also implies that $t_+ = \infty$). Thus $t_- > -\infty$ and since θ is strictly monotonic, it exhausts some interval (θ_0, ∞). We know that $\theta_0 \geq 0$ and by (20.20) it is clear that θ_0 can not be positive; θ_0 positive would imply $\theta(t) \to -\infty$ as $t \to \infty$.

Consider the possibility $\theta = 0$. Due to (20.19), we see that $\sigma_{ij} = 0$ so that (20.9) implies that n_{ij} is constant. Furthermore, due to (20.18), the only Bianchi types that can come into consideration are types I and VII_0 and the a_i are independent of t. In the case of Bianchi I, all the n_i are zero. In the case of Bianchi VII_0, two are nonzero, say n_2 and n_3, and in that case (20.18) implies that they have to equal. Due to Lemma 19.11, we see that the metric induced on $M_t = \{t\} \times G$ has zero Ricci tensor regardless of whether the Bianchi type under consideration is I or VII_0. Since M_t is 3-dimensional, this implies that the induced metric has zero Riemann curvature tensor. Since the induced metric is complete by an argument presented in the proof of Proposition 20.3, we conclude that the universal covering space of M_t with the induced metric is isometric to Euclidean space, cf. Proposition 23, p. 227 of [65]. □

It is significantly more difficult to compute the maximal existence interval and the range of the mean curvature in the case of Bianchi IX than in the case of the other Bianchi class A types. In fact, in order to be able to carry out this computation, it will be necessary to appeal to a result of Lin and Wald [56], the proof of which we shall need to devote an entire chapter to. Let us start by making a preliminary observation.

Lemma 20.8. *Consider a solution to* (20.7)–(20.11) *and* (20.14) *on an interval of the form* $I_+ = [t_0, t_+)$ *where* $t_+ < \infty$. *If* θ *is bounded on this interval, the solution can be extended beyond* t_+. *A similar statement holds for intervals of the form* $(t_-, t_0]$.

Proof. Let C_θ be an upper bound on $|\theta|$ on I_+ and assume $C_\theta \geq 1$. By (20.9), $(n_1 n_2 n_3)' = -\theta n_1 n_2 n_3$, so that $n_1 n_2 n_3$ is bounded on I_+ since $t_+ < \infty$. Let C_n be a constant such that $n_1 n_2 n_3 \leq C_n$ on I_+. We wish to prove that this implies that σ_{ij} is bounded on I_+. Let us consider two cases. At a given point in time, either $n_i \leq 2C_\theta$ for all i or there is an i such that $n_i \geq 2C_\theta$. Note that there is a numerical constant $C_\sigma > 0$ such that if $n_i \leq 2C_\theta$ at one point in time, then $|\sigma_{ij}| \leq C_\sigma$ at that point due to (20.18). Assume now that, at one point in time, there is an n_i such that $n_i \geq 2C_\theta$. Labeling the n_i in the right way, we can assume $n_3 \geq n_2 \geq n_1$. There are in principle two cases. Either $n_2 \leq n_3/10$ or $n_2 > n_3/10$. In the former case,

$$n_3^2 + (n_2 - n_1)^2 - 2n_3(n_2 + n_1) \geq n_3^2 - \frac{4}{10}n_3^2 \geq \frac{12}{5}C_\theta^2,$$

which contradicts (20.18). In the latter case,

$$n_1 n_2 \leq \frac{C_n}{n_3} \leq \frac{C_n}{2C_\theta} \leq \frac{1}{2}C_n, \quad n_1 n_3 \leq \frac{C_n}{n_2} \leq \frac{10C_n}{n_3} \leq 5C_n,$$

where we have used the fact that $C_\theta \geq 1$. Thus

$$\frac{1}{2}[n_1^2 + (n_2 - n_3)^2 - 2n_1(n_2 + n_3)] \geq -n_1(n_2 + n_3) \geq -6C_n.$$

Thus (20.18) implies that

$$\sigma_{ij}\sigma^{ij} \leq \frac{2}{3}\theta^2 + 6C_n.$$

Since the right-hand side is bounded on I_+, we conclude that σ_{ij} is bounded on I_+. By (20.9), we conclude that n_{ij} cannot grow faster than exponentially in I_+, i.e., they are bounded on I_+. Due to Lemma 56, p. 30 of [65], we conclude that the solution can be extended beyond t_+. The argument in the other case is similar. □

Lemma 20.9. *Consider a Bianchi IX development with $I = (t_-, t_+)$. Then there is a $t_0 \in I$ such that $\theta > 0$ in (t_-, t_0) and $\theta < 0$ in (t_0, t_+). Furthermore, θ exhausts the interval $(-\infty, \infty)$ and $\lim_{t \to t_\pm \mp} \theta(t) = \mp\infty$. Finally, $t_\pm$ are finite.*

Proof. Let us begin by proving that θ can be zero at most once. If $\theta(t_i) = 0$, $i = 1, 2$ and $t_1 < t_2$, then $\theta = 0$ in (t_1, t_2) since it is monotone by (20.17). Thus (20.17) implies $\sigma_{ij} = 0$ in (t_1, t_2) as well. Combining this fact with (20.18) and (20.8), we get $b_{ij} = 0$, which is impossible for a Bianchi IX solution. Assume θ is never zero. By a suitable choice of time orientation, we can assume that $\theta > 0$ on I. Let us prove that $t_+ = \infty$. Since θ is decreasing on $I_+ = [0, t_+)$ and non-negative on I it is bounded on I_+. Due to Lemma 20.8 and the fact that the solution restricted to I_+ is not extendible to the future, we conclude that $t_+ = \infty$. In order to use the arguments of Lin and Wald, we define

$$\begin{aligned}
\beta_i(t) &= \int_0^t \sigma_i(s)\, ds + \beta_i^0, \\
\alpha(t) &= \int_0^t \frac{1}{3}\theta(s)\, ds + \alpha_0, \\
\beta_+ &= -\frac{1}{2}\beta_3, \\
\beta_- &= \frac{1}{2\sqrt{3}}(\beta_1 - \beta_2),
\end{aligned}$$

where $2\beta_i^0 - \alpha_0 = \ln[n_i(0)]$ and $\sum_{i=1}^3 \beta_i^0 = 0$. Then

$$n_i = \exp(2\beta_i - \alpha).$$

Equations (20.14) and (20.17) then imply (20.21) and (20.22) below, and (20.8) implies (20.23) and (20.24):

$$\left(\frac{d\alpha}{dt}\right)^2 - \left(\frac{d\beta_+}{dt}\right)^2 - \left(\frac{d\beta_-}{dt}\right)^2 + \frac{1}{4}e^{-2\alpha}(1-V) = 0, \tag{20.21}$$

$$\frac{d^2\alpha}{dt^2} + \left(\frac{d\alpha}{dt}\right)^2 + 2\left[\left(\frac{d\beta_+}{dt}\right)^2 + \left(\frac{d\beta_-}{dt}\right)^2\right] = 0, \tag{20.22}$$

$$\frac{d^2\beta_+}{dt^2} + 3\frac{d\alpha}{dt}\frac{d\beta_+}{dt} + \frac{1}{8}e^{-2\alpha}\frac{\partial V}{\partial \beta_+} = 0, \tag{20.23}$$

$$\frac{d^2\beta_-}{dt^2} + 3\frac{d\alpha}{dt}\frac{d\beta_-}{dt} + \frac{1}{8}e^{-2\alpha}\frac{\partial V}{\partial \beta_-} = 0, \tag{20.24}$$

where

$$V(\beta_+, \beta_-) = 1 - \frac{4}{3}e^{-2\beta_+}\cosh(2\sqrt{3}\beta_-) + \frac{1}{3}e^{-8\beta_+} + \frac{2}{3}e^{4\beta_+}[\cosh(4\sqrt{3}\beta_-) - 1]. \tag{20.25}$$

By Theorem 21.1, which we shall prove below, we obtain a contradiction. In other words, there is a zero and since θ is decreasing it must be positive before the zero and negative after it. Due to (20.17), we conclude that θ blows up in finite time both to the future and to the past, assuming nothing else blows up first. However, the latter possibility is excluded by Lemma 20.8. Thus $\theta \to -\infty$ as $t \to t_+-$. Similarly $\theta \to \infty$ as $t \to t_-+$. The lemma follows. □

20.4 Causal geodesic completeness and incompleteness

The lemma concerning causal geodesic completeness will be based on the following estimate.

Lemma 20.10. *Consider a Bianchi class A development. Let $\gamma : (s_-, s_+) \to M$ be a future directed inextendible causal geodesic, and*

$$f_\nu(s) = \langle \gamma'(s), e_\nu|_{\gamma(s)} \rangle. \tag{20.26}$$

If $\theta = 0$ for the entire development, then f_0 is constant. Otherwise,

$$\frac{d}{ds}(f_0\theta) \geq \frac{2-\sqrt{2}}{3}\theta^2 f_0^2. \tag{20.27}$$

Remark 20.11. We consider functions of t as functions of s by evaluating them at $\pi_1[\gamma(s)]$, where $\pi_1 : M \to \mathbb{R}$ is the projection to the first factor.

Proof. Using the computations of Section B.2, we have

$$\frac{df_0}{ds} = \langle \gamma'(s), \nabla_{\gamma'(s)} e_0 \rangle = \sum_{k=1}^{3} \theta_k f_k^2,$$

where θ_k are the diagonal elements of θ_{ij}. If $\theta = 0$ for the entire development, then $\theta_k = 0$ for the entire development by Lemma 20.6 and Lemma 20.9, so that f_0 is constant. Compute, using Raychaudhuri's equation (20.17),

$$\frac{d}{ds}(f_0\theta) = \frac{1}{3}\theta^2 \sum_{k=1}^{3} f_k^2 + \sum_{k=1}^{3} \theta\sigma_k f_k^2 + f_0^2 \sum_{k=1}^{3} \sigma_k^2 + \frac{1}{3}\theta^2 f_0^2,$$

where σ_k are the diagonal elements of σ_{ij}. Let us assume that, at some point in time, $|\sigma_1| \leq |\sigma_2| \leq |\sigma_3|$. Then, due to the fact that σ is trace free, we have

$$3\sigma_3^2 = 2\sigma_3^2 + (\sigma_1 + \sigma_2)^2 = 2\sigma_3^2 + \sigma_1^2 + \sigma_2^2 + 2\sigma_1\sigma_2 \leq 2\sum_{k=1}^{3} \sigma_k^2.$$

Consequently,

$$\Big|\sum_{k=1}^{3} \sigma_k f_k^2\Big| \leq \Big(\frac{2}{3}\Big)^{1/2} \Big(\sum_{k=1}^{3} \sigma_k^2\Big)^{1/2} \sum_{k=1}^{3} f_k^2. \tag{20.28}$$

Let us consider the following three cases

- $\sum_{k=1}^{3} \sigma_k^2 \leq \theta^2/3$. In this case,

$$\frac{1}{3}\theta^2 \sum_{k=1}^{3} f_k^2 + \sum_{k=1}^{3} \theta\sigma_k f_k^2 \geq \frac{1-\sqrt{2}}{3}\theta^2 \sum_{k=1}^{3} f_k^2 \geq \frac{1-\sqrt{2}}{3}\theta^2 f_0^2,$$

  where we have used the causality of the curve and (20.28).

- $\theta^2/3 \leq \sum_{k=1}^{3} \sigma_k^2 \leq 2\theta^2/3$. In this case,

$$\frac{1}{3}\theta^2 \sum_{k=1}^{3} f_k^2 + \sum_{k=1}^{3} \theta\sigma_k f_k^2 + f_0^2 \sum_{k=1}^{3} \sigma_k^2 \geq -\frac{1}{3}\theta^2 \sum_{k=1}^{3} f_k^2 + \frac{1}{3}\theta^2 f_0^2 \geq 0,$$

  where we have used the causality of the curve and (20.28).

- $2\theta^2/3 \leq \sum_{k=1}^{3} \sigma_k^2$. In this case,

$$\begin{aligned}\sum_{k=1}^{3} \theta\sigma_k f_k^2 + f_0^2 \sum_{k=1}^{3} \sigma_k^2 \geq{}& -|\theta|\Big(\frac{2}{3}\Big)^{1/2} \Big(\sum_{k=1}^{3} \sigma_k^2\Big)^{1/2} \sum_{k=1}^{3} f_k^2 \\ &+ |\theta|\Big(\frac{2}{3}\Big)^{1/2} \Big(\sum_{k=1}^{3} \sigma_k^2\Big)^{1/2} f_0^2 \geq 0,\end{aligned}$$

  where we have used the causality of the curve and (20.28).

Thus (20.27) holds. □

Lemma 20.12. *Consider a Bianchi class A development for which θ is not identically zero (i.e., the development is not a quotient of Minkowski space), and let the existence interval be $I = (t_-, t_+)$. There are two possibilities:*

(1) *The development is not of type IX and, by a suitable choice of time orientation, $\theta > 0$. Then all inextendible causal geodesics are future complete and past incomplete. Furthermore, $t_- > -\infty$, $t_+ = \infty$ and θ exhausts the interval $(0, \infty)$.*

(2) *If the development is of type IX, then all inextendible causal geodesics are past and future incomplete. We also have $t_- > -\infty$, $t_+ < \infty$ and θ exhausts the interval $(-\infty, \infty)$.*

Proof. The only statements that remain to be proved are the ones concerning causal geodesic incompleteness. Let $\gamma\colon (s_-, s_+) \to M$ be a future directed inextendible causal geodesic and let f_ν be defined as in (20.26). Since every $M_t = \{t\} \times G, t \in I$, is a Cauchy surface by Proposition 20.3, $\pi_1[\gamma(s)]$ must cover the interval I as s runs through (s_-, s_+). Furthermore, $\pi_1[\gamma(s)]$ is monotone increasing so that

$$\pi_1[\gamma(s)] \to t_\pm \text{ as } s \to s_\pm. \tag{20.29}$$

Let $s_0 \in (s_-, s_+)$ and compute

$$\int_{s_0}^{s} -f_0(u)\,du = \pi_1[\gamma(s)] - \pi_1[\gamma(s_0)]. \tag{20.30}$$

Assume that the development is not of type IX and that $\theta > 0$. Since $f_0\theta$ is negative on $[s_0, s_+)$, its absolute value is bounded on that interval by (20.27). If s_+ were finite, θ would be bounded from below by a positive constant on $[s_0, s_+)$, since

$$\left|\frac{d\theta}{ds}\right| \leq -f_0\theta^2 \leq C\theta$$

on that interval for some $C > 0$, cf. (20.19) and the observations following that equation. Since $f_0\theta$ is bounded, we then deduce that f_0 is bounded on $[s_0, s_+)$. But then (20.29) and (20.30) cannot both hold, since $t_+ = \infty$ by Lemma 20.6. Thus, $s_+ = \infty$ and all inextendible causal geodesics must be future complete. Since $f_0\theta$ is negative on (s_-, s_+), (20.27) proves that this expression must blow up in finite s-time going backward, so that $s_- > -\infty$.

Consider the Bianchi IX case. By Proposition 20.3 and 20.9, we conclude the existence of an $s_0 \in (s_-, s_+)$ such that $f_0\theta$ is negative on (s_-, s_0) and positive on (s_0, s_+). By (20.27), $f_0\theta$ must blow up a finite s-time before s_0, and a finite s-time after s_0. Every inextendible causal geodesic is thus future and past incomplete. □

Corollary 20.13. *Given Bianchi class A initial data for Einstein's vacuum equations, the Bianchi class A development is the maximal globally hyperbolic development of the data.*

Remark 20.14. Note that our definition of unimodular Lie groups is such that the initial data are given on a connected manifold.

Proof. The statement is an immediate consequence of Lemma 20.12, Proposition 18.16 and the definition of the MGHD. □

21 Closed universe recollapse

That there is a maximal globally hyperbolic development, given initial data, is a fundamental result. However, this result does not provide any information concerning the geometric properties of the MGHD. One of the most fundamental questions to ask concerning a Lorentz manifold is if it is causally geodesically complete or incomplete. If all timelike geodesics in the MGHD are past and future incomplete, we shall say that it *recollapses*. There is a general conjecture relating the topology of the initial hypersurface and the issue of recollapse, and it goes under the name of *recollapse conjecture*, cf. e.g. [2]. The conjecture states that if the initial hypersurface admits a Riemannian metric of positive scalar curvature, then the MGHD recollapses. The conjecture does, of course, come with some caveats; it is not true regardless of the matter model and presence of a cosmological constant, as the example of de Sitter space illustrates. However, if one restricts one's attention to vacuum spacetimes, there are, as far as we know, no counterexamples. Since SU(2) certainly admits metrics, in fact left invariant metrics, of positive scalar curvature, we expect the MGHD of vacuum, left invariant initial data on SU(2) to recollapse, and the purpose of the present chapter is to provide a complete proof of this fact. The proof is due to Lin and Wald, cf. [56], a paper which also deals with certain types of matter models, and the argument we shall carry out below follows the presentation of [56] very closely.

As was demonstrated in the previous chapter, cf. Lemma 20.9, it is enough to prove the following theorem.

Theorem 21.1. *There is no solution to* (20.21)–(20.25) *on an interval of the form* $[t_0, \infty)$ *such that* $d\alpha/dt > 0$ *on the entire interval.*

We shall divide the proof of the theorem into several lemmas.

Lemma 21.2. *Let*

$$S = \{(x, y, z) \in \mathbb{R}^3 : x > 0, y > 0, z > 0, xyz = 1\},$$
$$f(x, y, z) = x^2 + y^2 + z^2 - 2(xy + yz + zx).$$

Then $f \geq -3$ *on* S. *Furthermore, if* $\gamma\colon [t_0, \infty) \to S$ *is a continuous map such that* $\lim_{t\to\infty} |\gamma(t)| = \infty$ *and* $f \circ \gamma$ *is bounded on* $[t_0, \infty)$, *then two of the components of* γ *have to converge to* ∞ *and one has to converge to zero.*

Remark 21.3. Even though it will be of no relevance in what follows, let us remark that finding the minimum of f on S corresponds to maximizing the scalar curvature of left invariant metrics on SU(2) while keeping the volume fixed. The maximum is attained by the standard metric on S^3 with the prescribed volume.

Proof. Let us first prove that f attains its minimum in a compact subset of S. Let $(x, y, z) \in S$ and assume, without loss of generality, that $x \geq y \geq z$. We have two exclusive possibilities. Either $y \leq x/10$, in which case

$$f(x, y, z) = x^2 + (y - z)^2 - 2x(y + z) \geq x^2 - \frac{4}{10}x^2 = \frac{3}{5}x^2, \tag{21.1}$$

or $y > x/10$, in which case

$$f(x, y, z) = (x - y)^2 + z^2 - 2z(x + y) \geq -\frac{2}{x} - \frac{2}{y} \geq -\frac{22}{x}.$$

Since $f(1, 1, 1) = -3$, we conclude that f attains its minimum on $S \cap K$, where $K = \{(x, y, z) : \max(|x|, |y|, |z|) \leq 10\}$. Thus

$$g(x, y) = x^2 + y^2 + \frac{1}{x^2 y^2} - 2xy - \frac{2}{x} - \frac{2}{y}$$

attains its minimum on $Q = \{(x, y) \in \mathbb{R}^2 : x > 0, y > 0\}$ in a compact subset of Q. Consequently, we have to have $\partial_x g = \partial_y g = 0$ at the minimum, which implies $x = y = 1$. The first statement of the lemma follows. Assume $\gamma : [t_0, \infty) \to S$ to be a continuous map such that $\lim_{t \to \infty} |\gamma(t)| = \infty$ and assume that there is a constant $C_0 \geq 1$ such that $|f \circ \gamma(t)| \leq C_0$ for $t \in [t_0, \infty)$. Let $T \in [t_0, \infty)$ be such that $\max\{\gamma_1(t), \gamma_2(t), \gamma_3(t)\} \geq 2\sqrt{C_0} + 10$ for $t \geq T$. Then the second largest component of γ has to be larger than a tenth of the largest component, since (21.1) would otherwise hold and imply $f \circ \gamma > C_0$ at the corresponding point in time, contradicting our assumptions. In order for the second largest to be equal to the smallest, we thus have to have

$$\gamma_1(t)\gamma_2(t)\gamma_3(t) \geq \frac{1}{100}(2C_0^{1/2} + 10)^3 > 1,$$

a contradiction. Thus, for $t \geq T$, the two largest will always remain the two largest and the smallest will always remain the smallest. Since the second largest always has to be greater than a tenth of the largest, we conclude that the two largest have to converge to infinity. Since the product of the components equal 1, the smallest has to converge to zero. □

Lemma 21.4. *Let V be defined as in (20.25). Then $V \geq 0$. Assume that there is a solution α, β_+, β_- to (20.21)–(20.25) defined on an interval of the form $[t_0, \infty)$, where $t_0 \in \mathbb{R}$. If $\xi(t) = [\beta_+(t), \beta_-(t)]$ has the property that $|\xi(t)| \to \infty$ as $t \to \infty$ and that $V \circ \xi$ is bounded on $[t_0, \infty)$, then we can assume, without loss of generality, that $\beta_+(t) \to \infty$ as $t \to \infty$.*

Proof. Let us use the notation of the proof of Lemma 20.9. One can compute that

$$3(V - 1)e^{-2\alpha} = n_1^2 + n_2^2 + n_3^2 - 2n_1n_2 - 2n_2n_3 - 2n_3n_1, \quad e^{-2\alpha} = (n_1n_2n_3)^{2/3}.$$

Introducing

$$x = \left(\frac{n_1^2}{n_2n_3}\right)^{1/3} = e^{2\beta_1}, \quad y = \left(\frac{n_2^2}{n_1n_3}\right)^{1/3} = e^{2\beta_2}, \quad z = \left(\frac{n_3^2}{n_1n_2}\right)^{1/3} = e^{2\beta_3},$$

we conclude that $xyz = 1$, $x, y, z > 0$, and

$$3(V - 1) = x^2 + y^2 + z^2 - 2xy - 2yz - 2zx.$$

Due to Lemma 21.2, we conclude that $V \geq 0$. Let ξ be a curve with the properties stated in the lemma. Note that the curve ξ determines a curve γ in S with components x, y, z given by the above relations; $\beta_\pm$ determine β_i, which in their turn determine x, y, z. Furthermore, the fact that $|\xi(t)| \to \infty$ as $t \to \infty$ implies that $|\gamma(t)| \to \infty$ as $t \to \infty$. Due to Lemma 21.2, we conclude that one of the components of γ has to converge to zero. Without loss of generality, we can assume this component to be the z-component. This corresponds to β_+ converging to ∞. □

Let us change time coordinate according to

$$\frac{d\tau}{dt} = e^{-\alpha}. \tag{21.2}$$

Note that time coordinates labeled t and τ appear in [56], but they correspond to the time coordinates τ and t, respectively, of our presentation. The equations (20.21)–(20.24) imply

$$\dot{\alpha}^2 - \dot{\beta}_+^2 - \dot{\beta}_-^2 + \frac{1}{4}(1 - V) = 0, \tag{21.3}$$

$$\ddot{\alpha} + 2(\dot{\beta}_+^2 + \dot{\beta}_-^2) = 0, \tag{21.4}$$

$$\ddot{\beta}_+ + 2\dot{\alpha}\dot{\beta}_+ + \frac{1}{8}\frac{\partial V}{\partial \beta_+} = 0, \tag{21.5}$$

$$\ddot{\beta}_- + 2\dot{\alpha}\dot{\beta}_- + \frac{1}{8}\frac{\partial V}{\partial \beta_-} = 0, \tag{21.6}$$

where the dot signifies differentiation with respect to τ. Note that (21.4)–(21.6) imply that if (21.3) is satisfied at one point in time, it is always satisfied.

Lemma 21.5. *Assume that we have a solution to* (20.21)–(20.24) *on* $[t_0, \infty)$. *Let* $\tau_0 \in \mathbb{R}$ *and define* τ *by* (21.2) *and the condition* $\tau(t_0) = \tau_0$. *Changing time coordinate to* τ, *we get a solution to* (21.3)–(21.6). *Furthermore, the existence interval in* τ*-time is* $[\tau_0, \infty)$.

Proof. The first statement follows by a computation. To prove the statement concerning the existence interval, note that $\ddot{\alpha} \leq 0$ due to (21.4). Thus

$$\alpha(\tau) \leq \dot{\alpha}_0(\tau - \tau_0) + \alpha_0,$$

for some constants $\dot{\alpha}_0 > 0$ and α_0 and $\tau \geq \tau_0$. By (21.2), we thus obtain

$$t \leq \frac{1}{\dot{\alpha}_0} e^{\dot{\alpha}_0(\tau - \tau_0) + \alpha_0} + t_0$$

for $\tau \geq \tau_0$. As $t \to \infty$, we thus have to have $\tau \to \infty$. □

Lemma 21.6. *Let* $\alpha_+, \beta_+, \beta_-$ *be a solution to* (21.3)–(21.6) *on an interval of the form* $[\tau_0, \infty)$ *such that* $\dot{\alpha} > 0$ *for all* $\tau \in [\tau_0, \infty)$. *Let* K *be a compact subset of the* $\beta_+\beta_-$*-plane. Then there is a* $\tau_1 \in [\tau_0, \infty)$ *such that* $[\beta_+(\tau), \beta_-(\tau)] \notin K$ *for all* $\tau \geq \tau_1$.

Remark 21.7. This is the lemma on page 3282 of [56].

Proof. Let

$$\mathcal{S} = \{(\beta_{+,0}, \beta_{-,0}, \beta_{+,1}, \beta_{-,1}) \in \mathbb{R}^4 : \beta_{+,1}^2 + \beta_{-,1}^2 + \tfrac{1}{4}[V(\beta_{+,0}, \beta_{-,0}) - 1] \geq 0.$$

Note that we can view $\mathcal{S}$ as the space of initial data for (21.3)–(21.6); α only appears in differentiated form, so that we can think of $\sigma = \dot{\alpha}$ as a new variable and the initial data for $\dot{\alpha}$ are given by the non-negative solution of (21.3). Note that $\mathcal{S}$ is closed. Given an $x \in \mathcal{S}$, we associate a solution to (21.3)–(21.6) to x by demanding that the initial data at τ_0 be x. We shall use the notation $\alpha(\cdot, x), \beta_+(\cdot, x), \beta_-(\cdot, x)$ to denote the solution for which $\alpha(\tau_0, x) = 0$. Let $\tau_+(x)$ be the maximal existence time to the future of the corresponding solution and define

$$F(x) = \int_{\tau_0}^{\tau_+(x)} [\dot{\beta}_+^2(\tau, x) + \dot{\beta}_-^2(\tau, x)]\, d\tau.$$

Let K' be a compact subset of $\mathcal{S}$. Let us prove that there is then an $\varepsilon > 0$ such that $F(x) \geq \varepsilon$ on K'. Note that by introducing new variables, we can consider (21.4)–(21.6) as a system of first order equations, i.e., $X' = G(X)$. By the properties of the flow of a vector field, we thus conclude that there is a $T > 0$ such that $\tau_+(x) \geq T + \tau_0$ for $x \in K'$. Furthermore,

$$F_T(x) = \int_{\tau_0}^{\tau_0+T} [\dot{\beta}_+^2(\tau, x) + \dot{\beta}_-^2(\tau, x)]\, d\tau$$

which is defined on K', is continuous, so that it attains its minimum, and bounded from above by $F(x)$. If the minimum is positive, we are done, so assume not. Then there is an $x \in K'$ such that $F_T(x) = 0$. Thus $\dot{\beta}_+^2(\cdot, x) + \dot{\beta}_-^2(\cdot, x) = 0$ on $[\tau_0, \tau_0 + T]$. By (21.5) and (21.6) we then conclude that $\partial V/\partial\beta_+ = \partial V/\partial\beta_- = 0$ on this interval. Since V has only one critical point, namely $\beta_+ = \beta_- = 0$, we conclude that $\beta_+(\cdot, x) = \beta_-(\cdot, x) = 0$ on $[\tau_0, \tau_0 + T]$. This contradicts (21.3) and the desired conclusion follows.

Due to (21.4) and the fact that $\dot{\alpha}$ is bounded from below, we conclude that $\dot{\beta}_+^2 + \dot{\beta}_-^2$ is integrable on $[\tau_0, \infty)$. Thus, if $x_0 \in \mathcal{S}$ are the initial data for the solution under consideration, we have $F(x_0) < \infty$. Thus

$$\lim_{\tau\to\infty} \int_\tau^\infty (\dot{\beta}_+^2 + \dot{\beta}_-^2)(s)\, ds = 0.$$

Let K be a compact subset of the $\beta_+\beta_-$-plane. Note that, due to (21.3) and the fact that $\dot{\alpha}$ decays,

$$\dot{\beta}_+^2 + \dot{\beta}_-^2 + \frac{1}{4}V \leq \dot{\alpha}_0^2 + \frac{1}{4},$$

where $\dot{\alpha}_0 = \dot{\alpha}(\tau_0)$. Define

$$K_1 = \{(a, b) \in \mathbb{R}^2 : a^2 + b^2 \leq \dot{\alpha}_0^2 + \tfrac{1}{4}\}, \quad K' = (K \times K_1) \cap \mathcal{S}.$$

As argued above, there is then an $\varepsilon > 0$ and a T such that $F_T(x) \geq \varepsilon$ for $x \in K'$. We wish to prove that there is a T_0 such that for $t \geq T_0$, the β_+, β_--components of the solution are always in the complement of K. In order to do so, we assume the opposite. Thus there is a sequence $\tau_k \to \infty$ such that the β_+, β_--components of the solution at τ_k are contained in K. By the above observations, the initial data for the solution corresponding to $\tau = \tau_k$ are then contained in K'. Thus we conclude that

$$\int_{\tau_k}^{\infty} (\dot\beta_+^2 + \dot\beta_-^2)(s)\, ds \geq \varepsilon, \quad \lim_{k\to\infty} \int_{\tau_k}^{\infty} (\dot\beta_+^2 + \dot\beta_-^2)(s)\, ds = 0,$$

where the first inequality holds for all k and we have used the fact that the system of differential equations (21.4)–(21.6) is autonomous. We obtain the desired contradiction. □

Proof of Theorem 21.1. Due to Lemma 21.6, we know that $\beta_+^2 + \beta_-^2$ tends to infinity. Due to the constraint (21.3) and the fact that $\dot\alpha$ is bounded to the future, we know that V is bounded to the future. By Lemma 21.4, we thus conclude that we, without loss of generality, can assume that $\beta_+(\tau) \to \infty$ as $\tau \to \infty$. We wish to prove that there is no solution with $\dot\alpha > 0$ and $\beta_+ \to \infty$. To this end, add five times (21.4) to (21.3) and (21.5) in order to obtain

$$\ddot\beta_+ + 5\ddot\alpha = -9(\dot\beta_+^2 + \dot\beta_-^2) - 2\dot\alpha\dot\beta_+ - \dot\alpha^2 - \frac{1}{8}\left[\frac{\partial V}{\partial\beta_+} + 2(1-V)\right]. \tag{21.7}$$

Note that

$$\begin{aligned}
-\frac{1}{8}\left[\frac{\partial V}{\partial\beta_+} + 2(1-V)\right] &= -\frac{2}{3}e^{-2\beta_+}\cosh(2\sqrt{3}\beta_-) \\
&\quad - \frac{1}{6}e^{4\beta_+}[\cosh(4\sqrt{3}\beta_-) - 1] + \frac{5}{12}e^{-8\beta_+} \\
&< -\frac{1}{3}e^{-2\beta_+},
\end{aligned} \tag{21.8}$$

assuming that $\beta_+ > 1$. Let $\tau_1 \in [\tau_0, \infty)$ be such that $\beta_+(\tau) > 1$ for $\tau \geq \tau_1$. From now on, we shall restrict our attention to the interval $[\tau_1, \infty)$. The first three terms on the right-hand side of (21.7) can be rewritten

$$\begin{aligned}
-9(\dot\beta_+^2 + \dot\beta_-^2) - 2\dot\alpha\dot\beta_+ - \dot\alpha^2 &= -(\dot\beta_+ + \dot\alpha)^2 - 8\dot\beta_+^2 - 9\dot\beta_-^2 \\
&= -\frac{6}{5}\dot\beta_+(\dot\beta_+ + 5\dot\alpha) - (2\dot\beta_+ - \dot\alpha)^2 - \frac{19}{5}\dot\beta_+^2 - 9\dot\beta_-^2.
\end{aligned} \tag{21.9}$$

Combining (21.7), (21.8) and (21.9), we get

$$\dot X < -\frac{1}{3}e^{-2\beta_+}, \tag{21.10}$$

$$\dot X < -\frac{6}{5}\dot\beta_+ X \tag{21.11}$$

for $\tau \geq \tau_1$, where we have introduced $X := \dot{\beta}_+ + 5\dot{\alpha}$. Note that (21.10) implies that $X > 0$ on $[\tau_1, \infty)$, since if it ever becomes zero, it has to be negative to the future. Since $\dot{\alpha} > 0$, this implies $\dot{\beta}_+ < 0$ to the future in contradiction with the fact that $\beta_+ \to \infty$. Since $X > 0$ and $\dot{\alpha} > 0$, we can multiply (21.10) with $2X$ and integrate in order to obtain, for $\tau_1 \leq \tau \leq \tau_2$,

$$\begin{aligned} X^2(\tau_2) - X^2(\tau) &< -\frac{2}{3}\int_\tau^{\tau_2} e^{-2\beta_+} X \, d\tau \\ &< -\frac{2}{3}\int_\tau^{\tau_2} e^{-2\beta_+} \dot{\beta}_+ \, d\tau \\ &= \frac{1}{3}[e^{-2\beta_+(\tau_2)} - e^{-2\beta_+(\tau)}], \end{aligned}$$

so that

$$\frac{1}{3}[e^{-2\beta_+(\tau)} - e^{-2\beta_+(\tau_2)}] < X^2(\tau).$$

Letting τ_2 tend to infinity, we obtain

$$\frac{1}{\sqrt{3}} e^{-\beta_+(\tau)} \leq X(\tau). \tag{21.12}$$

However, (21.11) implies

$$X(\tau) < c e^{-6\beta_+(\tau)/5}$$

for some constant $c > 0$ and $\tau \geq \tau_1$, which, recalling that $\beta_+(\tau) \to \infty$ as $\tau \to \infty$, contradicts (21.12). The theorem follows. □

21.1 Recollapse for an open set of initial data

Combining the recollapse result of Lin and Wald with Cauchy stability and Hawking's singularity theorem, we obtain the conclusion that there is an open set of initial data on the 3-sphere such that all the timelike geodesics in the corresponding MGHD's are future and past incomplete.

Theorem 21.8. *Let (G, g, k) be arbitrary Bianchi type* IX *vacuum initial data. Let $\|\cdot\|_{H^k}$ be a choice of Sobolev norms on G, cf. Definition* 15.1. *Then there is an $\varepsilon > 0$ such that if (G, ρ, κ) are initial data to Einstein's vacuum equations satisfying*

$$\|\rho - g\|_{H^4} + \|k - \kappa\|_{H^3} \leq \varepsilon, \tag{21.13}$$

then all timelike geodesics in the maximal globally hyperbolic development of the initial data (G, ρ, κ) are past and future incomplete.

Proof. Note that the Bianchi type IX development of the initial data (G, g, k) is a background solution, in the sense of Definition 15.8, if we define the scalar field to be zero and the potential to be zero. We wish to prove that there is an $\varepsilon > 0$ such that if

(21.13) holds, then the MGHD of the initial data (G, ρ, κ) has Cauchy hypersurfaces Σ_i, $i = 1, 2$, such that if κ_i is the mean curvature of Σ_i, then

$$\sup_{p\in\Sigma_1} \kappa_1(p) < 0, \qquad \inf_{p\in\Sigma_2} \kappa_2(p) > 0.$$

Due to Hawking's singularity theorem, Theorem 55A, p. 431 of [65], the existence of such hypersurfaces implies that timelike geodesics in the MGHD are past and future incomplete. In order to prove the existence of such hypersurfaces, let us assume that the above statement is false. Then there is a sequence of vacuum initial data (G, ρ_j, κ_j) converging to (G, g, k) with respect to the norms given in (21.13) such that the MGHD's corresponding to (G, ρ_j, κ_j) do not contain Cauchy hypersurfaces as above (and, therefore, nor does any other globally hyperbolic development of the data). However, this leads to a contradiction to Theorem 15.10, since the background solution has compact Cauchy hypersurfaces with strictly positive and with strictly negative mean curvature. The theorem follows. □

22 Asymptotic behaviour

In the previous two chapters, we established the basic properties of Bianchi class A developments. In the present chapter, we analyze some aspects of the asymptotic behaviour of the solutions to the corresponding ODE's. The main question of interest here is the asymptotic behaviour in the causally geodesically incomplete directions, the goal being to prove, with some exceptions, that the Kretschmann scalar $R_{\alpha\beta\gamma\delta}R^{\alpha\beta\gamma\delta}$ is unbounded in the incomplete directions of causal geodesics. Note that in these directions, the trace of the second fundamental form tends to either $+\infty$ or $-\infty$. Consequently, it is natural to normalize the variables to compensate for this. In a paper by Wainwright and Hsu, [84], such variables were introduced, based on the variables of Ellis and MacCallum which we have used in the previous chapters. Not only are the variables divided by the trace of the second fundamental form, the time coordinate is also changed in a similar fashion. In Section 22.1, we introduce the variables: N_i, $i = 1, 2, 3$ and $\Sigma_{\pm}$. The N_i are simply the n_i we have used earlier, divided by the trace of the second fundamental form, and the $\Sigma_{\pm}$ constitute information equivalent to the trace free part of the second fundamental form, similarly normalized. One advantage of these variables is that the state space becomes compact for the simpler Bianchi classes; in case of Bianchi class I, II and VI_0, this is an immediate consequence of the constraint. Furthermore, the Bianchi type I solutions become a circle of fixed points, referred to as the Kasner circle, with respect to the new variables. Finally, there is a natural hierarchy with the simpler Bianchi types constituting the boundary of the more complicated ones. Since there are monotone quantities which force the solutions toward the boundary, this means that analyzing the simpler models is of importance when considering the more complicated ones.

As was already mentioned, not only do we change the variables, we also introduce a new time coordinate, and in Section 22.2, we relate the different time coordinates. Before analyzing the asymptotics, we, in Section 22.3, introduce the basic terminology in terms of which the results will be phrased. We also introduce one very important tool for analyzing the asymptotics: the monotonicity principle. As we shall see in later chapters, some of the Taub type initial data are such that the corresponding MGHD's are extendible. It turns out that if one considers the asymptotic behaviour of the corresponding solutions with respect to the Wainwright–Hsu variables, then one can, without loss of generality, assume that (Σ_+, Σ_-) converges to $(-1, 0)$. Since the curvature cannot become unbounded as one approaches a singularity in a spacetime which is extendible beyond the singularity, it is clear that, if one wants to prove that the curvature generically blows up in the approach to a singularity, there could potentially be a problem if (Σ_+, Σ_-) converges to $(-1, 0)$. It is therefore of interest to characterize the solutions for which (Σ_+, Σ_-) converges to this value. This is done in Section 22.5. In the remaining sections, we then analyze the asymptotics, with the exception of Bianchi class VIII and IX for which we only state the relevant results. The main references for this chapter are [84], [68], [70], [71], and [85].

22.1 The Wainwright–Hsu variables

Before introducing the Wainwright–Hsu variables, we exclude the Bianchi class A developments isometric to Minkowski space, and in the case of Bianchi type IX, we consider the two halves of the development, corresponding to the mean curvature exhausting the intervals $(-\infty, 0)$ and $(0, \infty)$, separately.

In order to be able to define the Wainwright–Hsu variables, let us first define

$$\Sigma_{ij} = \sigma_{ij}/\theta, \quad N_{ij} = n_{ij}/\theta.$$

Furthermore, let τ be a function satisfying

$$\frac{dt}{d\tau} = \frac{3}{\theta}. \tag{22.1}$$

Let

$$\Sigma_+ = \frac{3}{2}(\Sigma_{22} + \Sigma_{33}) \quad \text{and} \quad \Sigma_- = \frac{\sqrt{3}}{2}(\Sigma_{22} - \Sigma_{33}). \tag{22.2}$$

If we let N_i be the diagonal elements of N_{ij}, equations (20.8), (20.17) and (20.9) imply

$$N_1' = (q - 4\Sigma_+)N_1, \tag{22.3}$$

$$N_2' = (q + 2\Sigma_+ + 2\sqrt{3}\Sigma_-)N_2, \tag{22.4}$$

$$N_3' = (q + 2\Sigma_+ - 2\sqrt{3}\Sigma_-)N_3, \tag{22.5}$$

$$\Sigma_+' = -(2-q)\Sigma_+ - 3S_+, \tag{22.6}$$

$$\Sigma_-' = -(2-q)\Sigma_- - 3S_-, \tag{22.7}$$

where the prime denotes derivative with respect to the time coordinate τ and

$$q = 2(\Sigma_+^2 + \Sigma_-^2), \tag{22.8}$$

$$S_+ = \frac{1}{2}[(N_2 - N_3)^2 - N_1(2N_1 - N_2 - N_3)], \tag{22.9}$$

$$S_- = \frac{\sqrt{3}}{2}(N_3 - N_2)(N_1 - N_2 - N_3). \tag{22.10}$$

The constraint, (20.14), becomes

$$\Sigma_+^2 + \Sigma_-^2 + \frac{3}{4}[N_1^2 + N_2^2 + N_3^2 - 2(N_1N_2 + N_2N_3 + N_1N_3)] = 1. \tag{22.11}$$

The equations (22.3)–(22.11) have certain symmetries described in [84]; by permuting N_1, N_2, N_3 arbitrarily, we get new solutions if we at the same time carry out appropriate combinations of rotations by integer multiples of $2\pi/3$ and reflections in the (Σ_+, Σ_-)-plane. We shall classify points $(\Sigma_+, \Sigma_-, N_1, N_2, N_3)$ according to the values of N_1, N_2, N_3 in the same way as in Table 19.1. Since the sets $N_i > 0$, $N_i < 0$ and $N_i = 0$ are invariant under the flow of the equation, we may classify solutions to (22.3)–(22.11) accordingly. The Raychaudhuri equation (20.17) takes the form

$$\theta' = -(1+q)\theta. \tag{22.12}$$

Definition 22.1. The *Kasner circle* is defined by the conditions $N_i = 0$ and the constraint (22.11). There are three points on this circle called *special*, defined by the $\Sigma_+\Sigma_-$-coordinates being $(\Sigma_+, \Sigma_-) = (-1, 0)$ and $(1/2, \pm\sqrt{3}/2)$. Define $\mathcal{K}_1$ to be the segment of the Kasner circle with $\Sigma_+ > 1/2$.

If the Bianchi class A development we started with was LRS (of Taub type), then we shall say that the corresponding solution to the equations of Wainwright and Hsu is an LRS (or a Taub type) solution. In particular, if $n_2 = n_3$ and $\sigma_2 = \sigma_3$, we have $N_2 = N_3$ and $\Sigma_- = 0$.

22.2 Relation between the time coordinates

It will be of interest to relate the time coordinate τ, which we shall refer to as Wainwright–Hsu time, to the time coordinate t.

Lemma 22.2. *Consider a Bianchi class A development which is not of type* IX *and for which θ is not identically zero (i.e., it is not a quotient of Minkowski space). Due to Lemma* 20.6*, we can then assume $\theta > 0$. If the existence interval is of the form $I = (t_-, t_+)$, then the corresponding solution to the equations of Wainwright and Hsu has existence interval $\mathbb{R}$, and $t \to t_\pm$ corresponds to $\tau \to \pm\infty$, i.e.,*

$$\theta \notin L^1[t_0, t_+), \quad \theta \notin L^1(t_-, t_0] \tag{22.13}$$

for $t_0 \in (t_-, t_+)$.

Proof. Due to Lemma 20.6, $t_+ = \infty$, and we have

$$\lim_{t\to\infty} \theta(t) = 0, \qquad \lim_{t\to t_-+} \theta(t) = \infty. \tag{22.14}$$

Due to the constraint, (22.11), $q \le 2$ for all the Bianchi types except IX. Combining this observation with (22.12) and (22.14), we conclude that

$$\lim_{t\to t_\pm\mp} \tau(t) = \pm\infty.$$

Since

$$\tau(t) - \tau(t_0) = \frac{1}{3}\int_{t_0}^{t} \theta(s)\,ds,$$

we conclude that (22.13) holds. □

Lemma 22.3. *Consider a Bianchi type* IX *development with $I = (t_-, t_+)$. According to Lemma* 20.9*, there is a $t_0 \in I$ such that $\theta > 0$ in $I_- = (t_-, t_0)$ and $\theta < 0$ in $I_+ = (t_0, t_+)$. The solution to the equations of Wainwright and Hsu corresponding to the interval I_- has existence interval $(-\infty, \tau_-)$ with $\tau_- < \infty$, and $t \to t_-$ corresponds to $\tau \to -\infty$. Similarly, I_+ corresponds to $(-\infty, \tau_+)$ with $\tau_+ < \infty$ and $t \to t_+$ corresponding to $\tau \to -\infty$. In particular,* (22.13) *holds.*

Proof. Let us relate the different time coordinates on I_-. According to equation (22.1), τ has to satisfy $dt/d\tau = 3/\theta$. Define

$$\tilde{\tau}(t) = \int_{t_1}^{t} \frac{\theta(s)}{3}\, ds,$$

where $t_1 \in I_-$. Then $\tilde{\tau}: I_- \to \tilde{\tau}(I_-)$ is a diffeomorphism and strictly monotone on I_-. Since θ is positive in I_-, $\tilde{\tau}$ increases with t. Since θ is continuous beyond t_0, it is clear that $\tilde{\tau}(t) \to \tau_- \in \mathbb{R}$ as $t \to t_0$. To prove that $t \to t_-$ corresponds to $\tau \to -\infty$, note that, due to Lemma 20.6, $\theta(t) \to \infty$ as $t \to t_-$. If $\tilde{\tau}$ were bounded from below on I_-, then the N_i would be bounded on $(t_-, t_0]$, since

$$N_1(\tau) = \exp\left\{ \int_0^{\tau} [q(s) - 4\Sigma_+(s)]\, ds \right\} N_1(0) \le e^{-2\tau} N_1(0)$$

for $\tau \le 0$ and similarly for N_2 and N_3. Consequently, q would be bounded on $(t_-, t_0]$ due to the constraint, (22.11), so that (22.12) would imply that θ is bounded on $(t_-, t_0]$, a contradiction. Similar arguments yield the same conclusion concerning I_+. □

22.3 Terminology, asymptotic behaviour

Before we discuss the asymptotics, let us introduce some terminology.

Definition 22.4. Let $f \in C^\infty(\mathbb{R}^n, \mathbb{R}^n)$, and consider a solution x to the equation

$$\frac{dx}{dt} = f \circ x, \quad x(0) = x_0,$$

with maximal existence interval (t_-, t_+). We call a point x_* an *α-limit point* of the solution x, if there is a sequence $t_k \to t_-$ with $x(t_k) \to x_*$. The *α-limit set* of x is the set of its α-limit points. The ω-limit set is defined similarly by replacing t_- with t_+.

Remark 22.5. If $t_- > -\infty$ then the α-limit set is empty and similarly for the ω-limit set, cf. Lemma 56, p. 30 of [65]. Furthermore, it is clear that the α- and ω-limit sets are closed and invariant under the flow of the vector field f.

The following lemma, which we shall refer to as *the monotonicity principle*, will be of central importance in the analysis of the asymptotics.

Lemma 22.6. *Consider*

$$\frac{dx}{dt} = f \circ x \tag{22.15}$$

where $f \in C^\infty(\mathbb{R}^n, \mathbb{R}^n)$. Let U be an open subset of $\mathbb{R}^n$ and M a closed subset invariant under the flow of the vector field f. Let $G: U \to \mathbb{R}$ be a continuous function such that $G \circ x$ is strictly monotone for any solution x of (22.15)*, as long as $x(t) \in U \cap M$. Then no solution of* (22.15) *whose image is contained in $U \cap M$ has an α- or ω-limit point in U.*

Proof. Suppose $p \in U$ is an α-limit point of a solution x whose image is contained in $U \cap M$. Then $G \circ x$ is strictly monotone and there is a sequence $t_n \to -\infty$ such that $x(t_n) \to p$ by our assumptions. Thus $G[x(t_n)] \to G(p)$. Since $G \circ x$ is monotone, we conclude that $G[x(t)] \to G(p)$. Thus $G(q) = G(p)$ for all α-limit points q of x. Since M is closed, $p \in M$. The solution $\bar{x}$ of (22.15) with initial value p is contained in M by the invariance property of M, and it consists of α-limit points of x so that $G[\bar{x}(t)] = G(p)$, which is constant. Furthermore, on an open set containing zero it takes values in U contradicting the assumptions of the lemma. The argument for the ω-limit set is similar. □

22.4 Criteria ensuring curvature blow up

One way to prove that a MGHD is inextendible is to prove that the curvature, e.g. the Kretschmann scalar, becomes unbounded in the incomplete directions of timelike geodesics, cf. Lemma 18.18. Let us formulate a result which relates the issue of blow up to the asymptotics of solutions to (22.3)–(22.11).

Lemma 22.7. *Consider a part of a Bianchi class A development such that θ ranges between 0 and ∞ (or between $-\infty$ and 0). If the corresponding solution to (22.3)–(22.11) has a non-special α-limit on the Kasner circle, then the Kretschmann scalar, $R_{\alpha\beta\gamma\delta}R^{\alpha\beta\gamma\delta}$, is unbounded along causal curves along which $\theta \to \infty$ (or $\theta \to -\infty$).*

Remark 22.8. The definition of "unbounded along causal curves" is similar in vein to the definition of "unbounded in the incomplete directions of timelike geodesics" given in Remark 18.19.

The proof is to be found in Section B.4.

22.5 Limit characterization of the Taub solutions

Due to Lemma 22.7, we see that the special points on the Kasner circle play an important role. Of course, the statement of the lemma is an implication, not an equivalence. Nevertheless, as we shall see in Chapter 23, if the Wainwright–Hsu variables of a Bianchi type I development is a special point on the Kasner circle, then the development has an open subset of Minkowski space as a universal covering space. Similarly, if the Wainwright–Hsu variables of a Bianchi type VII_0 development are given by $(\Sigma_+, \Sigma_-) = (-1, 0)$ and $N_2 = N_3$ (or if one of the permuted conditions hold), then the development has an open subset of Minkowski space as a universal covering space. In the above mentioned cases, it is thus clear that the Kretschmann scalar is identically zero. Consequently, even though we, at this stage, are not able to prove that the statement of Lemma 22.7 can be improved to an equivalence, it is clear that the special points on the Kasner circle play an essential role, and that it is important to characterize the solutions the $\Sigma_+\Sigma_-$-variables of which converge to, say, $(-1, 0)$. The purpose of the present section is to provide such a characterization. Finally, let us note that, in the end, it will be possible to improve Lemma 22.7 to an equivalence. The arguments of the present section are taken from [70], pp. 721–722.

Let us start by making an elementary observation that will be of importance in what follows.

Lemma 22.9. *Consider a positive function N such that $N' = hN$, where $h(\tau) \to \alpha \in \mathbb{R}$ as $\tau \to -\infty$. Then, for every $\varepsilon > 0$, there is a T such that $\tau \le T$ implies*

$$\exp[(\alpha+\varepsilon)\tau] \le N(\tau) \le \exp[(\alpha-\varepsilon)\tau].$$

The following statement is the main result of this section.

Proposition 22.10. *A solution to* (22.3)–(22.11) *satisfies*

$$\lim_{\tau\to-\infty} [\Sigma_+(\tau), \Sigma_-(\tau)] = (-1, 0) \tag{22.16}$$

only if it is contained in the invariant set $\Sigma_- = 0$ and $N_2 = N_3$.

Remark 22.11. The analogue for $(\Sigma_+, \Sigma_-) \to (1/2, \pm\sqrt{3}/2)$ is true due to the symmetries.

The proposition is a consequence of Lemma 22.13 and 22.14 below. We shall, in this section, assume that (Σ_+, Σ_-) converges to $(-1, 0)$. Consider

$$f = \frac{4}{3}\Sigma_-^2 + (N_2 - N_3)^2.$$

Note that f is either identically zero or always strictly positive due to the fact that $N_2 = N_3$, $\Sigma_- = 0$ is an invariant set. A related function, Z_{-1}, defined in (22.22), will be important in the analysis of the asymptotics of Bianchi VII_0 solutions. We shall prove an estimate of the form $f(T) \le g(\tau, T)$ for some T and $\tau \le T$ and then that $g(\tau, T) \to 0$ as $\tau \to -\infty$. Compute

$$f' = -(2-q)\,\frac{8}{3}\,\Sigma_-^2 + 4\sqrt{3}N_1(N_2 - N_3)\Sigma_- + 2(q + 2\Sigma_+)(N_2 - N_3)^2. \tag{22.17}$$

Lemma 22.12. *For every $\varepsilon > 0$ there is a T such that*

$$f(T) \le f(\tau)\exp[\varepsilon(T - \tau)] \tag{22.18}$$

for all $\tau \le T$.

Proof. Due to (22.16), (22.3) and Lemma 22.9, N_1 converges to zero as $\tau \to -\infty$. Furthermore, due to (22.16), $2 - q$ and $q + 2\Sigma_+$ converge to zero. Thus, for all $\varepsilon > 0$ there exists a T such that $\tau \le T$ implies $f' \le \varepsilon f$ by (22.17). The lemma follows. □

Lemma 22.13. *If there is a sequence $\tau_k \to -\infty$ as $k \to \infty$ such that $\Sigma_+(\tau_k) \le -1$, then $\Sigma_- = 0$ and $N_2 = N_3$.*

Proof. The constraint, (22.11) yields, at τ_k,

$$\frac{3}{4}f = \Sigma_-^2 + \frac{3}{4}(N_2 - N_3)^2 \leq \frac{3}{2}N_1(N_2 + N_3).$$

Applying Lemma 22.9 to N_1N_2 and N_1N_3, we conclude that for k large enough, $f(\tau_k) \leq \exp(2\tau_k)$. Choose a finite T corresponding to $\varepsilon = 1$ in (22.18). Then $f(T) \leq \exp(\tau_k + T)$. Letting $k \to \infty$, we obtain $f(T) = 0$, so that f is identically zero. □

Lemma 22.14. *If there is an S such that $\Sigma_+(\tau) \geq -1$ for all $\tau \leq S$, then $\Sigma_- = 0$ and $N_2 = N_3$.*

Proof. Using the constraint to express $2 - q$ in terms of the N_i, (22.6) yields

$$(\Sigma_+ + 1)' \leq \frac{9}{2}N_1^2 + \frac{9}{2}|N_1(N_2 + N_3)|,$$

assuming that $-1 \leq \Sigma_+ \leq 0$. Let T be such that the right-hand side $\leq 2e^{2\tau}$ and $-1 \leq \Sigma_+(\tau) \leq 0$ for all $\tau \leq T$. Integrating this inequality, using (22.16), we obtain

$$0 \leq \Sigma_+(\tau) + 1 \leq e^{2\tau}$$

for all $\tau \leq T$. But then the constraint yields

$$\frac{3}{4}f = -\frac{3}{4}N_1^2 + \frac{3}{2}N_1(N_2 + N_3) + (1 - \Sigma_+)(1 + \Sigma_+).$$

By the above argument we have control over the last term, and as in the previous lemma we have control over the first two terms. Thus, there exists an S' such that $\tau \leq S'$ implies $f(\tau) \leq e^{\tau}$. We deduce, using $\varepsilon = \frac{1}{2}$ in (22.18), that f is identically zero. □

22.6 Asymptotic behaviour, Bianchi type I and II

For reasons already mentioned, cf. Lemma 22.7, our main interest in the asymptotics is to prove that there are non-special α-limit points on the Kasner circle.

In the Wainwright–Hsu variables, the type I solutions are all fixed points, and, consequently, there is nothing to analyze. Let us turn to the Bianchi type II solutions. Before stating the result, recall that $\mathcal{K}_1$ is the segment of the Kasner circle with $\Sigma_+ > 1/2$.

Proposition 22.15. *A Bianchi type* II *vacuum solution of* (22.3)–(22.11) *with $N_1 > 0$ and $N_2 = N_3 = 0$ satisfies*

$$\lim_{\tau \to \pm\infty} N_1(\tau) = 0. \tag{22.19}$$

The ω-limit set is a point in $\mathcal{K}_1$ and the α-limit set is a point on the Kasner circle, in the complement of the closure of $\mathcal{K}_1$. Furthermore,

$$\lim_{\tau \to -\infty} \Sigma_-(\tau) = 0$$

if and only if $\Sigma_-(\tau) = 0$ for all τ.

Proof. Using the constraint, (22.11), we deduce that

$$\Sigma_+' = \frac{3}{2}N_1^2(2 - \Sigma_+).$$

We wish to apply the monotonicity principle, Lemma 22.6. There are three variables: N_1, Σ_+ and Σ_-. Let U be defined by $N_1 > 0$, M be defined by (22.11), and $G(\Sigma_+, \Sigma_-, N_1) = \Sigma_+$. To prove that (22.19) holds, assume there is a sequence $\tau_n \to \infty$ such that $N_1(\tau_n)$ is bounded from below by a positive constant. A subsequence yields an ω-limit point by (22.11). The monotonicity principle yields $N_1(\tau_{n_k}) \to 0$ for the subsequence, a contradiction. The argument for the α-limit set is similar, and (22.19) follows. Compute

$$\left(\frac{\Sigma_-}{2 - \Sigma_+}\right)' = 0. \tag{22.20}$$

Since Σ_+ is strictly monotonically increasing, this yields the remaining conclusions of the proposition; note in particular that the points on the Kasner circle with $\Sigma_+ = 1/2$ can neither be α- nor ω-limit points, since the lines in the $\Sigma_+\Sigma_-$-plane defined by (22.20) passing through these points do not intersect the interior of the Kasner circle. □

22.7 Type VI_0 solutions

Proposition 22.16. *Every Bianchi type* VI_0 *solution with* $N_1 = 0$ *converges to a point in* $\mathcal{K}_1$ *as* $\tau \to -\infty$.

Proof. Note that

$$\Sigma_+' = -(2 - q)(1 + \Sigma_+). \tag{22.21}$$

We wish to apply the monotonicity principle. With notation as in Lemma 22.6, let U be defined by $q < 2$. Note that, since $(N_2 - N_3)^2 > 0$, the image of the solution is contained in U. Let G be defined by $1 + \Sigma_+$, and M by the constraint (22.11). Since $q < 2$ in U, G evaluated on a solution is strictly monotone as long as the solution is contained in $U \cap M$. On the α-limit set, we thus have to $q = 2$. As a consequence, N_2 and N_3 converge to zero (note that, due to the constraint, the variables belong to a compact set). Due to (22.21), we also know that Σ_+ has to converge. Let σ_+ be the limit of Σ_+. In order to prove that Σ_- converges, assume that it does not. Then $\sigma_+^2 < 1$ and there are time sequences $t_k, s_k \to -\infty$ such that

$$\lim_{k\to\infty} \Sigma_-(t_k) = \sqrt{1 - \sigma_+^2}, \quad \lim_{k\to\infty} \Sigma_-(s_k) = -\sqrt{1 - \sigma_+^2}.$$

We can, without loss of generality, assume $t_k \le s_k$. For k large enough, we can extract a sequence $r_k \in (t_k, s_k)$ such that $\Sigma_-(r_k) = 0$. Since the variables are contained in a compact set, a subsequence of $\{r_k\}$ leads to an α-limit point with $\Sigma_- = 0$. Consequently, $\sigma_+^2 = 1$, contradicting our assumptions. Thus Σ_- has to converge to, say, σ_-. Note that, since N_2 and N_3 converge to zero, we have to have $\sigma_+ \ge 1/2$.

Furthermore, σ_+ cannot equal $1/2$, since Proposition 22.10 would then imply that either $N_2 = 0$ or $N_3 = 0$ for the entire solution. The proposition follows. □

22.8 Type VII$_0$ solutions

When speaking of Bianchi VII$_0$ solutions, we shall always assume $N_1 = 0$ and $N_2, N_3 > 0$.

Lemma 22.17. *Consider a type* VII$_0$ *solution such that* $N_2 = N_3$ *and* $\Sigma_- = 0$. *Then there are two possibilities. Either*

- $N_2 = N_3$ *is constant,* $\Sigma_+ = -1$ *and* $\Sigma_- = 0$, *or*
- $\Sigma_+ = 1$, *in which case*

$$\lim_{\tau\to-\infty} N_2(\tau) = \lim_{\tau\to-\infty} N_3(\tau) = 0.$$

Proof. Due to the constraint, (22.11), we have $\Sigma_+ = \pm 1$. Due to the equations, (22.3)–(22.11), the lemma follows. □

Proposition 22.18. *Every Bianchi type* VII$_0$ *solution with* $N_1 = 0$ *which is not of Taub type converges to a point in* $\mathcal{K}_1$ *as* $\tau \to -\infty$.

Before proving this statement, let us make some observations.

Lemma 22.19. *For a Bianchi type* VII$_0$ *solution with* $N_1 = 0$, *which is not of Taub type,* $(N_2, N_3)(-\infty, 0]$ *is contained in a compact set.*

Proof. Due to the assumptions,

$$Z_{-1} = \frac{\frac{4}{3}\Sigma_-^2 + (N_2 - N_3)^2}{N_2 N_3} \tag{22.22}$$

is never zero. This function occurs on p. 1429 of [84]. Note, however, that in [84], the work of Bogoyavlensky [7] is referred to as the source of this function, cf. the functions F_i on p. 63 of [7]. In [68], Z_{-1} was used to analyze the asymptotics. Compute

$$Z_{-1}' = -\frac{16}{3}\frac{\Sigma_-^2(1+\Sigma_+)}{\frac{4}{3}\Sigma_-^2 + (N_2 - N_3)^2} Z_{-1}. \tag{22.23}$$

The proof that the past dynamics are contained in a compact set is as in the work of Rendall, [68]. Let $\tau \le 0$. Then $Z_{-1}(\tau) \ge Z_{-1}(0)$, so that, using the constraint, (22.11),

$$(N_2 N_3)(\tau) \le \frac{4}{3Z_{-1}(0)}.$$

Combining this fact with the constraint, we see that all the variables are contained in a compact set during $(-\infty, 0]$. □

Lemma 22.20. *Every Bianchi type* VII_0 *solution with* $N_1 = 0$ *which is not of Taub type satisfies*

$$\lim_{\tau\to-\infty}(N_2N_3)(\tau) = 0.$$

Proof. Assume the contrary. Then we can use Lemma 22.19 to construct an α-limit point $(\sigma_+, \sigma_-, 0, n_2, n_3)$ where $n_2n_3 > 0$. We apply the monotonicity principle in order to arrive at a contradiction. With notation as in Lemma 22.6, let U be defined by $N_2 > 0$, $N_3 > 0$ and $\Sigma_-^2 + (N_2 - N_3)^2 > 0$. Let G be defined by Z_{-1}, and M by the constraint (22.11). We have to show that G, evaluated on a solution, is strictly monotone as long as the solution is contained in $U \cap M$. Consider (22.23). By the constraint (22.11), $\Sigma_-^2 + (N_2 - N_3)^2 > 0$ implies $\Sigma_+ > -1$. Furthermore, $Z_{-1} > 0$ on U. If $Z_{-1}' = 0$ in $U \cap M$, we thus have $\Sigma_- = 0$, but then $\Sigma_-' \neq 0$ since $\Sigma_-^2 + (N_2 - N_3)^2 > 0$ and $N_2 + N_3 > 0$. The α-limit point we have constructed cannot belong to U. On the other hand, $n_2, n_3 > 0$ and since Z_{-1} increases as we go backward, $\sigma_-^2 + (n_2 - n_3)^2$ cannot be zero. We have a contradiction. □

Proof of Proposition 22.18. Note that

$$\Sigma_+' = -(2-q)(1+\Sigma_+). \tag{22.24}$$

Again, we wish to apply the monotonicity principle. With notation as in Lemma 22.6, let U be defined by $N_2 + N_3 > 0$ and $\Sigma_-^2 + (N_2 - N_3)^2 > 0$. Note that, by assumption, this is fulfilled for the entire solution. Let G be defined by $1 + \Sigma_+$, and M by the constraint (22.11). We have to show that G evaluated on a solution is strictly monotone as long as the solution is contained in $U \cap M$. Consider (22.24). By the constraint (22.11), $\Sigma_-^2 + (N_2 - N_3)^2 > 0$ implies $\Sigma_+ > -1$. If $q = 2$, then $N_2 = N_3$, due to the constraint, and $\Sigma_- \neq 0$. As a consequence, $(N_2 - N_3)' \neq 0$. Thus G is strictly monotone. On the α-limit set, we thus have to have either $N_2 + N_3 = 0$ or $\Sigma_-^2 + (N_2 - N_3)^2 = 0$. Thus $q = 2$ on the α-limit set. Due to (22.24) we also know that Σ_+ has to converge to, say, σ_+. Note that $\sigma_+ > -1$. If $\sigma_+ = 1$, then $\Sigma_-, N_2, N_3 \to 0$. If $\sigma_+ \in (-1, 1)$, then Σ_- has to converge by the argument that was presented in the end of the proof of Proposition 22.16. If $\sigma_+ < 1/2$, either N_2 or N_3 has to converge to ∞, contradicting Lemma 22.19. Finally, σ_+ cannot equal $1/2$ since Proposition 22.10 would then imply that either $N_2 = 0$ or $N_3 = 0$ for the entire solution. The proposition follows. □

22.9 Bianchi type VIII and IX

Since the arguments needed to analyze the asymptotics in the case of Bianchi type VIII and IX are somewhat technical, we do not wish to write them down. Nevertheless, we wish to quote some of the results. The following result was proved in [70].

Theorem 22.21. *A solution of Bianchi type* VIII *or* IX *which is not of Taub type has at least two* α*-limit points on the Kasner circle, at least one of which is non-special.*

Remark 22.22. Note that, given the background material we have provided so far in these notes, the material presented on pp. 723–727 in [70] is essentially sufficient to prove this fact. Note also that one can obtain more information than this; in [71] it is, for instance, proved that Bianchi IX solutions which are not of Taub type have the property that

$$\lim_{\tau\to\infty}(N_1 N_2 + N_2 N_3 + N_3 N_1)(\tau) = 0,$$

even though none of the N_i converge to zero.

22.10 Curvature blow up

Let us express the results of the present chapter in terms of Bianchi class A initial data.

Proposition 22.23. *Let (G, g, k) be Bianchi class A initial data for Einstein's vacuum equations. Due to Corollary* 19.14, *there is a canonical basis e_i' of the Lie algebra, with associated commutator matrix ν, such that $k_{ij} = k(e_i', e_j')$ are the components of a diagonal matrix. Then the Kretschmann scalar is unbounded in the incomplete directions of causal geodesics in the corresponding MGHD's, with the following possible exceptions:*

- *the initial data are of Bianchi type* I, *$k_{11} \neq 0$ and $k_{22} = k_{33} = 0$ or one of the permuted conditions hold,*
- *the initial data are of Bianchi type* VII_0, *$k_{11} \neq 0$, $k_{22} = k_{33} = 0$ and $\nu_2 = \nu_3 \neq 0$,*
- *the initial data are locally rotationally symmetric initial data of Bianchi type* II, VIII *or* IX.

Remark 22.24. Note that the MGHD's of Bianchi type I and VII_0 vacuum initial data such that $\mathrm{tr}_g\, k = 0$ are causally geodesically complete, so that the Kretschmann scalar is unbounded in the incomplete directions of causal geodesics, since there are no incomplete geodesics.

Proof. The statement is an immediate consequence of the basic properties of Bianchi class A developments, Proposition 22.15, Proposition 22.16, Lemma 22.17, Proposition 22.18, Theorem 22.21 and Lemma 22.7, keeping in mind that $k_{11} \neq 0$, $k_{22} = k_{33} = 0$ and $\nu_2 = \nu_3$ corresponds to $\Sigma_+ = -1$ and $N_2 = N_3$. □

23 LRS Bianchi class A solutions

In Proposition 22.23 of the previous chapter, we demonstrated that, with a few exceptions, the MGHD's of left invariant vacuum initial data on 3-dimensional unimodular Lie groups are inextendible; in fact, the Kretschmann scalar is in most cases unbounded in the incomplete directions of causal geodesics. The main purpose of the present chapter is to prove that in the remaining cases, the metric of the MGHD can be rewritten in the form (17.1), cf. also the discussion in the adjacent text. This observation will form a basis for the construction of an extension. Note that, due to Proposition 22.23, all the left invariant initial data for which we do not already know the Kretschmann scalar to become unbounded in the incomplete directions of causal geodesics are of Taub type. Consequently, we shall in this chapter discuss the LRS solutions of the different Bianchi types one by one. In the context of that discussion, the following preliminary observation is of interest. If G and H are Lie groups with Lie algebras $\mathfrak{g}$ and $\mathfrak{h}$, and if there is a homomorphism $\phi\colon \mathfrak{g} \to \mathfrak{h}$, then there is, assuming G to be simply connected, a unique Lie group homomorphism $\psi\colon G \to H$ which induces the homomorphism ϕ on the Lie algebra level. This is a fundamental result concerning Lie groups, and those interested in a proof are referred to e.g. Warner's book [88], Theorem 3.27, p. 101 or, for a more recent reference, Lee's book [55], Theorem 20.15, p. 532. As a consequence of this result, two simply connected Lie groups with isomorphic Lie algebras are isomorphic. Thus, given a Lie algebra, it is sufficient to find one representative simply connected Lie group with that Lie algebra. The representatives we shall choose are as in [53].

Note that LRS developments have the property that (after having carried out a permutation if necessary) if we write the metric in the form (20.13), then $a_2 = a_3$, assuming that ξ^i are the duals of a canonical basis e_i' for the Lie algebra such that the associated commutator matrix ν has the property that $\nu_2 = \nu_3$.

23.1 Bianchi type I

Lemma 23.1. *Let (G, g, k) be Bianchi type* I *initial data for Einstein's vacuum equations with G simply connected. There are the following possibilities:*

- *if $\mathrm{tr}_g k = 0$, then the MGHD of (G, g, k) is isometric to Minkowski space;*
- *if $\mathrm{tr}_g k \neq 0$, then there are $p_i \in \mathbb{R}$ such that*

$$\sum_{i=1}^{3} p_i = 1, \quad \sum_{i=1}^{3} p_i^2 = 1, \tag{23.1}$$

and the MGHD of (G, g, k) is isometric to

$$g = -dt^2 + \sum_{i=1}^{3} t^{2p_i} dx^i \otimes dx^i \tag{23.2}$$

on $\mathbb{R}_+ \times \mathbb{R}^3$, *where* $\mathbb{R}_+ = (0, \infty)$. *In this case, the time orientation is determined by the sign of* $\mathrm{tr}_g\, k$; *if* $\mathrm{tr}_g\, k > 0$, ∂_t *is future oriented.*

Let e_i *be a canonical basis such that* $k_{ij} = k(e_i, e_j)$ *are the components of a diagonal matrix. If* $k_{22} = k_{33} = 0$ *and* $k_{11} \neq 0$ *or one of the permuted conditions hold, then the MGHD is isometric to*

$$g_K = -dt^2 + t^2 dx^2 + dy^2 + dz^2 \tag{23.3}$$

on $\mathbb{R}_+ \times \mathbb{R}^3$. *Note in particular that if we define*

$$\phi_K(t, x, y, z) = (t \cosh x, t \sinh x, y, z), \tag{23.4}$$

then $\phi_K \colon \mathbb{R}_+ \times \mathbb{R}^3 \to \mathbb{R}^4$ *is a diffeomorphism onto its image, and if we pull back the Minkowski metric on* $\mathbb{R}^4$ *using this map, we obtain the metric* (23.3). *Finally, changing the time coordinate to* $s = t^2/4$, *the metric* (23.3) *takes the form*

$$g_K = -\frac{1}{s} ds^2 + 4s dx^2 + dy^2 + dz^2. \tag{23.5}$$

Remark 23.2. The metric (23.3) is referred to as the *flat Kasner solution*. The reason for calling it the *flat* Kasner solution is that, due to the existence of the diffeomorphism ϕ_K, the Riemann curvature tensor of (23.3) is identically zero. In particular, the corresponding spacetime has a singularity in the sense of causal geodesic incompleteness which is not a singularity in the sense of curvature blow up. Metrics of the form (23.2) where the p_i are constant and satisfy (23.1), sometimes referred to as the *Kasner relations*, are called the *Kasner metrics*. Note that, in the Bianchi type I case, the conditions $k_{22} = k_{33} = 0$ and $k_{11} \neq 0$ exactly constitute the case in Proposition 22.23 for which we do not know the curvature to blow up in the incomplete directions of causal geodesics. For those interested in relating the form of the metric given in (23.2) to the variables of Wainwright and Hsu, let us note that

$$\Sigma_- = \frac{\sqrt{3}}{2}(p_2 - p_3), \quad \Sigma_+ = \frac{3}{2}\left(p_2 + p_3 - \frac{2}{3}\right). \tag{23.6}$$

Given p_i, $i = 1, 2, 3$, one can compute Σ_+ and Σ_- using (23.6), and if (23.1) is satisfied, $\Sigma_+^2 + \Sigma_-^2 = 1$. Given Σ_+ and Σ_-, one can compute p_i, $i = 1, 2, 3$, using (23.6) and the first equality in (23.1). If $\Sigma_+^2 + \Sigma_-^2 = 1$, one can then also check that the second equality in (23.1) holds.

Proof. Clearly, $\mathbb{R}^3$ is a simply connected Lie group of Bianchi type I, assuming that we define the product of two elements to be the sum. Thus, by the remarks made in the introduction to this chapter, we might as well assume G to be $\mathbb{R}^3$ with this product. The statement concerning the case $\mathrm{tr}_g\, k = 0$ follows from Lemma 20.6 and its proof. Thus, let us restrict our attention to the case $\mathrm{tr}_g\, k = \theta > 0$ (which we can always do by a suitable choice of time orientation). Since all the n_i are zero in this case, the relevant equations are (20.7), (20.8) and (20.14). Thus we have

$$\dot{\theta} = -\frac{3}{2}\sigma_{ij}\sigma^{ij}, \tag{23.7}$$

$$\dot{\sigma}_{ij} = -\theta\sigma_{ij}, \tag{23.8}$$

$$\frac{2}{3}\theta^2 = \sigma_{ij}\sigma^{ij}. \tag{23.9}$$

Combining (23.7) and (23.9), we conclude that

$$\dot{\theta} = -\theta^2, \tag{23.10}$$

which implies that we can choose a time coordinate such that $\theta(t) = 1/t$. Note, however, that this time coordinate is different from the one we used previously. Combining (23.10) with (23.8), we conclude that θ_i/θ is constant and we shall denote this constant by p_i. Note that (23.1) holds, where the first equality is a consequence of the definition and the second equality follows from (23.9). By choosing suitable constant multiples of ∂_i as a basis for the Lie algebra of $\mathbb{R}^3$, we can assume that $a_i(1) = 1$. Since $\dot{a}_i = \theta_i a_i$ (no summation), we obtain $a_i(t) = t^{p_i}$. By a suitable choice of basis of the Lie algebra and choice of time coordinate, we can thus write a Bianchi type I development, assuming that it is not Minkowski space, of simply connected initial data as (23.2). Finally, let us consider the case that, for the initial data, $k_{22} = k_{33} = 0$ and $k_{11} \neq 0$. Then $\theta_2 = \theta_3 = 0$ and $\theta_1 \neq 0$, so that $p_1 = 1$ and $p_2 = p_3 = 0$. The remaining statements of the lemma follow by straightforward computations. □

23.2 Bianchi VII$_0$

Lemma 23.3. *Let (G, g, k) be Bianchi type* VII$_0$ *initial data for Einstein's vacuum equations with G simply connected. Due to Corollary* 19.14, *there is a canonical basis e_i' of the Lie algebra, with associated commutator matrix ν, such that $k_{ij} = k(e_i', e_j')$ are the components of a diagonal matrix. If $k_{11} \neq 0$, $k_{22} = k_{33} = 0$ and $\nu_2 = \nu_3$, or one of the permuted conditions hold, then the MGHD of (G, g, k) is isometric to* (23.3).

Proof. Let us define a binary operation on $\mathbb{R}^3$ by

$$\begin{pmatrix} x_1 \\ y_1 \\ z_1 \end{pmatrix}\begin{pmatrix} x_2 \\ y_2 \\ z_2 \end{pmatrix} = \begin{pmatrix} x_2 + x_1 \\ z_2 \sin x_1 + y_2 \cos x_1 + y_1 \\ z_2 \cos x_1 - y_2 \sin x_1 + z_1 \end{pmatrix}. \tag{23.11}$$

One can check that this operation is associative, that there is an identity element (the origin) and that there is an inverse of every element given by

$$\begin{pmatrix} x \\ y \\ z \end{pmatrix}^{-1} = \begin{pmatrix} -x \\ z \sin x - y \cos x \\ -y \sin x - z \cos x \end{pmatrix}.$$

Furthermore, it is clear that the operation of taking the product and the operation of taking the inverse are smooth. Consequently, $\mathbb{R}^3$, with the product given by (23.11), is a simply connected Lie group. By left translation of ∂_x, $-\partial_y$ and ∂_z at the origin, one obtains a basis

$$e_1 = \partial_x,\ e_2 = -\cos x\partial_y + \sin x\partial_z,\ e_3 = \sin x\partial_y + \cos x\partial_z \tag{23.12}$$

of the Lie algebra such that if $[e_i, e_j] = \gamma_{ij}^k e_k$, then $\gamma_{ij}^k = \epsilon_{ijl}\nu^{kl}$, where ν is a diagonal matrix. If we let ν_i denote the diagonal entries of ν, then $\nu_1 = 0$, $\nu_2 = \nu_3 = 1$. Consequently, all simply connected Lie groups of Bianchi type VII_0 are isomorphic to $\mathbb{R}^3$ with the group structure defined by (23.11), and we can, without loss of generality, assume the e_i' mentioned in the statement of the lemma to be constant multiples of the e_i given by (23.12), the constant multiplying e_2 being equal to the constant multiplying e_3. Duals, say ξ^i, of the e_i are given by

$$\xi^1 = dx,\ \xi^2 = -\cos x dy + \sin x dz,\ \xi^3 = \sin x dy + \cos x dz. \tag{23.13}$$

The MGHD of the initial data given in the statement of the lemma can be written in the form (20.13) where the ξ^i are given in (23.13) and $a_2 = a_3$. Since

$$\xi^2 \otimes \xi^2 + \xi^3 \otimes \xi^3 = dy^2 + dz^2,$$

we see that an LRS Bianchi type VII_0 solution is actually an LRS Bianchi type I solution, something we have already discussed. The lemma follows. □

23.3 Bianchi type VI_0

For the sake of completeness, we shall write down the relevant Lie group, Sol, even though there are no LRS solutions in the case of Bianchi class VI_0. The underlying manifold is $\mathbb{R}^3$, but the group structure is given by

$$\begin{pmatrix} x_1 \\ y_1 \\ z_1 \end{pmatrix} \begin{pmatrix} x_2 \\ y_2 \\ z_2 \end{pmatrix} = \begin{pmatrix} x_1 + x_2 \\ y_1 + e^{x_1} y_2 \\ z_1 + e^{-x_1} z_2 \end{pmatrix}.$$

One can check that this product is associative, that there is an identity element (the origin) and that the inverse is given by

$$\begin{pmatrix} x \\ y \\ z \end{pmatrix}^{-1} = \begin{pmatrix} -x \\ -e^{-x} y \\ -e^{x} z \end{pmatrix}.$$

By left translation of ∂_x, $\partial_y + \partial_z$ and $-\partial_y + \partial_z$ at the origin, one obtains a basis

$$e_1 = \partial_x, \quad e_2 = e^x\partial_y + e^{-x}\partial_z, \quad e_3 = -e^x\partial_y + e^{-x}\partial_z.$$

of the Lie algebra such that if $[e_i, e_j] = \gamma_{ij}^k e_k$, then $\gamma_{ij}^k = \epsilon_{ijl}\nu^{kl}$, where ν is a diagonal matrix. If we let ν_i denote the diagonal entries of ν, then $\nu_1 = 0$, $\nu_2 = 1$ and $\nu_3 = -1$. There are clearly no LRS solutions.

23.4 Bianchi type II, VIII and IX

In the case of Bianchi type II, the relevant group is the Heisenberg group. In other words, the subgroup of $\mathrm{GL}(3, \mathbb{R})$ given by matrices of the form

$$\begin{pmatrix} 1 & x & y \\ 0 & 1 & z \\ 0 & 0 & 1 \end{pmatrix},$$

where $x, y, z \in \mathbb{R}$. Bianchi type VIII corresponds to the universal covering group of $\mathrm{SL}(2, \mathbb{R})$ and Bianchi type IX corresponds to $\mathrm{SU}(2)$.

Lemma 23.4. *Let (G, g, k) be locally rotationally symmetric vacuum initial data of Bianchi type* II, VIII *or* IX. *Due to Corollary* 19.14, *there is a canonical basis e'_i of the Lie algebra, with associated commutator matrix ν, such that $k_{ij} = k(e'_i, e'_j)$ are the components of a diagonal matrix. Furthermore, $k_{22} = k_{33}$ and $\nu_2 = \nu_3$. Let ξ^i be the duals of the e'_i. Then the MGHD of (G, g, k) is isometric to*

$$g = -\frac{L^2}{X} du^2 + X\xi^1 \otimes \xi^1 + Y^2(\xi^2 \otimes \xi^2 + \xi^3 \otimes \xi^3) \tag{23.14}$$

on $J \times G$, where J is an open interval to be specified below, L is a positive constant and Y is defined by the condition that it be positive and that

$$Y^2(u) = \alpha_2^2\left(u^2 + \frac{1}{4}\nu_1^2\right),$$

where α_2 is a positive constant. In the case of Bianchi type II, *the function X is defined by*

$$X(u)Y^2(u) = \alpha_{12}^2 u,$$

where $\alpha_{12} > 0$ is a constant. In the case of Bianchi type VIII *and* IX, *X is defined by*

$$X(u)Y^2(u) = -\alpha_2^4 \nu_1 \nu_2 u^2 + c_1 u + \frac{1}{4}\alpha_2^4 \nu_1^3 \nu_2, \tag{23.15}$$

where c_1 is a constant. The metric (23.14) *is defined on $J \times G$, where $J = (u_-, u_+)$ is an interval depending on the Bianchi type. We have the following cases:*

- *in the case of Bianchi* II, *$J = (0, \infty)$;*
- *in the case of Bianchi* VIII, *$J = (u_-, \infty)$, where u_- is the infimum of the set of η such that the polynomial on the right-hand side of* (23.15) *is positive on (η, ∞);*
- *in the case of Bianchi* IX, *J is the largest interval on which the right-hand side of* (23.15) *is positive.*

Remark 23.5. The form of the metric given in (23.14) should be compared with [37], p. 134.

Proof. For locally symmetric solutions, the relevant equations can be written

$$\dot{\theta}_1 + \theta\theta_1 + \frac{1}{2}n_1^2 = 0, \tag{23.16}$$

$$\dot{\theta}_2 + \theta\theta_2 - \frac{1}{2}n_1^2 + n_1 n_2 = 0, \tag{23.17}$$

$$\dot{n}_1 + (2\theta_2 - \theta_1)n_1 = 0, \tag{23.18}$$

$$\dot{n}_2 + \theta_1 n_2 = 0, \tag{23.19}$$

$$2\theta_2^2 + 4\theta_1\theta_2 - \frac{1}{2}n_1^2 + 2n_1 n_2 = 0, \tag{23.20}$$

cf. (20.5), (20.6) and (20.14). Using (23.17), (23.18) and (23.20), one can compute that

$$\frac{d}{dt}\left(\frac{\theta_2}{n_1}\right) = \frac{\theta_2^2 + \frac{1}{4}n_1^2}{n_1}. \tag{23.21}$$

In particular, $\nu_1\theta_2/n_1$, where $\nu_i = n_i(0)$, is strictly increasing, and we can use it as a time coordinate. Define

$$u := \nu_1 \frac{\theta_2}{n_1}.$$

Note that (23.20) can be reformulated to

$$4(\theta_1 + 2\theta_2)\theta_2 - 6\theta_2^2 - \frac{1}{2}n_1^2 + 2n_1 n_2 = 0.$$

In the case of Bianchi type II and VIII, we see that all the terms except the first one on the left-hand side are non-positive and that the third term on the left-hand side is negative. Since our conventions are such that $\theta = \theta_1 + 2\theta_2 > 0$, we see that θ_2 must always be strictly larger than zero. Thus, in the case of Bianchi type II and VIII, we have $u > u_+$ for some constant $u_+ \geq 0$. By (23.21), we have

$$\frac{dt}{du} = \frac{1}{\nu_1}\frac{n_1}{\theta_2^2 + \frac{1}{4}n_1^2} = \frac{\nu_1}{n_1}\frac{1}{u^2 + \frac{1}{4}\nu_1^2}. \tag{23.22}$$

Compute, using (20.12),

$$\frac{da_2^2}{du} = 2\theta_2 a_2^2 \frac{dt}{du} = 2\theta_2 a_2^2 \frac{\nu_1}{n_1}\frac{1}{u^2 + \frac{1}{4}\nu_1^2} = \frac{2u}{u^2 + \frac{1}{4}\nu_1^2}a_2^2.$$

Thus there is a constant $\alpha_2 > 0$ such that

$$a_2^2 = \alpha_2^2\left(u^2 + \frac{1}{4}\nu_1^2\right). \tag{23.23}$$

Observe that we use the notation $a_1^2 = X$ and $a_2 = Y$ in the statement of the lemma. Note that, cf. (20.2),

$$n_1 = \nu_1 \frac{a_1}{a_2^2}.$$

Thus, combining this with (23.22) and (23.23), we obtain

$$\frac{dt}{du} = \frac{1}{u^2 + \frac{1}{4}\nu_1^2} \frac{a_2^2}{a_1} = \frac{\alpha_2^2}{a_1}.$$

In particular,

$$-dt^2 = -\frac{\alpha_2^4}{a_1^2} du^2 = -\frac{L^2}{X} du^2,$$

if we let $L = \alpha_2^2$. Let us compute

$$\frac{d(a_1^2 a_2^2)}{du} = 2(\theta_1 + \theta_2) a_1^2 a_2^2 \frac{\nu_1}{n_1} \frac{1}{u^2 + \frac{1}{4}\nu_1^2} = 2 \frac{u + \frac{\nu_1 \theta_1}{n_1}}{u^2 + \frac{1}{4}\nu_1^2} a_1^2 a_2^2. \tag{23.24}$$

Note that, due to the constraint, (23.20), we have

$$2u^2 + 4u \frac{\nu_1 \theta_1}{n_1} - \frac{1}{2}\nu_1^2 + 2\nu_1^2 \frac{n_2}{n_1} = 0. \tag{23.25}$$

For $u \neq 0$, we thus have

$$\begin{aligned} 2\frac{u + \frac{\nu_1\theta_1}{n_1}}{u^2 + \frac{1}{4}\nu_1^2} &= \frac{4u^2 + 4u\frac{\nu_1\theta_1}{n_1}}{2u^3 + \frac{1}{2}\nu_1^2 u} \\ &= \frac{2u^2 + \frac{1}{2}\nu_1^2 - 2\frac{\nu_1^2 n_2}{n_1}}{2u^3 + \frac{1}{2}\nu_1^2 u} \\ &= \frac{1}{u} - \frac{u^2 + \frac{1}{4}\nu_1^2}{u} \frac{\frac{\nu_1^2 n_2}{n_1}}{(u^2 + \frac{1}{4}\nu_1^2)^2}. \end{aligned}$$

Furthermore, recall that $n_1 = \nu_1 a_1 / a_2^2$ and that $n_2 = \nu_2 / a_1$, cf. (20.2), so that

$$\frac{\frac{\nu_1^2 n_2}{n_1}}{(u^2 + \frac{1}{4}\nu_1^2)^2} a_1^2 a_2^2 = \frac{1}{(u^2 + \frac{1}{4}\nu_1^2)^2} \frac{\nu_1 \nu_2 a_2^2}{a_1^2} a_1^2 a_2^2 = \alpha_2^4 \nu_1 \nu_2, \tag{23.26}$$

where we have used (23.23). Combining the last two observations with (23.24), we obtain

$$\frac{d(a_1^2 a_2^2)}{du} = \frac{1}{u} a_1^2 a_2^2 - \alpha_2^4 \nu_1 \nu_2 \frac{u^2 + \frac{1}{4}\nu_1^2}{u},$$

so that

$$\frac{d}{du}\left(\frac{a_1^2 a_2^2}{u}\right) = -\alpha_2^4 \nu_1 \nu_2 \frac{u^2 + \frac{1}{4}\nu_1^2}{u^2}. \tag{23.27}$$

In particular, in the case of Bianchi type II, there is a constant $\alpha_{12} > 0$ such that

$$XY^2 = a_1^2 a_2^2 = \alpha_{12}^2 u.$$

In order to compute the second derivative of $a_1^2 a_2^2$, note that

$$\frac{d}{du}\left(\nu_1 \frac{\theta_1}{n_1}\right) = -2\frac{\left(\frac{\nu_1\theta_1}{n_1}\right)^2 + \frac{1}{4}\nu_1^2}{u^2 + \frac{1}{4}\nu_1^2}.$$

Combining this observation with (23.24), one can compute that

$$\frac{d^2(a_1^2 a_2^2)}{du^2} = \frac{4u\frac{\nu_1\theta_1}{n_1} + 2u^2 - \frac{1}{2}\nu_1^2}{\left(u^2 + \frac{1}{4}\nu_1^2\right)^2} a_1^2 a_2^2 = -2\frac{\frac{\nu_1^2 n_2}{n_1}}{\left(u^2 + \frac{1}{4}\nu_1^2\right)^2} a_1^2 a_2^2 = -2\alpha_2^4 \nu_1 \nu_2,$$

where we used (23.25) and (23.26). Thus there are constants c_1 and c_0 such that

$$a_1^2 a_2^2 = -\alpha_2^4 \nu_1 \nu_2 u^2 + c_1 u + c_0.$$

Consequently,

$$\frac{d}{du}\left(\frac{a_1^2 a_2^2}{u}\right) = -\alpha_2^4 \nu_1 \nu_2 - \frac{c_0}{u^2}.$$

Comparing this computation with (23.27), we obtain

$$c_0 = \frac{1}{4}\alpha_2^4 \nu_1^3 \nu_2.$$

Thus

$$XY^2 = a_1^2 a_2^2 = -\alpha_2^4 \nu_1 \nu_2 u^2 + c_1 u + \frac{1}{4}\alpha_2^4 \nu_1^3 \nu_2. \tag{23.28}$$

Note that the coefficient in front of u^2 has an opposite sign to the constant term coefficient in the Bianchi type VIII and IX cases. Consequently, if we let $u_0 < u_1$ denote the two zeros of the polynomial, we have $u_0 < 0 < u_1$.

Let us analyze the correspondence between the existence interval in t-time and the one in u-time. Due to Lemma 22.2 and 22.3 we know that (22.13) holds for Bianchi class A developments that are not quotients of Minkowski space. Since

$$\frac{da_1^2 a_2^4}{dt} = 2\theta a_1^2 a_2^4,$$

we conclude that, for all the Bianchi types except IX,

$$\lim_{t \to t_- +} a_1^2 a_2^4 = 0, \quad \lim_{t \to \infty} a_1^2 a_2^4 = \infty$$

and in the case of Bianchi IX,

$$\lim_{t \to t_\pm \mp} a_1^2 a_2^4 = 0.$$

Since we have, for Bianchi type II,

$$a_1^2 a_2^4 = \alpha_{12}^2 u \alpha_2^2 \left(u^2 + \frac{1}{4}\nu_1^2\right),$$

we conclude that the existence interval in u-time, corresponding to the existence interval (t_-, ∞) in t-time, is $(0, \infty)$. In the case of Bianchi VIII and IX, we have

$$a_1^2 a_2^4 = (-\alpha_2^4 \nu_1 \nu_2 u^2 + c_1 u + \frac{1}{4}\alpha_2^4 \nu_1^3 \nu_2)\alpha_2^2 \left(u^2 + \frac{1}{4}\nu_1^2\right).$$

Thus, for Bianchi VIII, the existence interval is (u_-, ∞) where $u_- = u_1$ and for Bianchi IX, it is (u_-, u_+), where $u_- = u_0$ and $u_+ = u_1$. □

24 Existence of extensions

In this chapter, we prove that the MGHD's (of left invariant vacuum initial data on 3-dimensional unimodular Lie groups) we do not already know to be inextendible, cf. Proposition 22.23, have extensions that solve Einstein's vacuum equations. In Section 24.1 we construct the embedding, the existence of which leads to the conclusion that there is an extension. In Section 24.2 we consider the question to what extent the extension is extendible. It turns out that there are examples for which the extension we construct is in its turn extendible, but that if the spatial topology is compact, it is not. Finally, in Section 24.3, we state a result amounting to strong cosmic censorship in the unimodular vacuum case.

24.1 Construction of an embedding

Combining Proposition 22.23, Lemma 23.1, 23.3 and 23.4, we conclude that all the MGHD's (of left invariant vacuum initial data on 3-dimensional unimodular Lie groups) for which we do not already know that the Kretschmann scalar becomes unbounded in the incomplete directions of causal geodesics can be written in the form

$$g_{\text{MGHD}} = -\frac{L^2}{X}du^2 + X\xi^1 \otimes \xi^1 + Y^2(\xi^2 \otimes \xi^2 + \xi^3 \otimes \xi^3). \tag{24.1}$$

In the case of Bianchi type II, VIII and IX, the objects appearing on the right-hand side are defined in the statement of Lemma 23.4; in particular, the ξ^i are the duals of a basis e_i of the Lie algebra with certain properties. In the remaining cases, $X(u) = 4u$, $Y(u) = 1$, $L = 2$, $\xi^1 = dx$, etc. and $e_1 = \partial_x$, etc., cf. (23.5). Let $J = (u_-, u_+)$ be the interval on which (24.1) is defined.

Definition 24.1. Define $h_\pm$ by

$$h_\pm(u) = \pm \int_{u_a}^{u} \frac{L}{X(s)}\, ds$$

for some $u_a \in J$. Let $\gamma\colon \mathbb{R} \to G$ be the smooth homomorphism which is an integral curve of e_1. Define two diffeomorphisms of $M = J \times G$ by

$$\phi_\pm(u, g) = (u, g\gamma[h_\pm(u)]). \tag{24.2}$$

Remark 24.2. Clearly, ϕ_+ and ϕ_- are each other's inverses.

Note that

$$\phi_{\pm *}\partial_u|_{(u,g)} = \partial_u|_{\phi_\pm(u,g)} \pm \frac{L}{X(u)} e_1|_{\phi_\pm(u,g)}, \quad \phi_{\pm *}e_1|_{(u,g)} = e_1|_{\phi_\pm(u,g)}. \tag{24.3}$$

Thus

$$\phi_\pm^* g_{\text{MGHD}}(\partial_u, \partial_u) = 0, \quad \phi_\pm^* g_{\text{MGHD}}(\partial_u, e_1) = \pm L, \quad \phi_\pm^* g_{\text{MGHD}}(e_1, e_1) = X.$$

Note that, in particular, in the $\partial_u e_1$-plane, the pulled-back metric remains perfectly Lorentzian even when X becomes zero. What remains to be done is to compute $\phi_{\pm *}e_i$ for $i = 2, 3$.

Lemma 24.3. *With the above notation,*

$$\begin{aligned}\phi_{\pm *}e_2|_{(u,g)} &= \cos[\nu_2 h_\pm(u)]e_2|_{\phi_\pm(u,g)} - \sin[\nu_2 h_\pm(u)]e_3|_{\phi_\pm(u,g)},\\ \phi_{\pm *}e_3|_{(u,g)} &= \sin[\nu_2 h_\pm(u)]e_2|_{\phi_\pm(u,g)} + \cos[\nu_2 h_\pm(u)]e_3|_{\phi_\pm(u,g)}.\end{aligned}$$

Proof. Let, for $a \in G$, $C_a : G \to G$ be defined by $C_a b = aba^{-1}$. Note that C_a is a smooth diffeomorphism and an isomorphism. We shall denote the pushforward of C_a by Ad_a. Let us define

$$\zeta_j(s) = \mathrm{Ad}_{\gamma(s)}\, e_j|_e$$

for $j = 2, 3$, where e is the identity element of G. Note that $\zeta_j(s) \in T_eG$ for all $s \in \mathbb{R}$. Since T_eG is a vector space, we shall also think of derivatives of the ζ_j as elements of T_eG. Note that the flow of the vector field e_1 is given by $\phi_t(g) = \phi(t, g) = g\gamma(t)$. Consequently, $\phi_{t*} = R_{\gamma(t)*}$, so that, for $j = 2, 3$,

$$\phi_{-t*}(e_j|_{\phi_t(e)}) = R_{\gamma(-t)*}L_{\gamma(t)*}e_j|_e = \mathrm{Ad}_{\gamma(t)}\, e_j|_e.$$

Due to Proposition 58, p. 31 of [65], we thus conclude that

$$\zeta_2'(0) = [e_1, e_2]|_e = \nu_2 e_3|_e, \quad \zeta_3'(0) = [e_1, e_3]|_e = -\nu_2 e_2|_e.$$

Since Ad and γ are homomorphisms, we have

$$\begin{aligned}\zeta_2'(s) &= \lim_{h\to 0} \frac{\mathrm{Ad}_{\gamma(s+h)}\, e_2|_e - \mathrm{Ad}_{\gamma(s)}\, e_2|_e}{h}\\ &= \lim_{h\to 0} \mathrm{Ad}_{\gamma(s)} \frac{\mathrm{Ad}_{\gamma(h)}\, e_2|_e - e_2|_e}{h} = \mathrm{Ad}_{\gamma(s)}\, \zeta_2'(0) = \nu_2\, \mathrm{Ad}_{\gamma(s)}\, e_3|_e\\ &= \nu_2\zeta_3(s).\end{aligned}$$

Similarly, $\zeta_3'(s) = -\nu_2\zeta_2(s)$. Introducing

$$\begin{aligned}\varrho_2(s) &= \cos(\nu_2 s)\zeta_2(s) - \sin(\nu_2 s)\zeta_3(s),\\ \varrho_3(s) &= \sin(\nu_2 s)\zeta_2(s) + \cos(\nu_2 s)\zeta_3(s),\end{aligned}$$

one can compute that $\varrho_j'(s) = 0$ for $j = 2, 3$. Since $\varrho_j(0) = e_j|_e$ for $j = 2, 3$, we obtain

$$\begin{aligned}\zeta_2(s) &= \cos(\nu_2 s)e_2|_e + \sin(\nu_2 s)e_3|_e,\\ \zeta_3(s) &= -\sin(\nu_2 s)e_2|_e + \cos(\nu_2 s)e_3|_e.\end{aligned}$$

Let $\psi_\pm = R_{\gamma[h_\pm(u)]}$. Then

$$\psi_{\pm *}e_2|_g = L_{g\gamma[h_\pm(u)]*}\zeta_2[-h_\pm(u)],$$

and similarly for $\psi_{\pm *}e_3|_g$. Combining the above observations, the statement of the lemma follows. □

Corollary 24.4. *Let g_{MGHD} be defined as in* (24.1) *and $\phi_\pm$ be defined as in* (24.2). *Then*

$$\phi_\pm^* g_{\mathrm{MGHD}} = \pm L du \otimes \xi^1 \pm L\xi^1 \otimes du + X\xi^1 \otimes \xi^1 + Y^2(\xi^2 \otimes \xi^2 + \xi^3 \otimes \xi^3).$$

As a consequence of this corollary, we conclude that there is an extension of metrics of the form (24.1).

Definition 24.5. With conventions as above, define

$$g_{\mathrm{EXT},\pm} := \pm L du \otimes \xi^1 \pm L\xi^1 \otimes du + X\xi^1 \otimes \xi^1 + Y^2(\xi^2 \otimes \xi^2 + \xi^3 \otimes \xi^3) \quad (24.4)$$

on $\mathbb{R} \times G$.

24.2 Basic properties of the extensions

Proposition 24.6. *The Lorentz metric $g_{\mathrm{EXT},\pm}$ defined in* (24.4) *is a solution to the Einstein vacuum equations.*

Proof. Note that if we compute the components of the Ricci tensor with respect to the basis defined by $e_0 = \partial_u$ and e_i, then they will be quotients of polynomials, the polynomial in the denominator being strictly positive. Due to Corollary 24.4, we know that on an open subinterval of $\mathbb{R}$, all the components of the Ricci tensor are zero. This means that for a given component of the Ricci tensor, the polynomial in the numerator has to be zero on an open subinterval of $\mathbb{R}$, and consequently, the same has to be true of its derivatives. Thus, the polynomial in the numerator is zero, and the Ricci tensor is identically zero. Thus the metric solves the Einstein vacuum equations. □

It is natural to ask if the extension we have constructed is extendible. For LRS Bianchi type IX solutions, the case of interest when proving the existence of inequivalent maximal extensions, the answer turns out to be no. However, in the case of Bianchi type I, the answer is sometimes yes. Let $\phi_\pm$ be defined as in (24.2) with $u_a = 1$. Then

$$\phi_\pm(t,x,y,z) = \left(t, x \pm \frac{1}{2}\ln t, y, z\right).$$

Recall the embedding ϕ_K, defined in (23.4), of the flat Kasner solution into Minkowski space and define

$$\phi_{CT}(s,x,y,z) = (2\sqrt{s}, x, y, z).$$

Then $\phi_K \circ \phi_{CT} \circ \phi_\pm$ defines isometries from $\mathbb{R}_+ \times \mathbb{R}^3$ with the metric $g_{\mathrm{EXT},\pm}$ to an open subset of Minkowski space. Note that they are given by

$$\phi_K \circ \phi_{CT} \circ \phi_+(u,x,y,z) = (ue^x + e^{-x}, ue^x - e^{-x}, y, z),$$
$$\phi_K \circ \phi_{CT} \circ \phi_-(u,x,y,z) = (e^x + ue^{-x}, e^x - ue^{-x}, y, z).$$

Both of these maps can, in an obvious way, be extended to isometries from $(\mathbb{R}^4, g_{\mathrm{EXT},\pm})$ to the open subsets

$$\{(t,x,y,z) \in \mathbb{R}^4 : t \mp x > 0\}$$

respectively, endowed with the Minkowski metric. In particular, it is clear that $(\mathbb{R}^4, g_{\mathrm{EXT},\pm})$ are both extendible.

The extendibility of the extensions $(\mathbb{R} \times G, g_{\mathrm{EXT},\pm})$ in the Bianchi type I case is related to the non-compactness of the spatial hypersurfaces. If we had considered Bianchi type I initial data on the 3-torus instead, the extension would have been inextendible, as we shall see below. One way to ensure spatial compactness is to demand that the Lie group under consideration be compact. Another is to consider quotients of the initial manifold by co-compact subgroups of the isometry group of the initial data. When considering the second possibility, it is of interest to analyze to what extent taking the quotient is compatible with the construction of the extension. In preparation for this analysis, let us make the following definition.

Definition 24.7. Let (G, g, k) be locally rotationally symmetric vacuum initial data of Bianchi type I, II, VII_0, VIII or IX, where G is simply connected. Due to Corollary 19.14, there is a canonical basis e_i of the Lie algebra, with associated commutator matrix ν, such that $k_{ij} = k(e_i, e_j)$ are the components of a diagonal matrix. Furthermore, we can assume $k_{22} = k_{33}$ and $\nu_2 = \nu_3$. Let $\mathcal{D}_{\mathrm{LRS}}$ be the subgroup of the diffeomorphism group of G generated by

- the left translations,
- the isomorphisms of G arising from the Lie algebra isomorphisms given by rotations in the e_2e_3-plane.

It is natural to ask if $\mathcal{D}_{\mathrm{LRS}}$ coincides with the isometry group of the initial data. That this is not the case is easily seen by noting that the isomorphisms of G arising from the Lie algebra isomorphisms that change the sign of two of the elements in the basis e_i without changing the sign of the third element are not (all) included.

Lemma 24.8. *Let (G, g, k) be locally rotationally symmetric vacuum initial data of Bianchi type* I, II, VII_0, VIII *or IX, where G is simply connected, such that the associated MGHD is of the form* (24.1) *on $J \times G$, where J is an open interval. Let $\psi \in \mathcal{D}_{\mathrm{LRS}}$ and extend ψ to a diffeomorphism of $J \times G$ by $\psi(t,h) = [t, \psi(h)]$. Then $\psi \circ \phi_\pm = \phi_\pm \circ \psi$.*

Remark 24.9. It is of interest to note that the proof demonstrates that $\psi_* e_1 = e_1$ for all $\psi \in \mathcal{D}_{\mathrm{LRS}}$.

Proof. It is enough to prove the statement for the generators of $\mathcal{D}_{\mathrm{LRS}}$. Since

$$L_{h_1} \circ \phi_\pm(u, h_2) = L_{h_1}(u, h_2\gamma[h_\pm(u)]) = (u, h_1h_2\gamma[h_\pm(u)]) = \phi_\pm \circ L_{h_1}(u, h_2),$$

it is clear that left translations commute with $\phi_\pm$. Let ψ be an isomorphism of G arising from a Lie algebra isomorphism given by a rotation in the e_2e_3-plane by an angle, say θ. Let $\gamma\colon \mathbb{R} \to G$ be the smooth homomorphism which is an integral curve of e_1. Then $\psi \circ \gamma(0) = e$, where e is the identity element of G. Furthermore, the derivative of $\psi \circ \gamma$ at the origin is e_1. Finally, since $\psi \circ \gamma(t+h) = L_{\psi\circ\gamma(t)}\psi \circ \gamma(h)$, we conclude that the derivative of $\psi \circ \gamma$ at t is simply $e_1|_{\psi\circ\gamma(t)}$. Thus $\psi \circ \gamma(t) = \gamma(t)$, so that

$$\psi \circ \phi_\pm(u,h) = \psi(u, h\gamma[h_\pm(u)]) = (u, \psi(h)\gamma[h_\pm(u)]) = \phi_\pm \circ \psi(u,h).$$

The lemma follows. □

Note that if Γ is a properly discontinuous subgroup of $\mathfrak{D}_{\mathrm{LRS}}$ acting freely on G, then (24.1) defines a solution to Einstein's vacuum equations on the quotient $J \times G/\Gamma$. Furthermore, as a consequence of Lemma 24.8, this solution can be extended to $\mathbb{R}\times G/\Gamma$ with the metric induced by (24.4), the embedding being induced by ϕ_+.

Lemma 24.10. *Let $\mathfrak{D}_{\mathrm{LRS}}$ be defined as in Definition* 24.7 *and the Lorentz metric $g_{\mathrm{EXT},\pm}$ be defined as in* (24.4). *Let Γ be a properly discontinuous subgroup of $\mathfrak{D}_{\mathrm{LRS}}$ acting freely on G and assume $\Sigma := G/\Gamma$ to be compact. Then $g_{\mathrm{EXT},\pm}$ defines a Lorentz metric on $\mathbb{R} \times \Sigma$ which is a C^2-inextendible solution to Einstein's vacuum equations.*

Remark 24.11. In the case of Bianchi IX, G is compact in itself, so that the extension is always C^2-inextendible in that case.

Proof. The only thing that remains to be proved is the C^2-inextendibility. In preparation for the proof of this, let us consider the timelike geodesics with respect to $g_{\mathrm{EXT},\pm}$ on $\mathbb{R} \times G$. To simplify the notation, let us consider metrics of the form

$$g = c_{01}du \otimes \xi^1 + c_{01}\xi^1 \otimes du + X\xi^1 \otimes \xi^1 + Y^2(\xi^2 \otimes \xi^2 + \xi^3 \otimes \xi^3), \tag{24.5}$$

where c_{01} is a constant. Assume γ to be a timelike geodesic and let us introduce the notation $\dot{\gamma} = v^\mu e_\mu$. Then the geodesic equation, $\ddot{\gamma} = 0$, implies that

$$\dot{v}^0 + \frac{X_u}{c_{01}} v^0 v^1 + \frac{X}{2c_{01}^2}[X_u v^1 v^1 + 2YY_u(v^2v^2 + v^3v^3)] = 0, \tag{24.6}$$

$$\dot{v}^1 - \frac{1}{2c_{01}}[X_u v^1 v^1 + 2YY_u(v^2v^2 + v^3v^3)] = 0. \tag{24.7}$$

As a consequence, one can compute that

$$\frac{d}{ds}\left(v^0 + \frac{X}{c_{01}}v^1\right) = 0. \tag{24.8}$$

Thus the quantity inside the parenthesis is conserved. Due to causality, we also have

$$2c_{01}v^0v^1 + Xv^1v^1 + Y^2(v^2v^2 + v^3v^3) = -\beta_0, \tag{24.9}$$

for some constant $\beta_0 > 0$.

To prove C^2-inextendibility, assume (M, g), where $M = \mathbb{R} \times \Sigma$ and $g = g_{\mathrm{EXT},\pm}$, is C^2-extendible and let $(\widehat{M}, \hat{g})$ be an extension with embedding $i : M \to \widehat{M}$. Just as in the proof of Lemma 18.18, we can assume that there is a timelike geodesic γ intersecting both $i(M)$ and $\widehat{M} - i(M)$. We can thus assume $\gamma : (s_-, s_+) \to \widehat{M}$ to be such that $\gamma(s) \in i(M)$ for $s \in (s_-, s_0)$, but $\gamma(s_0) \in \widehat{M} - i(M)$, where $s_0 \in (s_-, s_+)$. We shall below restrict our attention to $\gamma|_{(s_-,s_0)}$ and consider it to be a timelike geodesic in M. It is natural to divide the analysis into several different cases.

Bianchi type I and II. There are two possibilities for the geodesic. The u-coordinate of γ has to converge to ∞ or to $-\infty$ as $s \to s_0-$. The reason this is true is that γ has to leave every compact subset of $M = \mathbb{R} \times \Sigma$, i.e., it has to leave $K \times \Sigma$ for every compact subset K of $\mathbb{R}$. Due to the fact that the Bianchi class A developments of type I and II initial data are future causally geodesically complete, convergence to ∞ is not possible. Let us therefore assume the u-coordinate converges to $-\infty$. Note that in this case

$$X(u) = \frac{x_1 u}{x_2 u^2 + x_0}$$

for non-negative constants x_i, $i = 1, 2, 3$, such that $x_1, x_0 > 0$ and $x_2 > 0$ in the case of Bianchi II and $x_2 = 0$ in the case of Bianchi I. Let γ^0 denote the u-coordinate of γ. Then $v^0 = \dot{\gamma}^0$. Since γ^0 tends to $-\infty$ in finite time, there is a sequence $s_k \to s_0-$ such that

$$\gamma^0(s_k) \to -\infty \quad \text{and} \quad \dot{\gamma}^0(s_k)/\gamma^0(s_k) \to \infty. \tag{24.10}$$

Due to (24.8), there is a constant C such that

$$v^0 + \frac{X}{c_{01}} v^1 = C. \tag{24.11}$$

Since X is negative and bounded by a constant times $|\gamma^0| + 1$ and since (24.10) holds, this leads to the conclusion that $v^1(s_k)/c_{01} \to -\infty$. However, combining (24.9) and (24.11), we obtain

$$c_{01} v^0 v^1 + c_{01} v^1 C + Y^2 (v^2 v^2 + v^3 v^3) = -\beta_0.$$

Note that the first term converges to ∞ along s_k. Since the quotient between the second and the first term converges to zero, we conclude that the left-hand side converges to ∞. This contradicts the fact that the right-hand side is negative.

Bianchi VIII. Just as in the Bianchi type I and II cases, there are two possibilities for the geodesic: the u-coordinate of γ has to converge to ∞ or to $-\infty$ as $s \to s_0-$, the reason being the same as before. In the case of Bianchi type VIII, there is an interval (u_-, u_+) with $u_\pm \in \mathbb{R}$ such that $X(u) > 0$ for $u > u_+$ and for $u < u_-$. Restricting the metrics to one of the intervals (u_+, ∞) and $(-\infty, u_-)$, we obtain metrics isometric to a Bianchi type VIII development. Keeping track of the time orientation of these developments and recalling that they are causally geodesically complete in one time direction due to Lemma 20.12, we conclude that it is not possible to extend the spacetime.

Bianchi IX. In this case, an argument similar to the one presented in the case of Bianchi type I and II yields the conclusion that the spacetime is inextendible. □

24.3 SCC, unimodular vacuum case

Theorem 24.12. *Consider Bianchi class A initial data (G, g, k) for Einstein's vacuum equations. In the case of Bianchi* IX, *the maximal globally hyperbolic development (MGHD) is past and future causally geodesically incomplete, and if the initial data are of type* I *or* VII_0 *with* $\mathrm{tr}_g\, k = 0$, *then the MGHD is causally geodesically complete. In the remaining cases, the MGHD is causally geodesically complete in one direction and incomplete in the other, and in these cases we shall below assume the time orientation to be such that the development is future causally geodesically complete. There is the following division:*

- *If the initial data are of type* I *or* VII_0, *then there are the following cases:*
  - *If* $\mathrm{tr}_g\, k = 0$, *then the MGHD is a quotient of Minkowski space and is thus inextendible.*
  - *There is a canonical basis e_i of the Lie algebra, with associated commutator matrix ν, such that k is diagonal with respect to this basis, with diagonal components k_i. Furthermore, $k_1 \neq 0$, $k_2 = k_3 = 0$ and $\nu_2 = \nu_3$. Then the MGHD is a quotient of the flat Kasner solution,* (23.3), *and consequently, it has a smooth extension which solves the Einstein vacuum equations.*
  - *In all the remaining cases, the Kretschmann scalar, $R_{\alpha\beta\gamma\delta}R^{\alpha\beta\gamma\delta}$, is unbounded along past inextendible causal curves in the MGHD. In particular, the MGHD is C^2-inextendible.*
- *If the initial data are of type* VI_0, *the Kretschmann scalar is unbounded along past inextendible causal curves in the MGHD. In particular, the MGHD is C^2-inextendible.*
- *If the initial data are of type* II, VIII *or* IX, *there are the following cases:*
  - *If the initial data are locally rotationally symmetric, the maximal globally hyperbolic development has a smooth extension which solves the Einstein vacuum equations.*
  - *If the initial data are not rotationally symmetric, the Kretschmann scalar is unbounded along past inextendible causal curves in the maximal globally hyperbolic development. In the case of Bianchi* IX *initial data, it is also unbounded along future inextendible causal curves. In particular, the MGHD is C^2-inextendible.*

Remark 24.13. As a consequence of the above theorem, strong cosmic censorship holds in the class of left invariant vacuum initial data on 3-dimensional unimodular Lie groups. One can of course ask similar questions concerning quotients of the initial data: if Γ is a free and properly discontinuous group of isometries of the initial data, what are the extendibility properties of the MGHD of the induced initial data on M/Γ? A partial answer to this question is given in connection with Lemma 24.8. Nevertheless, it is important to note that some groups Γ are such that only locally rotationally symmetric initial data are invariant under them. Consequently, for some groups Γ, all the initial data on G/Γ that arise from Bianchi class A initial data on G are such that the corresponding MGHD's have smooth extensions that solve the Einstein vacuum equations.

Proof. The statement is an immediate consequence of statements already made. □

25 Existence of inequivalent extensions

The purpose of this chapter is to demonstrate that there are vacuum initial data, the MGHD's of which allow inequivalent maximal extensions. The argument, which is taken from [23], see also [62], is based on a study of the null geodesics in the $\partial_u e_1$-plane. As a sketch of the proof was already presented in the introduction to the present part of these notes, let us proceed to the details.

Lemma 25.1. *Consider a spacetime* $(\mathbb{R} \times G, g_{\mathrm{EXT},\pm})$*, where* $g_{\mathrm{EXT},\pm}$ *is given by* (24.4) *and* G *is a Bianchi type* IX *Lie group. Define a time orientation by demanding that* ∂_u *be future oriented. Let* $J = (u_-, u_+)$ *be the interval on which* X *is positive. Let* γ *be a null geodesic starting at a point of* $J \times G$ *with initial velocity in the plane spanned by* ∂_u *and* e_1*. Then either* γ *is future and past incomplete or it is complete. The complete geodesics are the ones parallel to* ∂_u*. Finally, let* γ *be a future directed timelike curve such that* $\gamma^0(s_0) = u_+$ *or* $\gamma^0(s_0) = u_-$*, where* γ^0 *is the* u*-coordinate of* γ*. Then* $\dot{\gamma}^0(s_0) > 0$.

Proof. For any curve γ, define v^μ by

$$\dot{\gamma} = v^\mu e_\mu,$$

where $e_0 = \partial_u$. Considering the geodesic equation, one can then see that if $v^A = 0$ for $A = 2, 3$ originally, then they remain zero. Consequently, it is meaningful to consider null geodesics, the tangent vectors of which remain in the $e_0 e_1$-plane. The relevant equations are, using the notation (24.5) for the metric $g_{\mathrm{EXT},\pm}$, given by

$$\dot{v}^0 + \frac{X_u}{c_{01}} v^0 v^1 + \frac{1}{2c_{01}^2} X X_u v^1 v^1 = 0, \tag{25.1}$$

$$\dot{v}^1 - \frac{X_u}{2c_{01}} v^1 v^1 = 0, \tag{25.2}$$

cf. (24.6)–(24.7). As a consequence of the second equation, we see that if $v^1 = 0$ at some point, then it is always zero. The resulting geodesics have the property that $\dot{v}^0 = 0$ due to (25.1). Consequently, v^0 is conserved, so that the geodesic is complete, since if γ^0 is the u-coordinate of γ, then $\dot{\gamma}^0 = v^0$. Consider a $u_0 \in \mathbb{R}$ such that $X(u_0) = 0$ and assume γ passes through $\{u_0\} \times G$. Assume $\gamma^0(s_0) = u_0$. Due to the form of the metric, we have to have $v^0(s_0) = 0$ or $v^1(s_0) = 0$. Since we have already considered the latter case, let us assume $v^0(s_0) = 0$. Then, due to (25.1), $v^0 = 0$ for the entire solution. Consequently, the geodesic is contained in the hypersurface $\{u_0\} \times G$. As a consequence, the only null geodesics (with tangent vectors in the $e_0 e_1$-plane) that can pass through a hypersurface $\{u_0\} \times G$, where $X(u_0) = 0$, are the null geodesics with $v^1 = 0$. A null geodesic (with tangent vectors in the $e_0 e_1$-plane) which starts in $J \times G$ and does not pass through either $\{u_-\} \times G$ or $\{u_+\} \times G$ is contained in $J \times G$ and can thus be considered to be a null geodesic in the Bianchi class A development of initial data. Consequently, it is both future and past incomplete.

To prove the last statement, let γ be a future oriented timelike curve and let v^μ be as above. Let s_0 be as in the statement of the lemma. Since γ is timelike and future oriented, we have to have, for $s = s_0$,

$$2c_{01}v^0v^1 + Y^2(v^2v^2 + v^3v^3) < 0, \quad c_{01}v^1 < 0,$$

where the latter inequality is a consequence of $\langle \partial_u, \dot{\gamma} \rangle < 0$. Combining these two observations, we conclude that $\dot{\gamma}^0(s_0) = v^0(s_0) > 0$. □

Before proving the main theorem, we need a technical lemma.

Lemma 25.2. *Let G be a Bianchi type* IX *Lie group and let e_i be a basis of the Lie algebra such that the associated commutator matrix ν is diagonal with positive diagonal components. Then the integral curves of the elements of the basis through the identity element of G are periodic.*

Remark 25.3. One could of course make an explicit choice of basis for the Lie algebra of SU(2) and carry out a less abstract argument than the one presented below. However, we here wish to avoid explicit constructions on particular representatives of the Lie groups under consideration if possible.

Proof. Let γ_i denote the integral curve of e_i through the identity element of G. Let $\widetilde{G}$ be the universal covering group of G and let π be the covering map (which is also a homomorphism), cf. Theorem 2.13, p. 43 of [55]. Let $\tilde{e}_i$ be the element of the Lie algebra of $\widetilde{G}$ such that $\pi_* \tilde{e}_i|_{\tilde{e}} = e_i|_e$, where $\tilde{e}$ and e denote the identity elements of $\widetilde{G}$ and G respectively, and let $\tilde{\gamma}_i$ be the integral curve of $\tilde{e}_i$ through the identity element of $\widetilde{G}$. Note that $\pi \circ \tilde{\gamma}_i = \gamma_i$. Thus, if we can prove $\tilde{\gamma}_i$ to be periodic, the same follows for γ_i. In other words, we might as well assume G to be simply connected. Rescaling the e_i, we obtain a basis e_i' so that the corresponding matrix ν is the identity matrix. Since if the integral curve of e_i' through the identity is periodic, the same is true of γ_i, we might as well assume the matrix ν associated with the e_i to be the identity matrix. Define a metric on G by demanding that the basis e_i be orthonormal. Then, due to Lemma 19.11 and the fact that G is 3-dimensional, g is a metric of constant positive curvature. Since g is a left invariant metric, we also know it to be complete, cf. the proof of Proposition 20.3. Thus (G, g) is a simply connected and geodesically complete Riemannian manifold of constant positive curvature. Due to Proposition 23, p. 227 of [65], we conclude that (G, g) is isometric to the 3-sphere with the appropriate rescaling of the standard metric. Consequently, the geodesics of (G, g) are periodic. All that remains to be proved is thus that γ_i is a geodesic of (G, g). To this end, note that

$$\dot{\gamma}_i(t) = e_i|_{\gamma_i(t)}, \quad \ddot{\gamma}_i(t) = \nabla_{\dot{\gamma}_i(t)} e_i$$

(no summation), where ∇ is the Levi-Civita connection associated with g. Since $\nabla_{e_i} e_i = 0$ (no summation) due to the fact that ν is diagonal, we conclude that γ_i is a geodesic of the 3-sphere. The lemma follows. □

Considering (24.3), we see that $\phi_\pm$ sends null geodesics, the tangent vectors of which are parallel to

$$\partial_u \mp \frac{L}{X}e_1 \quad \text{and} \quad \partial_u \pm \frac{L}{X}e_1$$

respectively, to null geodesics whose tangent vectors are parallel to ∂_u and $\partial_u \pm 2Le_1/X$ respectively. Thus, due to Lemma 25.1, the image of a null geodesic falling into the first category under $\phi_\pm$ is part of a complete null geodesic. However, the image of a null geodesic falling into the second category under $\phi_\pm$ cannot be extended, and is thus an incomplete null geodesic even with respect to $g_{\text{EXT},\pm}$.

Theorem 25.4. *The maximal globally hyperbolic development of locally rotationally symmetric Bianchi type* IX *vacuum initial data has two non-isometric extensions which solve the Einstein vacuum equations and are C^2-inextendible.*

Remark 25.5. We emphasize the C^2-inextendibility of the extensions since it is of course trivial to construct an infinite number of non-isometric extensions of the MGHD; starting with $(\mathbb{R} \times G, g_{\text{EXT},\pm})$, where $g_{\text{EXT},\pm}$ is defined in (24.4), and removing suitable subsets of $(u_+, \infty) \times G$ yields the desired extensions, the inequivalence being a direct consequence of topological inequivalence. The general philosophy concerning the existence of non-isometric maximal extensions is that if the Cauchy horizon, i.e., the hypersurfaces $u = u_+$ and $u = u_-$ in the case of the metric $g_{\text{EXT},\pm}$, has more than one component, then there are non-isometric extensions. In fact, the results of [23] demonstrate that there are spacetimes with a Cauchy horizon consisting of an arbitrarily large number of components, and in that case, the authors construct an arbitrarily large number of inequivalent maximal extensions, cf. the last sentence of Section V on p. 1626 of [23].

Proof. Let us define the two extensions. The first one is simply $M_1 = \mathbb{R} \times G$, $g_1 = g_{\text{EXT},+}$. The second one is defined through identifications. To start with, let $I = (u_-, u_+)$ be the interval on which X is positive and

$$N_a = (-\infty, u_+) \times G, \quad N_b = (u_-, \infty) \times G.$$

Consider N_a to be a Lorentz manifold by giving it the metric $g_a = g_{\text{EXT},-}$ and N_b to be a Lorentz manifold by giving it the metric $g_b = g_{\text{EXT},+}$. Let N_2 be the disjoint union of N_a and N_b. We define an equivalence relation on N_2 by saying that $q_1 \sim q_2$ if one of the following holds,

- $q_1, q_2 \in N_a$ and $q_1 = q_2$,
- $q_1, q_2 \in N_b$ and $q_1 = q_2$,
- $q_1 \in N_a$, $q_2 \in N_b$, and if $q_i = (u_i, p_i)$ for $i = 1, 2$, then $u_1 = u_2 \in (u_-, u_+)$ and $\phi_- \circ \phi_-(q_1) = q_2$,
- $q_1 \in N_b$, $q_2 \in N_a$, and if $q_i = (u_i, p_i)$ for $i = 1, 2$, then $u_1 = u_2 \in (u_-, u_+)$ and $\phi_- \circ \phi_-(q_2) = q_1$.

Define M_2 to be the quotient manifold. Since $\phi_- \circ \phi_-$ is an isometry between the relevant subsets of N_a and N_b, we get a Lorentz metric on M_2, which we shall denote by g_2. Note that both M_1 and M_2 are time oriented. An orientation can be fixed by demanding that ∂_u be future oriented in the case of M_1, N_a and N_b. Note, furthermore, that (M_1, g_1) is already known to be inextendible and that (M_2, g_2) can be proven to be inextendible by arguments similar to the proof of Lemma 24.10. Finally, (M_i, g_i), $i = 1, 2$, both solve Einstein's vacuum equations.

Let C_2 be the region defined as the quotient of the disjoint union of the subset $[u_-, u_+) \times G$ of N_a and the subset $(u_-, u_+] \times G$ of N_b under the above equivalence relation. Then a statement similar to the one made in Lemma 25.1 holds concerning future directed timelike curves passing through the boundary of C_2. Let $C_1 = [u_-, u_+] \times G$. Assume there is an isometry, say ψ, between (M_1, g_1) and (M_2, g_2). To start with we wish to prove that ψ has to map C_1 to C_2. Assume there is a $q \in C_1$ such that $\psi(q) \notin C_2$. Then, either $\psi(q) \in N_a$ with u-coordinate $\psi^0(q) < u_-$, or $\psi(q) \in N_b$ with u-coordinate $\psi^0(q) > u_+$. In both cases, $X[\psi^0(q)] < 0$. Thus, the integral curve of e_1 through $\psi(q)$ is timelike. Due to Lemma 25.2, there is thus a closed timelike curve passing through $\psi(q)$, say γ. Then $\psi^{-1} \circ \gamma$ is a closed timelike curve passing through q. This is not possible due to Lemma 25.1 and the global hyperbolicity of $g_{\text{EXT},\pm}$ restricted to $(u_-, u_+) \times G$. Similarly, ψ^{-1} has to map C_2 to C_1. Consequently, ψ restricts to an isometry of the interior of C_1 to the interior of C_2. Note that the interior of C_1 and the interior of C_2 are isometric to $M_3 = J \times G$ with the metric $g_3 = g_{\text{MGHD}}$, where $J = (u_-, u_+)$. We can thus view ψ, restricted to the interior of C_1, as an isometry from (M_3, g_3) to itself. Let us denote this isometry by χ. Note that all the isometries involved, except possibly for ψ and χ, are such that they map tangent vectors in the $\partial_u e_1$-plane to tangent vectors in the same plane. Furthermore, if we consider a null-geodesic in (M_1, g_1), starting in the interior of C_1 with initial velocity in the $\partial_u e_1$-plane, then it is either future complete and past complete, or future and past incomplete due to Lemma 25.1. On the other hand, if we consider a null-geodesic in (M_2, g_2), starting in the interior of C_2 with initial velocity in the $\partial_u e_1$-plane, then it is either future complete and past incomplete, or future incomplete and past complete due to the observations preceding the statement of the theorem and the definition of (M_2, g_2). Consequently, if we can prove that χ maps the $\partial_u e_1$-plane to itself, we obtain the desired contradiction to the assumption that there is an isometry between M_1 and M_2.

Note that (M_3, g_3) is time oriented, globally hyperbolic and foliated by compact constant mean curvature hypersurfaces that are also Cauchy hypersurfaces. Furthermore, it satisfies the timelike convergence condition since it satisfies Einstein's vacuum equations. Due to Corollary 18.14, we conclude that χ has to map the compact spacelike hypersurfaces of constant mean curvature $\kappa \neq 0$ to themselves or to compact spacelike hypersurfaces of constant mean curvature $-\kappa \neq 0$, if χ reverses the time orientation. Since the hypersurface of constant mean curvature 0 can be characterized as the complement of the hypersurfaces of constant mean curvature different from zero, we conclude that χ respects the foliation by constant mean curvature hypersurfaces.

Since ∂_u is orthogonal to the CMC hypersurfaces, $\chi_*\partial_u$ has to be some function times ∂_u (in fact it has to be ∂_u or $-\partial_u$).

To prove that e_1 is mapped to some function times e_1, let us make the following preliminary observations. Up till now, e_i has denoted the basis dual to the ξ^i appearing in (24.4). In the below argument, it will be convenient to rescale the e_i so that they become orthonormal. We shall denote the rescaled basis by e_i as well. As a consequence, the commutator matrix changes and we shall denote the resulting diagonal components by n_i. Note that the n_i depend on the time coordinate. Since we already know that $\chi_* e_1$ is orthogonal to ∂_u, it is enough to prove that it is also orthogonal to e_2 and e_3. Consider

$$f_j(p) = \langle \chi_* e_1|_{\chi^{-1}(p)}, e_j|_p \rangle$$

for $j = 2, 3$. Note that the f_j are smooth functions. Consequently, if f_j is zero on a dense subset of $J \times G$, it is zero. Consider the set of u's such that $n_1 = n_2$ (note that we always have $n_2 = n_3$). Due to (23.18)–(23.19), we see that if $n_2 = n_1$ and the derivative of the quotient is zero, then $n_1 = n_2 = n_3$ and $\theta_1 = \theta_2 = \theta_3$, contradicting the constraint (23.20). Consequently, the points at which $n_1 = n_2$ are isolated, and it is enough to prove that f_j is zero on $\{u\} \times G$, where u is such that $n_1(u) \neq n_2(u)$ and $n_1(u') \neq n_2(u')$, where u' is the u coordinate of the hypersurface $\chi^{-1}(\{u\} \times G)$; note that χ (and χ^{-1}) has to map hypersurfaces of the form $\{u\} \times G$ to hypersurfaces of the same form, since they are characterized as CMC hypersurfaces. Assuming u and u' to fulfill these criteria, note that χ restricts to an isometry from $\Sigma' = \{u'\} \times G$ to $\Sigma = \{u\} \times G$ with the induced metrics. Let $p \in \Sigma$. We can consider the Ricci tensor Ric of Σ to be a linear operator, say $\mathcal{R}$, from $T_p\Sigma$ to itself, by defining

$$\mathcal{R}v = \sum_{i=1}^{3} \mathrm{Ric}(v, E_i) E_i,$$

where $v \in T_p\Sigma$ and E_i is any orthonormal basis of $T_p\Sigma$. Note that $\mathcal{R}$ is symmetric and that, due to Lemma 19.11 and the fact that χ restricts to an isometry from Σ' to Σ, $\chi_* e_1|_{\chi^{-1}(p)}$ is an eigenvector with eigenvalue λ_1', say, and $\chi_* e_j|_{\chi^{-1}(p)}$, $j = 2, 3$, are eigenvectors with eigenvalue λ_2'. Since $\lambda_2' \neq \lambda_1'$, the corresponding eigenspaces are orthogonal. Analogously, $e_1|_p$ is an eigenvector with eigenvalue λ_1, say, and $e_j|_p$, $j = 2, 3$, are eigenvectors with eigenvalue λ_2. Again, $\lambda_2 \neq \lambda_1$ and the corresponding eigenspaces are orthogonal. As a consequence, $\chi_* e_1|_{\chi^{-1}(p)}$ has to be orthogonal to $e_j|_p$, $j = 2, 3$, and the theorem follows. □

Part V

Appendices

Appendix A

A.1 Conventions

If $(S, \mathcal{A}, m)$ is a measure space and $A \in \mathcal{A}$, we shall denote the characteristic function of A by χ_A, i.e., $\chi_A(x) = 1$ if $x \in A$ and $\chi_A(x) = 0$ if $x \notin A$.

To fix notation, if X is a Banach space, we shall use $\|\cdot\|_X$ to denote the associated norm. In cases where the space we are dealing with is clear from the context, we shall simply write $\|\cdot\|$. Furthermore, we shall use X^* to denote the dual of X. In the case of a Hilbert space, we shall use $(\cdot,\cdot)_H$ to denote the associated inner product, and $(\cdot,\cdot)$ when the space is clear from the context. We shall also assume that, in the case of complex Hilbert spaces, the inner product is complex linear in the first argument and that $(x, y)_H = \overline{(y, x)_H}$, where the bar denotes complex conjugation. If $z, w \in \mathbb{C}^n$, we shall write

$$z \cdot w = \sum_{i=1}^{n} z^i w^i,$$

where z^i and w^i are the components of z and w respectively. The convention for elements of $\mathbb{R}^n$ can be considered to be a special case of this. We write $\bar{z}$ for the component wise conjugate of z and

$$(z, w) = z \cdot \bar{w}, \quad |z| = (z \cdot \bar{z})^{1/2}.$$

Again the convention for elements of $\mathbb{R}^n$ can be considered to be a special case.

For $x \in \mathbb{R}^n$, we shall write x^i, $i = 1, \dots, n$ for the coordinates. We shall always write an element of $\mathbb{R}^{n+1}$ as (t, x) (assuming that the n is clear from the context), where t is the first variable and x the n last. We shall use x^i, $i = 1, \dots, n$ to index the last n variables and we shall sometimes write $x^0 = t$. If we speak of the x-direction, we shall mean in the direction of the last n variables. Concerning derivatives, we shall write

$$\partial_i = \frac{\partial}{\partial x^i}, \quad \partial_0 = \partial_t = \frac{\partial}{\partial t}.$$

If f is differentiable, we shall write $\partial_\mu f$, $\mu = 0, \dots, n$. In other words, Greek indices are assumed to range from 0 to n and Latin indices are assumed to range from 1 to n. A *multiindex* is a vector $\alpha = (\alpha_1, \dots, \alpha_n)$ such that all the α_i are non-negative integers. If α is a multiindex and f is sufficiently differentiable, we shall write

$$|\alpha| = \alpha_1 + \dots + \alpha_n, \quad \partial^\alpha f = \partial_1^{\alpha_1} \dots \partial_n^{\alpha_n} f.$$

Furthermore,

$$\binom{\alpha}{\beta} = \binom{\alpha_1}{\beta_1} \cdots \binom{\alpha_n}{\beta_n}$$

and $\alpha \le \beta$ means $\alpha_i \le \beta_i$, $i = 1, \dots, n$. If u is a function on $\mathbb{R}^{n+1}$ or $[0, T] \times \mathbb{R}^n$, where the n is clear from the context, and we write $\partial^\alpha u$, we shall take for granted that

α is an n-dimensional multiindex and that all the derivatives hit the last n variables. To conclude, if we write $\partial^\alpha u$, we assume α is a multiindex and that we differentiate with respect to the last n variables and if we write $\partial_\mu u$, we assume μ is an integer between 0 and n.

We shall use the Einstein summation convention, meaning that we sum over repeated upstairs and downstairs indices. What range we sum over should be clear from the above conventions or the context. If there is to be no summation, we state it explicitly.

If A is an $n \times n$-matrix, we shall write $A \geq c$, where c is a real number, if and only if A is real-valued, symmetric and

$$v^t A v \geq c|v|^2$$

for all $v \in \mathbb{R}^n$. We define $A > c$ similarly.

Concerning function spaces, we take it for granted that the members of the spaces take values in $\mathbb{R}$ unless otherwise stated (with two exceptions). When we write $C_0^\infty(\mathbb{R}^n)$, we thus have functions from $\mathbb{R}^n$ to $\mathbb{R}$ in mind. The exceptions are $H_{(s)}(\mathbb{R}^n)$ and $\mathcal{S}(\mathbb{R}^n)$, where we take it for granted that the functions take values in $\mathbb{C}$.

A.2 Different notions of measurability

The proof of the equivalence of weak and strong measurability assuming that the measure space is σ-finite and the Banach space is separable is taken from Yosida [86]. Note that [86] does not require the measure space to be σ-finite. Reasonably, this is tacitly assumed, since if X is a separable Banach space and $(S, \mathcal{A}, m)$ is a measure space which is not σ-finite, we can let $x\colon S \to X$ be a constant function, where the constant is different from zero. Then x is clearly weakly measurable. If the result holds in general, we conclude that x is strongly measurable, so that there is a sequence x_n of finitely-valued functions such that $x_n \to x$ a.e. Let A_0 be the set on which x_n does not converge to x and let A_n be the set on which x_n is non zero. Then A_0 has measure zero, A_n has finite measure and S can be written as the union of $A_n, n = 1, \dots$ and A_0. We conclude that S is σ-finite.

Theorem A.1. *Let X be a separable Banach space and let $(S, \mathcal{A}, m)$ be a σ-finite measure space. A function $x\colon S \to X$ is strongly measurable if and only if it is weakly measurable.*

Remark A.2. Recall that a Banach space is said to be *separable* if there is a countable dense subset.

Proof. Assume x is strongly measurable and that x_n is a sequence of finitely-valued functions converging pointwise to x a.e. Let $f \in X^*$. Then $f \circ x_n$ are measurable functions converging a.e. to $f \circ x$. By standard measure and integration theory, $f \circ x$ is measurable. Thus x is weakly measurable.

Assume x is weakly measurable. Let $U^* = \{f \in X^* : \|f\| \leq 1\}$. Let us start by proving that there exists a sequence $\{f_n\} \subseteq U^*$ such that for any $f_0 \in U^*$, there is a subsequence $\{f_{n_k}\}$ of $\{f_n\}$ such that $f_{n_k}(\xi) \to f_0(\xi)$ for all $\xi \in X$. Let the sequence $\{\xi_n\}$ be dense in X. Define a mapping $\phi_n \colon U^* \to \mathbb{C}^n$ (or $\mathbb{R}^n$ if the Banach space is real) by $\phi_n(f) = [f(\xi_1), \dots, f(\xi_n)]$. Since $\mathbb{C}^n$ is separable, there is, for every n, a sequence $\{f_{n,k}\}$ such that $\{\phi_n(f_{n,k})\}$ is dense in $\phi_n(U^*)$. Then $\{f_{n,k}\}$ is the desired sequence. Fix $f_0 \in U^*$. For every n, there is an f_{n,m_n} such that $|f_{n,m_n}(\xi_i) - f_0(\xi_i)| \leq 1/n$ for all $i = 1, \dots, n$. Thus $f_{n,m_n}(\xi_i) \to f_0(\xi_i)$ as $n \to \infty$ for all i. By a standard argument, $f_{n,m_n}(\xi) \to f_0(\xi)$ as $n \to \infty$ for all $\xi \in X$, since the sequence $\{\xi_i\}$ is dense and $f_{n,k}, f_0 \in U^*$.

Let $\{f_n\}$ be a sequence of the form proved to exist above. Let us prove that $\|x\|$ is measurable. Let a be a real number and define $A = \{s : \|x(s)\| \leq a\}$ and $A_f = \{s : |f[x(s)]| \leq a\}$, where $f \in U^*$. If we can show that $A = \bigcap_j A_{f_j}$, then it is clear that A is measurable due to the weak measurability of x, and thus $\|x\|$ is measurable. We have $A = \bigcap_{\|f\| \leq 1} A_f$, where one inclusion is trivial and the other is due to the Hahn–Banach theorem, cf. also Theorem 5.20 of [79]. Due to the properties of the sequence $\{f_n\}$, we obtain $\bigcap_j A_{f_j} = \bigcap_{\|f\| \leq 1} A_f$.

Since X is separable, there is a countable number of open balls $S_{j,n}$ $j = 1, \dots$ of radius $< 1/n$ whose union covers X. Let $x_{j,n}$ be the center of the ball $S_{j,n}$. Since $\|x - x_{j,n}\|$ is measurable, the set $B_{j,n} = \{s \in S : x(s) \in S_{j,n}\}$ is measurable. Furthermore, $\bigcup_j B_{j,n} = S$. Define

$$x_n(s) = x_{i,n} \quad \text{if} \quad s \in B'_{i,n} = B_{i,n} - \bigcup_{j=1}^{i-1} B_{j,n}.$$

Then, since $\bigcup_i B'_{i,n} = S$, x_n is defined on all of S. Furthermore, $\|x(s) - x_n(s)\| \leq 1/n$ for all $s \in S$. Let us define y_k as follows. If $s \notin \bigcup_{i,n \leq k} B'_{i,n}$, then $y_k(s) = 0$. If $s \in \bigcup_{i \leq k} B'_{i,k}$, let $y_k(s) = x_k(s)$. If $s \in \bigcup_{i \leq k} B'_{i,k-1} - \bigcup_{i \leq k} B'_{i,k}$, let $y_k(s) = x_{k-1}(s)$. We continue in this fashion, the last step being $y_k(s) = x_1(s)$ for

$$s \in \bigcup_{i \leq k} B'_{i,1} - \bigcup_{i \leq k, 2 \leq j \leq k} B'_{i,j}.$$

Then, if $i, n \leq k$, $\|y_k(s) - x(s)\| \leq 1/n$ on $B'_{i,n}$. Let $s \in S$ and $\varepsilon > 0$. Let n be large enough that $\varepsilon > 1/n$ and let $N \geq n$ be large enough that there is an $i \leq N$ with $s \in B'_{i,n}$. Then $\|y_k(s) - x(s)\| < \varepsilon$ for all $k \geq N$. Thus y_k converges pointwise to x. Note that, unfortunately, it is not clear that the y_k are finitely-valued. In order to remedy this, let $A_j \in \mathcal{A}$ be a countable sequence of sets of finite measure whose union is S. Assume furthermore that $A_j \subseteq A_{j+1}$ for all j. Define $y'_k = y_k \chi_{A_k}$, where for any set $A \in \mathcal{A}$, we use χ_A to denote the characteristic function of A ($\chi_A(s) = 1$ for $s \in A$ and $\chi_A(s) = 0$ for $s \notin A$). Then y'_k is finitely-valued and it converges pointwise everywhere to x. Thus x is strongly measurable. □

A.3 Separability

We shall need to know that the L^p spaces are separable.

Lemma A.3. *The Banach spaces $L^p(\mathbb{R}^n, \mathbb{R})$ are separable for $1 \le p < \infty$.*

Remark A.4. The same is of course true for $L^p(\Omega, \mathbb{R}^m)$ and $L^p(\Omega, \mathbb{C}^m)$ for $1 \le p < \infty$ and any measurable $\Omega \subseteq \mathbb{R}^n$.

Proof. Note first of all that the set of continuous functions with compact support is dense in the mentioned spaces. For the purpose of the present proof, let us define the *elementary sets* to be sets of the form

$$E = \{x \in \mathbb{R}^n : \alpha_i \le x_i < \beta_i\},$$

where α_i and β_i are rational numbers. We also define the *elementary functions of order m* to be functions of the form

$$f = \sum_{i=1}^{m} \gamma_i \chi_{E_i},$$

where $\gamma_1, \dots, \gamma_m \in \mathbb{Q}$, where $\mathbb{Q}$ are the rational numbers, E_i are elementary sets and

$$\chi_{E_i}(x) = \begin{cases} 1, & x \in E_i, \\ 0, & x \notin E_i. \end{cases}$$

We denote the set of elementary functions of order m by F_m and define the set of elementary functions F to be the union of F_m for all m. Note that each elementary set can be considered to be an element of $\mathbb{Q}^{2n}$, so that F_m can be considered to be a subset of $\mathbb{Q}^{m+2nm}$. Consequently F_m is countable, cf. Theorem 2.8 and Theorem 2.13 of [78]. Thus F is a countable union of countable sets and is thus countable by Theorem 2.12 of [78]. Let $u \in L^p(\mathbb{R}^n, \mathbb{R})$. Then there is a continuous function ϕ with compact support such that $\|u - \phi\|_p < \varepsilon/2$. Since ϕ has compact support, there is an integer $N > 0$ such that $\phi = 0$ outside of $[-N, N]^n$. Furthermore, ϕ is uniformly continuous so that for every $\xi > 0$ there is a δ such that $|x - y| \le \delta$ implies $|\phi(x) - \phi(y)| < \xi$. For a given $\delta > 0$, we can divide $[-N, N]^n$ into a finite number of elementary sets such that their diameter is strictly less than δ. For each such elementary set E, we choose a rational number γ such that $|\phi(x) - \gamma| \le \xi$ on E. This yields an elementary function f such that $|f(x) - \phi(x)| \le \xi$ for all $x \in \mathbb{R}^n$, excepting possibly the boundary points of the elementary sets, whose union is a set of measure zero and is therefore of no relevance to the present discussion. Compute

$$\|\phi - f\|_p \le \xi 2^{n/p} N^{n/p}.$$

Assuming $\xi = \varepsilon 2^{-1} 2^{-n/p} N^{-n/p}$, we obtain $\|f - u\|_p < \varepsilon$. We conclude that the set of elementary functions is dense, so that $L^p(\mathbb{R}^n, \mathbb{R})$ has a countable dense subset. □

A.4 Measurability

Let $I \subseteq \mathbb{R}$ be an open interval and assume that

$$u \in L^\infty[I, H_{(k)}(\mathbb{R}^n, \mathbb{C}^N)], \quad v \in L^1[I, H_{(-k)}(\mathbb{R}^n, \mathbb{C}^N)].$$

Then $\langle v, u\rangle$ can be defined as in (5.12) and if $\phi \in C_0^\infty(I \times \mathbb{R}^n, \mathbb{C}^N)$, then ϕ can be considered to be an element of $L^1[I, H_{(-k)}(\mathbb{R}^n, \mathbb{C}^N)]$. The reason is as follows. Since ϕ can be considered to be a continuous function from I into $H_{(-k)}(\mathbb{R}^n, \mathbb{C}^N)$ it is weakly and thus strongly measurable. That the relevant integral is bounded is immediate.

Lemma A.5. *Let $I \subseteq \mathbb{R}$ be an open interval and let $u \in L^\infty[I, H_{(k)}(\mathbb{R}^n, \mathbb{C}^N)]$, where k is a non-negative integer. Then there is a $U \in L^2_{\text{loc}}(I \times \mathbb{R}^n, \mathbb{C}^N)$ such that for any $\phi \in C_0^\infty(I \times \mathbb{R}^n, \mathbb{C}^N)$*

$$\langle \phi, u\rangle = \int_{I\times\mathbb{R}^n} \phi \cdot \bar{U}\, dt dx. \tag{A.1}$$

Furthermore, U is k times weakly differentiable with respect to x with derivatives in L^2_{loc} in the sense that for any multiindex $\alpha = (\alpha_1, \dots, \alpha_n)$ with $|\alpha| \leq k$, there is a $U_\alpha \in L^2_{\text{loc}}(I \times \mathbb{R}^n, \mathbb{C}^N)$ such that for all $\phi \in C_0^\infty(I \times \mathbb{R}^n, \mathbb{C}^N)$

$$\int_{I\times\mathbb{R}^n} \partial^\alpha \phi \cdot \bar{U}\, dt dx = (-1)^{|\alpha|} \int_{I\times\mathbb{R}^n} \phi \cdot \bar{U}_\alpha\, dt dx. \tag{A.2}$$

Proof. Let us start by assuming that $k = 0$ and let us denote the Lebesgue measure on $\mathbb{R}^n$ by μ_n. Identify u with a representative of its equivalence class. By definition there is a sequence u_l of finitely-valued functions converging strongly a.e. to u. Thus, for every l, there is a finite number of μ_1-measurable sets $A_{l,j} \subseteq I$, $j = 1, \dots, k_l$ with finite measure and a finite number of $f_{l,j} \in L^2(\mathbb{R}^n, \mathbb{C}^N)$, $j = 1, \dots, k_l$ such that

$$u_l = \sum_{j=1}^{k_l} f_{l,j} \chi_{A_{l,j}}.$$

This function can be considered to be a $\mu_1 \times \mu_n$-measurable function from $I \times \mathbb{R}^n$ to $\mathbb{C}^N$, and since μ_{n+1} is the completion of $\mu_1 \times \mu_n$, it is μ_{n+1}-measurable. It will be convenient to modify this function somewhat. Let B_l be the subset of I on which $\|u_l(t)\|_2 \leq 2\|u(t)\|_2$ and $u_l' = u_l \chi_{B_l}$. Note that u_l' converges μ_1-a.e. to u with respect to the L^2 norm and that it is μ_{n+1}-measurable. Let K be any compact subset of $I \times \mathbb{R}^n$, K_1 be the projection of K into I and $U_l = u_l' \chi_K$. Note that U_l is $\mu_1 \times \mu_n$-measurable and that

$$\left(\int_{\mathbb{R}^n} |U_l(t,x) - U_m(t,x)|^2\, dx\right)^{1/2} \leq \chi_{K_1}(t)\|u_l'(t) - u_m'(t)\|_2. \tag{A.3}$$

Let us consider

$$\|u_l'(t) - u_m'(t)\|_2 \leq \|u_l'(t) - u(t)\|_2 + \|u(t) - u_m'(t)\|_2.$$

Consider the first term on the right-hand side. It converges to zero pointwise a.e. as $l \to \infty$ and it is bounded by a function which is in $L^\infty(I, \mathbb{R})$. Since a similar statement is true concerning the second term, we can use Lebesgue's dominated convergence theorem to conclude that the L^2-norm of the right-hand side of (A.3) converges to zero as $n, m \to \infty$. Thus U_l defines a Cauchy sequence in $L^2(K, \mathbb{C}^N)$. This yields, for every compact set K, an element U_K of $L^2(I \times \mathbb{R}^n, \mathbb{C}^N)$. Letting K_l be an increasing sequence of compact sets whose union is $I \times \mathbb{R}^n$, we obtain a measurable function U from $I \times \mathbb{R}^n$ into $\mathbb{C}^N$ with the property that $|U|^2$ is locally integrable. Then (A.1) follows by observing that

$$\langle \phi, u \rangle = \lim_{l \to \infty} \langle \phi, u_l' \rangle = \int_{I \times \mathbb{R}^n} \phi \cdot \bar{U} \, d\mu$$

for all $\phi \in C_0^\infty(I \times \mathbb{R}^n, \mathbb{C}^N)$. Since $\partial^\alpha u \in L^\infty[I, H_{(0)}(\mathbb{R}^n, \mathbb{C}^N)]$ for all α such that $|\alpha| \le k$, we get, by the same argument, a $U_\alpha \in L^2_{\text{loc}}(I \times \mathbb{R}^n, \mathbb{C}^N)$ such that

$$\langle \phi, \partial^\alpha u \rangle = \int_{I \times \mathbb{R}^n} \phi \cdot \bar{U}_\alpha \, d\mu$$

for all $\phi \in C_0^\infty(I \times \mathbb{R}^n, \mathbb{C}^N)$. Due to (5.13), we obtain (A.2). $\square$

A.5 Hilbert spaces

For Hilbert spaces it is convenient to have an orthonormal (ON) basis.

Definition A.6. A set S of vectors u_α in a Hilbert space H, where α runs through some index set A is called *orthonormal* if $(u_\alpha, u_\beta)_H = 0$ if $\alpha \neq \beta$ and $\|u_\alpha\|_H = 1$. If the closure of the set of all finite linear combinations of elements in S equals H, we say that S is an *orthonormal basis* or a *complete orthonormal set*. A *countable orthonormal basis* is an orthonormal basis for which the index set A can be taken to be the positive integers.

Lemma A.7. *A real or complex infinite dimensional Hilbert space H is separable if and only if it has a countable orthonormal basis. If e_j, $j = 1, 2, \ldots$ is a countable ON basis and $x \in H$, then*

$$x_n = \sum_{j=1}^{n} (x, e_j)_H e_j \tag{A.4}$$

has the property that $x_n \to x$.

Proof. Let us assume that H is separable and let x_j, $j = 1, 2, \ldots$ be a countable dense subset. By taking away some of the x_j if necessary, we can ensure that x_{n+1} is linearly independent of $x_1, \ldots, x_n$ for every n. The usual Gram-Schmidt procedure produces an orthonormal set S consisting of e_j, $j = 1, 2, \ldots$ where $\{e_1, \ldots, e_n\}$ has the same linear span as $\{x_1, \ldots, x_n\}$ for all n. Let P be the set of all finite linear combinations of elements in S. Then the entire original dense sequence $\{x_j\}$ is a subset of P, so that P is dense. Consequently S is a countable ON basis. Assume now that H has a countable ON basis consisting of the vectors e_j, $j = 1, 2, \ldots$ and define the elements of a set M to be vectors of the form

$$x = \sum_{j=1}^{n} (\alpha_j + i\beta_j) e_j$$

where $\alpha_j, \beta_j \in \mathbb{Q}$. Then M is a countable set and for every $x \in H$ and $\varepsilon > 0$ there is a $y \in M$ such that $\|x - y\|_H < \varepsilon$. Finally, let us assume we have a countable ON basis $\{e_j\}$ and let $x \in H$. We know that there is a sequence $y_k \to x$ such that y_k is in the linear span of $\{e_1, \ldots, e_{n_k}\}$. Thus

$$y_k = \sum_{n=1}^{n_k} \alpha_{k,n} e_n,$$

where $\alpha_{k,n} \in \mathbb{C}$. Let us compare this with x_{n_k} defined in (A.4). We have

$$(x - y_k, x - y_k)_H = \|x\|_H^2 - \|x_{n_k}\|_H^2 + \|x_{n_k} - y_k\|_H^2.$$

Consequently, $\|x - x_{n_k}\|_H \leq \|x - y_k\|_H$. Thus $x_{n_k} \to x$, but by the same argument, x_N is a better approximation to x than x_{n_k} for any $N \geq n_k$ so that $x_n \to x$. □

A.6 Smooth functions with compact support

Let us start by introducing some terminology.

Definition A.8. Let $\Omega \subseteq \mathbb{R}^n$ be an open set. The set of smooth (infinitely differentiable) $f : \Omega \to \mathbb{R}^k$ will be denoted $C^\infty(\Omega, \mathbb{R}^k)$. The set of $f \in C^\infty(\Omega, \mathbb{R}^k)$ such that there is a compact set $K \subset \Omega$ with $f = 0$ for $f \notin K$ is denoted $C_0^\infty(\Omega, \mathbb{R}^k)$. We shall also use the notation $C^\infty(\Omega, \mathbb{R}) = C^\infty(\Omega)$ and $C_0^\infty(\Omega, \mathbb{R}) = C_0^\infty(\Omega)$. An element $f \in C_0^\infty(\Omega, \mathbb{R}^k)$ is called a *smooth function with compact support*. Similarly, if $m \geq 0$, we denote the set of m times differentiable functions by $C^m(\Omega, \mathbb{R}^k)$. The notation $C_0^m(\Omega, \mathbb{R}^k)$, $C_0^m(\Omega)$ and $C^m(\Omega)$ is analogous to the above. Finally, we shall use the notation $C^0(\Omega) = C(\Omega)$, $C_0^0(\Omega) = C_0(\Omega)$, etc.

Let us prove that, given an open set $\Omega \subseteq \mathbb{R}^n$ and a compact set $K \subset \Omega$, there is a function $\phi \in C_0^\infty(\Omega)$ such that $\phi = 1$ on K. We shall do so in several steps.

Lemma A.9. *Define* $f : \mathbb{R} \to \mathbb{R}$ *by*

$$f(t) = \begin{cases} e^{-1/t}, & t > 0, \\ 0, & t \leq 0. \end{cases} \tag{A.5}$$

Then f *is smooth.*

Proof. Note that f is smooth for $t > 0$ and for $t < 0$. The only problem is $t = 0$. Note that for $t > 0$ and any integer $m \geq 0$,

$$e^{-1/t} = \frac{1}{e^{1/t}} = \left(\sum_{k=0}^{\infty} \frac{t^{-k}}{k!}\right)^{-1} \leq \left(\frac{t^{-m}}{m!}\right)^{-1} = m!t^m.$$

Since $f(t) = 0$ for $t \leq 0$, we obtain

$$|f(t)| \leq m!|t|^m \tag{A.6}$$

for all integers $m \geq 0$ and $t \in \mathbb{R}$. As a consequence, it is clear that f is continuous at 0. Let us turn to the differentiability of f at 0. Estimate, for $h \neq 0$,

$$\left|\frac{f(h) - f(0)}{h}\right| = \left|\frac{f(h)}{h}\right| \leq m!|h|^{m-1}, \tag{A.7}$$

where we have used (A.6). We conclude that f is differentiable at zero and the derivative is zero. We leave it as an exercise to prove that f is continuously differentiable. By an induction argument, one can prove that f is smooth. □

Lemma A.10. *Let $\varepsilon > 0$. There is a $g \in C^\infty(\mathbb{R})$ such that $0 \leq g(t) \leq 1$ for all $t \in \mathbb{R}$, $g(t) = 0$ for $t \leq 0$ and $g(t) = 1$ for $t \geq \varepsilon$.*

Proof. Define

$$\phi(t) = f(t)f(\varepsilon - t),$$

where f is as in (A.5). Note that $\phi \in C^\infty(\mathbb{R})$, $\phi \geq 0$, $\phi(\varepsilon/2) > 0$, and that $\phi(t) = 0$ for $t \leq 0$ and for $t \geq \varepsilon$. Define

$$g(t) = \int_0^t \phi(s)\,ds \left[\int_0^\varepsilon \phi(s)\,ds\right]^{-1}.$$

Note that the integral in the denominator is non-zero. Then g has the desired properties. □

Lemma A.11. *Let $x_0 \in \mathbb{R}^n$ and $0 < r_1 < r_2$. Then there is a $\phi \in C^\infty(\mathbb{R}^n)$ such that $\phi(x) = 1$ for $x \in \overline{B}_{r_1}(x_0)$, $\phi(x) = 0$ for $x \notin B_{r_2}(x_0)$ and $0 \leq \phi \leq 1$.*

Proof. Let f be as in (A.5) and define

$$\phi_1(x) = f[r_2^2 - |x - x_0|^2].$$

Then $\phi_1 \in C^\infty(\mathbb{R}^n)$. Furthermore, $\phi_1(x) = 0$ if $x \notin B_{r_2}(x_0)$ and $\phi_1(x) \geq f(r_2^2 - r_1^2)$ for $x \in \overline{B}_{r_1}(x_0)$, since f is monotonically increasing. Let $\varepsilon = f(r_2^2 - r_1^2) > 0$ and let g be a function as in Lemma A.10. Define

$$\phi(x) = g[\phi_1(x)].$$

Then ϕ has the desired properties. □

Proposition A.12. *Let $\Omega \subseteq \mathbb{R}^n$ be open and let $K \subset \Omega$ be compact. Then there is a $\phi \in C_0^\infty(\Omega)$ such that $\phi(x) = 1$ for $x \in K$ and $0 \le \phi \le 1$.*

Proof. For every $x \in K$, let $r_x > 0$ be such that $\bar{B}_{2r_x}(x) \subset \Omega$. There is such an $r_x > 0$ since Ω is open. The collection of sets $B_{r_x}(x)$ for $x \in K$ is an open covering of K. By compactness, there is a finite subcovering $B_{r_i}(x_i)$ $i = 1, \dots, m$, where $r_i = r_{x_i}$. Let K_1 be the union of the $\bar{B}_{2r_i}(x_i)$. Since K_1 is a union of a finite number of compact sets, it is compact. Furthermore, it is contained in Ω by construction. Let $\phi_i \in C_0^\infty(\Omega)$ satisfy $\phi_i(x) = 1$ for $x \in \bar{B}_{r_i}(x_i)$ and $\phi_i = 0$ for $x \notin B_{2r_i}(x_i)$. That there are such functions follows from Lemma A.11. Define

$$\psi = \sum_{i=1}^m \phi_i.$$

Note that $\psi \in C^\infty(\Omega)$, $\psi(x) = 0$ for $x \notin K_1$ and $\psi(x) \ge 1$ for $x \in K$ (since for each $x \in K$, there is at least one i such that $x \in B_{r_i}(x_i)$). Let $\varepsilon = 1$ and let g be the function one obtains as a result of Lemma A.10. Define

$$\phi(x) = g[\psi(x)].$$

Then ϕ has the desired properties. □

A.7 Differentiability in the infinite dimensional case

Let X and Y be real vector spaces with norms $|\cdot|_X$ and $|\cdot|_Y$ respectively. Let $B(X, Y)$ denote the set of bounded linear transformations from X to Y, cf. Definition 5.3, p. 96 of [79]. Note that $B(X, Y)$ is a real vector space with a norm. For $f: X \to Y$, we shall say that f is differentiable at $x \in X$ if there is a $T \in B(X, Y)$ such that

$$\lim_{h\to 0} \frac{|f(x+h) - f(x) - Th|_Y}{|h|_X} = 0$$

Note that T, if it exists, is unique, and we shall call it the derivative of f at x, denoted $(Df)(x)$. Note that it is possible to define the derivative of a function under more general circumstances than the above, cf. e.g. p. 6 of [54]. However, the above situation will do for our purposes. If f is differentiable at every $x \in X$, we thus get a map $Df: X \to B(X, Y)$. If Df is continuous, we say that f is continuously differentiable. Iterating this procedure, we can speak of C^k functions, as well as C^∞ functions, $f: X \to Y$. We shall denote the kth derivative by $D^k f$. Let us now assume Y is an inner product space, i.e., there is an inner product $\langle \cdot, \cdot \rangle$ on Y such that $|y|_Y^2 = \langle y, y\rangle$ for all $y \in Y$. Assume $f, g: X \to Y$ are differentiable at $x \in X$ and consider $\phi(x) = \langle f(x), g(x)\rangle$. Then, by a straightforward argument, ϕ is differentiable at x and

$$[(D\phi)(x)]h = \langle [(Df)(x)]h, g(x)\rangle + \langle f(x), [(Dg)(x)]h\rangle.$$

If $X = \mathbb{R}^n$ in the above setting, we shall use the notation $\partial_j f$ to denote the function whose value at $x \in X$ is given by $(Df)(x)e_j$, where e_j is the vector in $\mathbb{R}^n$ all of whose components are zero except for the jth component which is given by 1. Thus

$$\partial_j \phi(x) = \langle \partial_j f(x), g(x) \rangle + \langle f(x), \partial_j g(x) \rangle.$$

Note also that if $f : \mathbb{R}^n \to Y$ is smooth, then all the partial derivatives exist and are smooth; in fact, we have

$$(\partial_j f)(x) = [(Df)(x)]e_j, \quad (\partial_k \partial_j f)(x) = \{[(D^2 f)(x)]e_k\}e_j,$$

etc.

Appendix B

B.1 Identities concerning permutation symbols

Let ε_{ijk} be antisymmetric in all of its indices and such that $\varepsilon_{123} = 1$. We shall raise and lower the indices of this object with the Kronecker delta. In other words $\varepsilon^{ijk} = \varepsilon_{ijk}$. We have

$$\varepsilon^{ijk}\varepsilon_{lmn} = \delta^i_l\delta^j_m\delta^k_n + \delta^i_m\delta^j_n\delta^k_l + \delta^i_n\delta^j_l\delta^k_m - \delta^i_l\delta^j_n\delta^k_m - \delta^i_n\delta^j_m\delta^k_l - \delta^i_m\delta^j_l\delta^k_n. \tag{B.1}$$

In order to prove this statement, let us make the following observations. First of all, both sides are antisymmetric in i, j, k and antisymmetric in l, m, n. If $\{i, j, k\} \neq \{1, 2, 3\}$ or if $\{l, m, n\} \neq \{1, 2, 3\}$, both sides of the equation are thus zero, and to check the equality, all we need to do is to check that we have equality for $i = l = 1$, $j = m = 2$ and $k = n = 3$. Then the left-hand side is 1 as well as the right-hand side. The conclusion follows. Taking $i = l$ and using Einstein's summation convention, i.e., that we sum over repeated upstairs and downstairs indices, yields the conclusion that

$$\varepsilon^{ijk}\varepsilon_{imn} = \delta^j_m\delta^k_n - \delta^j_n\delta^k_m. \tag{B.2}$$

Setting $j = m$, we then get

$$\varepsilon^{ijk}\varepsilon_{ijn} = 2\delta^k_n. \tag{B.3}$$

Proof of Lemma 19.11. Let G be a Bianchi class A Lie group, let g be a left invariant metric on G, let e_i be an orthonormal basis of $\mathfrak{g}$ and let ν be the associated commutator matrix. Below, we shall raise and lower indices using the Kronecker delta, i.e., there is no difference between upstairs and downstairs indices beyond the aesthetical. Let D be the Levi-Civita connection associated with the metric g and define Γ^i_{jk} by the condition that

$$D_{e_j}e_k = \Gamma^i_{jk}e_i.$$

Due to the Koszul formula, we have

$$\begin{aligned}\Gamma^i_{jk} = \langle D_{e_j}e_k, e_i\rangle &= \frac{1}{2}\{-\langle e_j, [e_k, e_i]\rangle + \langle e_k, [e_i, e_j]\rangle + \langle e_i, [e_j, e_k]\rangle\} \\ &= \frac{1}{2}\{-\gamma^j_{ki} + \gamma^k_{ij} + \gamma^i_{jk}\}.\end{aligned} \tag{B.4}$$

In order to be able to compute the Ricci curvature of G, let us compute

$$\begin{aligned}\langle R_{e_ie_j}e_j, e_m\rangle &= \langle D_{e_i}D_{e_j}e_j - D_{e_j}D_{e_i}e_j - D_{[e_i,e_j]}e_j, e_m\rangle \\ &= \langle D_{e_i}(\Gamma^k_{jj}e_k) - D_{e_j}(\Gamma^k_{ij}e_k) - \gamma^k_{ij}D_{e_k}e_j, e_m\rangle.\end{aligned}$$

Note that $e_i(\Gamma^m_{kl}) = 0$ and that $\Gamma^i_{ik} = \Gamma^i_{ki} = \Gamma^k_{ii} = 0$, assuming that one sums over the index i, due to the symmetries of the structure constants and the fact that the group is unimodular. Thus

$$\mathrm{Ric}(e_i, e_m) = \langle R_{e_ie_j}e_j, e_m\rangle = -\Gamma^k_{ij}\Gamma^m_{jk} - \gamma^k_{ij}\Gamma^m_{kj},$$

where it is understood that we sum over j and k. Below, it will be of interest to note that, by using the symmetries of the curvature tensor and carrying out a computation similar to the one above, we also have

$$\mathrm{Ric}(e_l, e_m) = \langle R_{e_j e_l} e_m, e_j \rangle = -\Gamma^i_{jm}\Gamma^j_{li} - \gamma^i_{jl}\Gamma^j_{im}. \tag{B.5}$$

Consider

$$\Gamma^k_{ij}\Gamma^m_{jk} = \frac{1}{4}(-\gamma^i_{jk} + \gamma^j_{ki} + \gamma^k_{ij})(-\gamma^j_{km} + \gamma^k_{mj} + \gamma^m_{jk}).$$

Consider the expression inside the first parenthesis. If one interchanges k and i in the second term, one sees that the last two terms taken together form an expression which is antisymmetric in k and j. The first term is also antisymmetric. Concerning the expression in the second parenthesis, one can argue similarly in order to conclude that the first two terms form a symmetric expression and the last term an antisymmetric expression in j and k. Since the contraction between an object which is antisymmetric in k and j with an object which is symmetric in the same indices yields zero as a result, we conclude that

$$\begin{aligned}\Gamma^k_{ij}\Gamma^m_{jk} = \frac{1}{4}(-\gamma^i_{jk} + \gamma^j_{ki} + \gamma^k_{ij})\gamma^m_{jk} &= -\frac{1}{4}\gamma^i_{jk}\gamma^m_{jk} + \frac{1}{4}(-\gamma^j_{ik} + \gamma^k_{ij})\gamma^m_{jk} \\ &= -\frac{1}{4}\gamma^i_{jk}\gamma^m_{jk} - \frac{1}{2}\gamma^j_{ik}\gamma^m_{jk}.\end{aligned}$$

Compute

$$\gamma^i_{jk}\gamma^m_{jk} = \varepsilon_{jkp}\varepsilon_{jkq}\nu^{ip}\nu^{mq} = 2\delta_{pq}\nu^{ip}\nu^{mq} = 2\nu^i{}_j\nu^{jm},$$

where we have used (B.3), and

$$\gamma^j_{ik}\gamma^m_{jk} = \varepsilon_{ikp}\varepsilon_{jkq}\nu^{pj}\nu^{qm} = (\delta_{ij}\delta_{pq} - \delta_{iq}\delta_{pj})\nu^{pj}\nu^{qm} = \nu_{ij}\nu^{jm} - (\mathrm{tr}\,\nu)\nu_i{}^m.$$

Thus

$$-\Gamma^k_{ij}\Gamma^m_{jk} = \nu_{ij}\nu^{jm} - \frac{1}{2}(\mathrm{tr}\,\nu)\nu_i{}^m. \tag{B.6}$$

Consider

$$\gamma^k_{ij}\Gamma^m_{kj} = \frac{1}{2}\gamma^k_{ij}(-\gamma^k_{jm} + \gamma^j_{mk} + \gamma^m_{kj}) = \frac{1}{2}\gamma^k_{ji}\gamma^k_{jm} + \frac{1}{2}\gamma^k_{ij}\gamma^j_{mk} + \frac{1}{2}\gamma^k_{ij}\gamma^m_{kj}.$$

By computations similar to ones given above, we have

$$\begin{aligned}\gamma^k_{ji}\gamma^k_{jm} &= \nu_{kl}\nu^{kl}\delta_{im} - \nu_i{}^l\nu_{lm}, \\ \gamma^k_{ij}\gamma^j_{mk} &= \nu_{kl}\nu^{kl}\delta_{im} + 2(\mathrm{tr}\,\nu)\nu_{im} - 2\nu_{ik}\nu^k{}_m - (\mathrm{tr}\,\nu)^2\delta_{im}, \\ \gamma^k_{ij}\gamma^m_{kj} &= \nu_{ik}\nu^{km} - (\mathrm{tr}\,\nu)\nu_i{}^m,\end{aligned}$$

where we used (B.2) in the first and last equation and (B.1) in the middle equation. Thus

$$-\gamma^k_{ij}\Gamma^m_{kj} = -\nu_{kl}\nu^{kl}\delta^m_i + \nu_{ik}\nu^{km} - \frac{1}{2}(\mathrm{tr}\,\nu)\nu_i{}^m + \frac{1}{2}(\mathrm{tr}\,\nu)^2\delta^m_i. \tag{B.7}$$

Combining (B.6) and (B.7), we obtain (19.5), and taking the trace of this equation, we obtain (19.6). □

B.2 Proof of Lemma 20.1

We wish to compute the Ricci curvature of a metric of the form (20.1). By the Koszul formula, we have

$$\theta_{ij} = \frac{1}{2}(-\langle e_i, [e_0, e_j]\rangle + \langle e_0, [e_j, e_i]\rangle + \langle e_j, [e_i, e_0]\rangle).$$

Note that

$$[e_0, e_i] = -a_i^{-1} e_0(a_i) e_i \tag{B.8}$$

(no summation), so that

$$\theta_{ij} = \frac{1}{2}[a_j^{-1} e_0(a_j)\delta_{ij} + a_i^{-1} e_0(a_i)\delta_{ij}]$$

(no summation). As a consequence, θ_{ij} is zero for $i \neq j$ and

$$\theta_{ii} = a_i^{-1} e_0(a_i) \tag{B.9}$$

(no summation).

B.2.1 Connection coefficients. Let us define $\Gamma^{\alpha}_{\beta\delta}$ by

$$\nabla_{e_\beta} e_\delta = \Gamma^{\alpha}_{\beta\delta} e_\alpha.$$

Note that

$$\Gamma^0_{\alpha\beta} = -\langle \nabla_{e_\alpha} e_\beta, e_0\rangle.$$

Due to the Koszul formula, we obtain

$$\Gamma^0_{\alpha\beta} = -\frac{1}{2}(-\langle e_\alpha, [e_\beta, e_0]\rangle + \langle e_\beta, [e_0, e_\alpha]\rangle + \langle e_0, [e_\alpha, e_\beta]\rangle).$$

Due to (B.8), $\langle e_0, [e_0, e_i]\rangle = 0$ and we conclude that $\Gamma^0_{0i} = \Gamma^0_{i0} = 0$. It is also clear that $\Gamma^0_{00} = 0$. Compute

$$\Gamma^0_{ij} = -\langle \nabla_{e_i} e_j, e_0\rangle = \langle \nabla_{e_i} e_0, e_j\rangle = \theta_{ij}. \tag{B.10}$$

Let us consider $\Gamma^i_{\alpha\beta}$. In the particular case that $\beta = 0$, we have

$$\Gamma^i_{\alpha 0} = \langle \nabla_{e_\alpha} e_0, e_i\rangle = -\langle e_0, \nabla_{e_\alpha} e_i\rangle = \Gamma^0_{\alpha i}.$$

Thus $\Gamma^i_{00} = 0$ and

$$\Gamma^i_{j0} = \theta_{ij}. \tag{B.11}$$

In case $\beta = l$, the Koszul formula yields

$$\Gamma^i_{\alpha l} = \frac{1}{2}(-\langle e_\alpha, [e_l, e_i]\rangle + \langle e_l, [e_i, e_\alpha]\rangle + \langle e_i, [e_\alpha, e_l]\rangle).$$

Due to (B.8), one can see that $\Gamma^i_{0l} = 0$. What remains is

$$\Gamma^i_{jl} = \frac{1}{2}(-\langle e_j, [e_l, e_i]\rangle + \langle e_l, [e_i, e_j]\rangle + \langle e_i, [e_j, e_l]\rangle) = \frac{1}{2}(-\gamma^j_{li} + \gamma^l_{ij} + \gamma^i_{jl}). \tag{B.12}$$

We see that the only non-zero connection coefficients $\Gamma^{\alpha}_{\beta\delta}$ are given by (B.10)–(B.12).

B.2.2 Commutators. Let the commutators $\gamma^{\mu}_{\alpha\beta}$ be defined by

$$[e_\alpha, e_\beta] = \gamma^{\mu}_{\alpha\beta} e_\mu.$$

Then

$$\gamma^{0}_{\alpha\beta} = -\langle [e_\alpha, e_\beta], e_0 \rangle = 0$$

and

$$\gamma^{i}_{\alpha\beta} = \langle [e_\alpha, e_\beta], e_i \rangle.$$

If $\alpha = \beta = 0$, we get zero as a result and if $\alpha = j$ and $\beta = k$, we get the same γ^i_{jk} that we would obtain by restricting ourselves to a hypersurface $\{t\} \times G$. Due to the fact that θ_{ij} are the components of a diagonal matrix with diagonal elements given by (B.9) and the fact that (B.8) holds, we conclude that

$$\gamma^{i}_{0j} = -\gamma^{i}_{j0} = -\theta_{ij}. \tag{B.13}$$

B.2.3 Ricci curvature. In order to compute the Ricci tensor, let $s_\mu = -1$ if $\mu = 0$ and $s_i = 1$ for $i = 1, 2, 3$, and compute (summation over μ being understood)

$$\begin{aligned}
\mathrm{Ric}(e_\alpha, e_\beta) &= s_\mu \langle \nabla_{e_\mu} \nabla_{e_\alpha} e_\beta - \nabla_{e_\alpha} \nabla_{e_\mu} e_\beta - \nabla_{[e_\mu, e_\alpha]} e_\beta, e_\mu \rangle \\
&= s_\mu \langle \nabla_{e_\mu} (\Gamma^{\nu}_{\alpha\beta} e_\nu) - \nabla_{e_\alpha} (\Gamma^{\nu}_{\mu\beta} e_\nu) - \gamma^{\nu}_{\mu\alpha} \Gamma^{\kappa}_{\nu\beta} e_\kappa, e_\mu \rangle \\
&= s_\mu [e_\mu (\Gamma^{\nu}_{\alpha\beta}) \eta_{\mu\nu} - e_\alpha (\Gamma^{\nu}_{\mu\beta}) \eta_{\mu\nu} + \Gamma^{\nu}_{\alpha\beta} \Gamma^{\kappa}_{\mu\nu} \eta_{\kappa\mu} \\
&\quad - \Gamma^{\nu}_{\mu\beta} \Gamma^{\kappa}_{\alpha\nu} \eta_{\kappa\mu} - \gamma^{\nu}_{\mu\alpha} \Gamma^{\kappa}_{\nu\beta} \eta_{\kappa\mu}] \\
&= e_\mu (\Gamma^{\mu}_{\alpha\beta}) - e_\alpha (\Gamma^{\mu}_{\mu\beta}) + \Gamma^{\nu}_{\alpha\beta} \Gamma^{\mu}_{\mu\nu} - \Gamma^{\nu}_{\mu\beta} \Gamma^{\mu}_{\alpha\nu} - \gamma^{\nu}_{\mu\alpha} \Gamma^{\mu}_{\nu\beta},
\end{aligned}$$

where $\eta_{\mu\nu}$ are the components of the Minkowski metric, i.e., $\eta_{00} = -1$, $\eta_{ii} = 1$ (no summation) and the remaining components are zero. This formula should be compared with Lemma 52, p. 87 of [65]. Note, however, the convention used for curvature in [65]. Using the fact that $e_\mu(\Gamma^{\nu}_{\alpha\beta}) = 0$ unless $\mu = 0$ and the fact that $\Gamma^{0}_{0\beta} = 0$, we obtain

$$\mathrm{Ric}(e_\alpha, e_\beta) = e_0 (\Gamma^{0}_{\alpha\beta}) - e_\alpha (\Gamma^{i}_{i\beta}) + \Gamma^{\nu}_{\alpha\beta} \Gamma^{\mu}_{\mu\nu} - \Gamma^{\nu}_{\mu\beta} \Gamma^{\mu}_{\alpha\nu} - \gamma^{\nu}_{\mu\alpha} \Gamma^{\mu}_{\nu\beta}.$$

Due to the fact that the Lie group is unimodular and the antisymmetries of γ^i_{jk}, we conclude that $\Gamma^{i}_{ij} = \Gamma^{j}_{ii} = \Gamma^{i}_{ji} = 0$ (where we sum over i), cf. Section B.1. As a consequence, $\Gamma^{i}_{i\beta} = \delta^{ij}\theta_{ij}$ if $\beta = 0$ and it equals zero if $\beta \neq 0$. In what follows, we shall abuse notation and write θ instead of $\delta^{ij}\theta_{ij}$. Note that

$$\Gamma^{\mu}_{\mu\nu} = \Gamma^{0}_{0\nu} + \Gamma^{i}_{i\nu} = \Gamma^{i}_{i\nu}.$$

By the above observations we thus conclude that

$$\Gamma^{\nu}_{\alpha\beta} \Gamma^{\mu}_{\mu\nu} = \theta \Gamma^{0}_{\alpha\beta}.$$

Thus

$$\mathrm{Ric}(e_\alpha, e_\beta) = e_0(\Gamma^0_{\alpha\beta}) - e_\alpha(\Gamma^i_{i\beta}) + \theta\Gamma^0_{\alpha\beta} - \Gamma^\nu_{\mu\beta}\Gamma^\mu_{\alpha\nu} - \gamma^\nu_{\mu\alpha}\Gamma^\mu_{\nu\beta}.$$

Let us consider the second to last term in this expression. Due to the fact that the only non-zero connection coefficients are given by (B.10)–(B.12), we have

$$\Gamma^\nu_{\mu\beta}\Gamma^\mu_{\alpha\nu} = \Gamma^0_{i\beta}\Gamma^i_{\alpha 0} + \Gamma^i_{j\beta}\Gamma^j_{\alpha i}.$$

Since $\Gamma^\mu_{0\beta} = 0$ regardless of what μ and β are, we have

$$\gamma^\nu_{\mu\alpha}\Gamma^\mu_{\nu\beta} = \gamma^i_{0\alpha}\Gamma^0_{i\beta} + \gamma^i_{j\alpha}\Gamma^j_{i\beta}.$$

To conclude

$$\mathrm{Ric}(e_\alpha, e_\beta) = e_0(\Gamma^0_{\alpha\beta}) - e_\alpha(\Gamma^i_{i\beta}) + \theta\Gamma^0_{\alpha\beta} - \Gamma^0_{i\beta}\Gamma^i_{\alpha 0} - \Gamma^i_{j\beta}\Gamma^j_{\alpha i} - \gamma^i_{0\alpha}\Gamma^0_{i\beta} - \gamma^i_{j\alpha}\Gamma^j_{i\beta}.$$

Let us compute

$$\mathrm{Ric}(e_0, e_0) = -e_0(\theta) - \gamma^i_{j0}\Gamma^j_{i0} = -\dot{\theta} - \theta^{ij}\theta_{ij},$$

so that (20.3) holds. Furthermore,

$$\mathrm{Ric}(e_0, e_m) = -\gamma^i_{j0}\Gamma^j_{im} = -\frac{1}{2}\theta_{ij}(-\gamma^i_{mj} + \gamma^m_{ji} + \gamma^j_{im}) = \theta_{ij}\gamma^i_{mj} = \varepsilon_{mjl}n^{li}\theta_{ij},$$

where we have used the fact that $\theta_{ij} = \theta_{ji}$. Thus (20.4) holds. Finally, consider

$$\begin{aligned}\mathrm{Ric}(e_l, e_m) &= \dot{\theta}_{lm} + \theta\theta_{lm} - \theta_{li}\theta^i{}_m - \Gamma^i_{jm}\Gamma^j_{li} + \theta_{li}\theta^i{}_m - \gamma^i_{jl}\Gamma^j_{im} \\ &= \dot{\theta}_{lm} + \theta\theta_{lm} - \Gamma^i_{jm}\Gamma^j_{li} - \gamma^i_{jl}\Gamma^j_{im}.\end{aligned}$$

Since Γ^i_{jk} and γ^i_{jk} are the same in the current situation as in Section B.1, assuming that we replace ν by n, and since (B.5) holds, we conclude that the last two terms are simply the Ricci curvature of the spatial hypersurface $\{t\} \times G$ evaluated at (e_l, e_m). Thus (20.5) follows from the above and (19.5).

B.3 Proof of Lemma 20.2

Due to the Jacobi identities, we have to have

$$[e_\alpha, [e_\beta, e_\delta]] + [e_\beta, [e_\delta, e_\alpha]] + [e_\delta, [e_\alpha, e_\beta]] = 0.$$

Reformulating this in terms of the $\gamma^\alpha_{\mu\nu}$'s, we obtain

$$[e_\alpha, \gamma^\mu_{\beta\delta}e_\mu] + [e_\beta, \gamma^\mu_{\delta\alpha}e_\mu] + [e_\delta, \gamma^\mu_{\alpha\beta}e_\mu] = 0.$$

Thus

$$e_\alpha(\gamma^\nu_{\beta\delta}) + e_\beta(\gamma^\nu_{\delta\alpha}) + e_\delta(\gamma^\nu_{\alpha\beta}) + \gamma^\mu_{\beta\delta}\gamma^\nu_{\alpha\mu} + \gamma^\mu_{\delta\alpha}\gamma^\nu_{\beta\mu} + \gamma^\mu_{\alpha\beta}\gamma^\nu_{\delta\mu} = 0.$$

In order for us to get something which is not automatically zero on the left-hand side, we have to have $\nu = i$. Thus

$$e_\alpha(\gamma^i_{\beta\delta}) + e_\beta(\gamma^i_{\delta\alpha}) + e_\delta(\gamma^i_{\alpha\beta}) + \gamma^j_{\beta\delta}\gamma^i_{\alpha j} + \gamma^j_{\delta\alpha}\gamma^i_{\beta j} + \gamma^j_{\alpha\beta}\gamma^i_{\delta j} = 0.$$

If all of α, β and δ are zero, the left-hand side is automatically zero. If all of α, β and δ are between 1 and 3, we get zero due to the Jacobi identities on $\{t\} \times G$. If $\alpha = l$, $\beta = \delta = 0$, we get

$$e_0(\gamma^i_{0l}) + e_0(\gamma^i_{l0}) + \gamma^j_{0l}\gamma^i_{0j} + \gamma^j_{l0}\gamma^i_{0j} = 0,$$

an equation which contains no information. Finally, let $\alpha = l$, $\beta = m$ and $\delta = 0$. We get

$$e_0(\gamma^i_{lm}) + \gamma^j_{m0}\gamma^i_{lj} + \gamma^j_{0l}\gamma^i_{mj} + \gamma^j_{lm}\gamma^i_{0j} = 0.$$

Noting that, due to (B.2) and (B.3),

$$\begin{aligned}
\varepsilon^{lmk}\gamma^i_{lm} &= 2n^{ki},\\
\varepsilon^{lmk}\gamma^i_{lj} &= n^{ki}\delta^m_j - n^{mi}\delta^k_j,\\
\varepsilon^{mkl}\gamma^i_{mj} &= n^{li}\delta^k_j - n^{ki}\delta^l_j,\\
\varepsilon^{lmk}\gamma^j_{lm} &= 2n^{kj},
\end{aligned}$$

we obtain, cf. (B.13),

$$2\dot{n}^{ik} - 2\theta^k{}_l n^{li} - 2\theta^i{}_j n^{jk} + 2\theta n^{ki} = 0.$$

The lemma follows.

B.4 Proof of Lemma 22.7

Using the notation and results of Section B.2 together with (20.7)–(20.11), one concludes that

$$\langle R_{e_\alpha e_\beta} e_\gamma, e_\delta\rangle \tag{B.14}$$

can be expressed as a homogeneous polynomial in the n_i and the θ_i of order 2. Dividing by θ^2, one obtains a, typically inhomogenous, polynomial in Σ_+, Σ_- and the N_i. Consequently, the Kretschmann scalar divided by θ^4 is a polynomial in the same variables. Since $\theta \to \pm\infty$ corresponds to $\tau \to -\infty$ with respect to the Wainwright–Hsu time, cf. Lemma 22.2 and 22.3, the existence of an α-limit point with respect to the Wainwright–Hsu variables such that this polynomial is non-zero implies that the Kretschmann scalar is unbounded in the direction of interest. If one is interested in the behaviour along a time sequence corresponding to an α-limit point on the Kasner circle, it is enough to compute (B.14) in a situation where all the n_i are zero in order to compute the limit of the Kretschmann scalar divided by θ^4 along the time sequence corresponding to the α-limit point. For this reason, we shall, from now on, assume $n_i = 0$ for $i = 1, 2, 3$. Due to the computations carried out in Section B.2, we have

$$\Gamma^0_{ij} = \theta_{ij}, \quad \Gamma^i_{j0} = \theta_{ij},$$

cf. (B.10) and (B.11), where θ_{ij} are the components of a diagonal matrix with diagonal elements θ_i, and all the remaining connection coefficients are zero. Concerning the commutators, we have

$$\gamma^i_{j0} = -\gamma^i_{0j} = \theta_{ij},$$

cf. (B.13), and the remaining commutators are zero. Compute

$$\begin{aligned}\langle R_{e_\alpha e_\beta} e_\gamma, e_\delta\rangle &= \langle \nabla_{e_\alpha}\nabla_{e_\beta} e_\gamma - \nabla_{e_\beta}\nabla_{e_\alpha} e_\gamma - \nabla_{[e_\alpha, e_\beta]} e_\gamma, e_\delta\rangle \\ &= \langle e_\alpha(\Gamma^\mu_{\beta\gamma}) e_\mu + \Gamma^\mu_{\beta\gamma}\Gamma^\nu_{\alpha\mu} e_\nu - e_\beta(\Gamma^\mu_{\alpha\gamma}) e_\mu \\ &\quad - \Gamma^\mu_{\alpha\gamma}\Gamma^\nu_{\beta\mu} e_\nu - \gamma^\mu_{\alpha\beta}\Gamma^\nu_{\mu\gamma} e_\nu, e_\delta\rangle.\end{aligned}$$

Consider the case when both of α and β are spacelike. Then

$$e_\alpha(\Gamma^\mu_{\beta\gamma}) = e_\beta(\Gamma^\mu_{\alpha\gamma}) = \gamma^\mu_{\alpha\beta}\Gamma^\nu_{\mu\gamma} = 0.$$

In order to get a non-zero result, $\alpha = i$ and $\beta = j$ have to be different. The relevant expression to consider is

$$\Gamma^\mu_{j\gamma}\Gamma^\delta_{i\mu} - \Gamma^\mu_{i\gamma}\Gamma^\delta_{j\mu}.$$

If $\delta = 0$, we have to have $\mu = i$ in the first term in order for the second factor of the first term to be non-zero. However, then the first factor of the first term is zero. In other words, if $\delta = 0$, then the first term is zero. For similar reasons, the second term has to equal zero. If δ is spatial, it has to equal either i or j in order for us to get a non-zero result. If it equals i, the second term is zero and we have to have $\gamma = j$ and $\mu = 0$ in order that the first term be non-zero. We then get

$$\langle R_{e_i e_j} e_j, e_i\rangle = \theta_i\theta_j$$

(no summation) for $j \neq i$. By the symmetries of the curvature tensor, we of course also have

$$\langle R_{e_i e_j} e_i, e_j\rangle = -\theta_i\theta_j$$

(no summation) for $j \neq i$. Now consider the case $\alpha = 0$ and $\beta = i$. If both of γ and δ are spacelike, we can use the symmetries of the curvature tensor together with the above arguments in order to conclude that the result is zero. Without loss of generality, it is thus enough to consider

$$\langle R_{e_0 e_i} e_0, e_j\rangle = e_0(\Gamma^j_{i0}) + \Gamma^\mu_{i0}\Gamma^j_{0\mu} - \Gamma^\mu_{00}\Gamma^j_{i\mu} - \gamma^\mu_{0i}\Gamma^j_{\mu 0}$$

If $i \neq j$, we obtain zero. If $i = j$, we obtain that

$$\langle R_{e_0 e_i} e_0, e_i\rangle = \dot\theta_i + \theta_i^2 = -\theta\theta_i + \theta_i^2,$$

cf. (20.5). Since indices are raised and lowered with the Minkowski metric, we get

$$\kappa = R_{\alpha\beta\gamma\delta} R^{\alpha\beta\gamma\delta} = 4\sum_i (\langle R_{e_0 e_i} e_0, e_i \rangle)^2 + 2\sum_{i,j} (\langle R_{e_i e_j} e_i, e_j \rangle)^2.$$

In other words, in order for κ/θ^4 to be zero, two of θ_i/θ have to be zero, and the remaining one non-zero, since the sum of the θ_i/θ equals one. Without loss of generality, we may assume $\theta_2 = \theta_3 = 0$ and $\theta_1 = \theta$. Then $\Sigma_2 = \Sigma_3 = -1/3$ and $\Sigma_1 = 2/3$. Thus $\Sigma_+ = -1$ and $\Sigma_- = 0$. In other words, the only possibility for the limit of κ/θ^4 to be different from zero along a sequence corresponding to an α-limit point on the Kasner circle is that the point on the Kasner circle is a special point.

Bibliography

[1] R. A. Adams, *Sobolev spaces*. Academic Press, New York, 1975.

[2] M. T. Anderson, On long-time evolution in general relativity and geometrization of 3-manifolds. *Commun. Math. Phys.* **222** (3) (2001), 533–567.

[3] L. Andersson and V. Moncrief, Future complete vacuum spacetimes. In *The Einstein equations and the large scale behavior of gravitational fields*, Birkhäuser, Basel, 2004.

[4] A. N. Bernal and M. Sánchez, On smooth Cauchy surfaces and Geroch's splitting theorem. *Commun. Math. Phys.* **243** (2003), 461–470.

[5] A. N. Bernal and M. Sánchez, Smoothness of time functions and the metric splitting of globally hyperbolic spacetimes. *Commun. Math. Phys.* **257** (2005), 43–50.

[6] A. N. Bernal and M. Sánchez, Further results on the smoothability of Cauchy hypersurfaces and Cauchy time functions. *Lett. Math. Phys.* **77** (2) (2006), 183–197.

[7] O. I. Bogoyavlensky, *Methods in the qualitative theory of dynamical systems in astrophysics and gas dynamics*. Springer Ser. Soviet Math., Springer-Verlag, New York, 1985.

[8] G. E. Bredon, *Topology and geometry*. Grad. Texts in Math. 139, Springer-Verlag, New York, 1995.

[9] Y. Choquet-Bruhat, *General relativity and the Einstein equations*. Oxford Math. Monogr., Oxford University Press, Oxford, 2009.

[10] Y. Choquet-Bruhat and R. Geroch, Global aspects of the Cauchy problem in general relativity. *Commun. Math. Phys.* **14** (1969), 329–335.

[11] D. Christodoulou, Global existence of generalised solutions of the spherically symmetric Einstein-scalar equations in the large. *Commun. Math. Phys.* **106** (1986), 587–621.

[12] D. Christodoulou, The problem of a self-gravitating scalar field. *Commun. Math. Phys.* **105** (1986), 337–361.

[13] D. Christodoulou, A mathematical theory of gravitational collapse. *Commun. Math. Phys.* **109** (1987), 613–647.

[14] D. Christodoulou, The structure and uniqueness of generalised solutions of the spherically symmetric Einstein-scalar equations. *Commun. Math. Phys.* **109** (1987), 591–611.

[15] D. Christodoulou, The formation of black holes and singularities in spherically symmetric gravitational collapse. *Commun. Pure Appl. Math.* **44** (1991), 339–373.

[16] D. Christodoulou, Bounded variation solutions of the spherically symmetric Einstein-scalar field equations. *Commun. Pure Appl. Math.* **46** (1993), 1131–1220.

[17] D. Christodoulou, Examples of naked singularity formation in the gravitational collapse of a scalar field. *Ann. Math.* **140** (1994), 607–653.

[18] D. Christodoulou, The instability of naked singularities in the gravitational collapse of a scalar field. *Ann. Math.* (2) **149** (1999), 183–217.

[19] D. Christodoulou, *Mathematical problems of general relativity* I. Zur. Lect. Adv. Math., EMS Publ. House, Zürich, 2008.

[20] D. Christodoulou and S. Klainerman, *The global nonlinear stability of the Minkowski space*. Princeton University Press, Princeton, N.J., 1993.

[21] P. T. Chruściel, J. A. Isenberg and V. Moncrief, Strong cosmic censorship in polarised Gowdy spacetimes. *Class. Quantum Grav.* **7** (1990), 1671–1680.

[22] P. T. Chruściel, *On uniqueness in the large of solutions of Einstein's equations ("strong cosmic censorship")*. Proc. Centre Math. Appl. Austral. Nat. Univ. 27, National University 27, Canberra, Australia, 1991.

[23] P. T. Chruściel and J. Isenberg, Non-isometric vacuum extensions of vacuum maximal globally hyperbolic spacetimes. *Phys. Rev. D* **48** (1993), 1616–1628.

[24] P. T. Chruściel and A. D. Rendall, Strong cosmic censorship in vacuum space-times with compact, locally homogeneous Cauchy surfaces. *Ann. Physics* **242** (2) (1995), 349–385.

[25] W. D. Curtis and F. R. Miller, *Differential manifolds and theoretical physics*. Academic Press, Orlando, 1985.

[26] M. Dafermos, Stability and instability of the Cauchy horizon for the spherically symmetric Einstein-Maxwell-scalar field equations. *Ann. Math.* **158** (2003), 875–928.

[27] M. Dafermos, The interior of charged black holes and the problem of uniqueness in general relativity. *Comm. Pure Appl. Math.* **58** (4) (2005), 445–504.

[28] M. Dafermos and I. Rodnianski, A proof of Price's law for the collapse of a self-gravitating scalar field. *Invent. Math.* **162** (2) (2005), 381–457.

[29] M. Dafermos and A. D. Rendall, Inextendibility of expanding cosmological models with symmetry. *Class. Quantum Grav.* **22** (23) (2005), L143–L147.

[30] M. Dafermos and A. D. Rendall, Strong cosmic censorship for $\mathbb{T}^2$-symmetric cosmological spacetimes with collisionless matter. Preprint, arXiv:gr-qc/0610075.

[31] M. Dafermos and A. D. Rendall, Strong cosmic censorship for surface-symmetric cosmological spacetimes with collisionless matter. Preprint, arXiv:gr-qc/0701034.

[32] D. M. Eardley and V. Moncrief, The global existence problem and cosmic censorship in general relativity. *Gen. Rel. Grav.* **13** (1981), 887–892.

[33] D. M. Eardley and V. Moncrief, The global existence of Yang-Mills-Higgs fields in 4-dimensional Minkowski space. I. Local existence and smoothness properties. *Commun. Math. Phys.* **83** (2) (1982), 171–191.

[34] D. M. Eardley and V. Moncrief, The global existence of Yang-Mills-Higgs fields in 4-dimensional Minkowski space. II. Completion of proof. *Commun. Math. Phys.* **83** (2) (1982), 193–212.

[35] A. Einstein, Zur allgemeinen Relativitätstheorie. *Preuss. Akad. Wiss. Berlin, Sitzber.* (1915), 778–786.

[36] A. Einstein, Die Feldgleichungen der Gravitation. *Preuss. Akad. Wiss. Berlin, Sitzber.* (1915), 844–847.

[37] G. F. R. Ellis and M. A. H. MacCallum, A class of homogeneous cosmological models. *Commun. Math. Phys.* **12** (1969), 108–141.

[38] A. E. Fischer and J. E. Marsden, The Einstein evolution equations as a first-order quasi-linear symmetric hyperbolic system, I. *Commun. Math. Phys.* **28** (1972), 1–38.

[39] Y. Fourès-Bruhat, Théorème d'existence pour certains systèmes d'équations aux derivées partielles non linéaires. *Acta Math.* **88** (1952), 141–225.

[40] H. Friedrich, On the existence of n-geodesically complete or future complete solutions of Einstein's field equations with smooth asymptotic structure. *Commun. Math. Phys.* **107** (1986), 587–609.

[41] H. Friedrich and A. D. Rendall, The Cauchy problem for the Einstein equations. In *Einstein's field equations and their physical implications*, Lecture Notes in Phys. 540, Springer-Verlag, Berlin, 2000, 127–223.

[42] E. Gagliardo, Ulterio proprietà de alcune classi de funzioni in più variabili. *Ricerche Mat.* **8** (1959), 24–51.

[43] R. Geroch, Domain of dependence. *J. Math. Phys.* **11** (1970), 437–439.

[44] D. Gilbarg and N. S. Trudinger, *Elliptic partial differential equations of second order.* Classics Math., Springer-Verlag, Berlin, 2001.

[45] P. B. Gilkey, *Invariance theory, the heat equation, and the Atiyah-Singer index theorem.* Second edition, Stud. Adv. Math., CRC Press, Boca Raton, FL, 1995.

[46] S. W. Hawking, The occurrence of singularities in cosmology. III. Causality and singularities. *Proc. Roy. Soc. Lond.* **A300** (1967), 182–201.

[47] S. W. Hawking and R. Penrose, The singularities of gravitational collapse and cosmology. *Proc. Roy. Soc. Lond.* **A314** (1970), 529–548.

[48] S. W. Hawking and G. F. R. Ellis, *The large scale structure of space time.* Cambridge Monogr. Math. Phys. 1, Cambridge University Press, Cambridge, 1973.

[49] L. Hörmander, *The analysis of linear partial differential operators* I. Grundlehren Math. Wiss. 256, Springer-Verlag, Berlin, 1990.

[50] L. Hörmander, *Lectures on nonlinear hyperbolic differential equations.* Math. Appl. (Berlin) 26, Springer-Verlag, Berlin, 1997.

[51] S. Klainerman and M. Machedon, Finite energy solutions of the Yang-Mills equations in $\mathbb{R}^{3+1}$. *Ann. of Math.* (2) **142** (1) (1995), 39–119.

[52] S. Klainerman and I. Rodnianski, Rough solutions of the Einstein-vacuum equations. *Ann. of Math.* (2) **161** (3) (2005), 1143–1193.

[53] T. Koike, M. Tanimoto and A. Hosoya, Compact homogeneous universes. *J. Math. Phys.* **35** (9) (1994), 4855–4888.

[54] S. Lang, *Differential and Riemannian manifolds.* Grad. Texts in Math. 160, Springer-Verlag, New York, 1995.

[55] J. M. Lee, *Introduction to smooth manifolds.* Grad. Texts in Math. 218, Springer-Verlag, New York, 2003.

[56] X.-F. Lin and R. Wald, Proof of the closed-universe-recollapse conjecture for diagonal Bianchi type-IX cosmologies. *Phys. Rev. D.* **40** (1989), 3280–3286.

[57] H. Lindblad and I. Rodnianski, Global existence for the Einstein vacuum equations in wave coordinates. *Commun. Math. Phys.* **256** (1) (2005), 43–110.

[58] H. Lindblad and I. Rodnianski, The global stability of Minkowski space-time in harmonic gauge. *Ann. of Math.*, to appear.

[59] A. Majda, *Compressible fluid flow and systems of conservation laws in several space variables.* Appl. Math. Sci. 53, Springer-Verlag, New York, 1984.

[60] J. E. Marsden and F. J. Tipler, Maximal hypersurfaces and foliations of constant mean curvature in general relativity. *Phys. Rep.* **66** (1980), 109–139.

[61] J. Milnor, Curvatures of left invariant metrics on Lie groups. *Adv. Math.* **21** (3) (1976), 293–329.

[62] C. Misner and A. Taub, A singularity-free empty universe. *Sov. Phys. JETP* **28** (1969), 122–133.

[63] J. Moser, A rapidly convergent iteration method and nonlinear partial differential equations, I. *Ann. Sc. Norm. Sup. Pisa* **20** (1966), 265–315.

[64] L. Nirenberg, On elliptic partial differential equations. *Ann. Sc. Norm. Sup. Pisa* **13** (1959), 116–162.

[65] B. O'Neill, *Semi-Riemannian geometry*. Pure Appl. Math. 103, Academic Press, Orlando, 1983.

[66] R. Penrose, Gravitational collapse and space-time singularities. *Phys. Rev. Lett.* **14** (1965), 57–59.

[67] R. Penrose, Gravitational collapse – the role of general relativity. *Riv. del Nuovo Cim.* **1** (numero speziale) (1969), 252–276.

[68] A. D. Rendall, Global dynamics of the mixmaster model. *Class. Quantum Grav.* **14** (1997), 2341–2356.

[69] A. D. Rendall, *Partial differential equations in general relativity*. Oxford University Press, Oxford, 2008.

[70] H. Ringström, Curvature blow up in Bianchi VIII and IX vacuum spacetimes. *Class. Quantum Grav.* **17** (2000), 713–731.

[71] H. Ringström, The Bianchi IX attractor. *Ann. Henri Poincaré* **2** (2001), 405–500.

[72] H. Ringström, On Gowdy vacuum spacetimes. *Math. Proc. Cambridge. Philos. Soc.* **136** (2004), 485–512.

[73] H. Ringström, Asymptotic expansions close to the singularity in Gowdy spacetimes. In *A spacetime safari: essays in honour of Vincent Moncrief* (ed. by B. Berger and J. Isenberg), *Class. Quantum Grav.* **21** (2004), S305–S322.

[74] H. Ringström, On a wave map equation arising in general relativity. *Comm. Pure Appl. Math.* **57** (2004), 657–703.

[75] H. Ringström, Existence of an asymptotic velocity and implications for the asymptotic behaviour in the direction of the singularity in T^3-Gowdy. *Comm. Pure Appl. Math.* **59** (2006), 977–1041.

[76] H. Ringström, Strong cosmic censorship in T^3-Gowdy. *Ann. of Math*, to appear.

[77] H. Ringström, Future stability of the Einstein-non-linear scalar field system. *Invent. Math.* **173** (2008), 123–208.

[78] W. Rudin, *Principles of mathematical analysis*. McGraw Hill, Singapore, 1976.

[79] W. Rudin, *Real and complex analysis*. McGraw Hill, Singapore, 1987.

[80] W. Rudin, *Functional analysis*. McGraw Hill, New York, 1991.

[81] R. Schoen and S.-T. Yau, Proof of the positive mass theorem. II. *Commun. Math. Phys.* **79** (1981), 231–260.

[82] C. D. Sogge, *Lectures on nonlinear wave equations*. International Press, Boston, 1995.

[83] M. E. Taylor, *Partial differential equations*. III. *Nonlinear equations*. Appl. Math. Sci. 117, Springer-Verlag, New York, 1996.

[84] J. Wainwright and L. Hsu, A dynamical systems approach to Bianchi cosmologies: orthogonal models of class A. *Class. Quantum Grav.* **6** (1989), 1409–1431.

[85] J. Wainwright and G. F. R. Ellis, editors, *Dynamical systems in cosmology*. Cambridge University Press, Cambridge, 1997.

[86] K. Yosida, *Functional analysis*. Grundlehren Math. Wiss. 123, Springer-Verlag, Berlin, 1965.

[87] R. M. Wald, *General relativity*. University of Chicago Press, Chicago, 1984.

[88] F. W. Warner, *Foundations of differentiable manifolds and Lie groups*. Scott, Foresman and Co., Glenview, Ill.-London, 1971.

Index